Load and Resistance Factor Design
of Steel Structures

PRENTICE HALL INTERNATIONAL SERIES
IN CIVIL ENGINEERING AND ENGINEERING MECHANICS
William J. Hall, Editor

Au and Christiano, *Fundamentals of Structural Analysis*
Au and Christiano, *Structural Analysis*
Barson and Rolfe, *Fracture and Fatigue Control in Structures, 2/e*
Bathe, *Finite Element Procedures in Engineering Analysis*
Berg, *Elements of Structural Dynamics*
Biggs, *Introduction to Structural Engineering*
Chajes, *Structural Analysis, 2/e*
Collins and Mitchell, *Prestressed Concrete Structures*
Cooper and Chen, *Designing Steel Structures*
Cording et al., *The Art and Science of Geotechnical Engineering*
Gallagher, *Finite Element Analysis*
Geschwindner, Disque, and Bjorhovde, *Load and Resistance Factor Design of Steel Structures*
Hendrickson and Au, *Project Management for Construction*
Higdon et al., *Engineering Mechanics, 2nd Vector Edition*
Holtz and Kovacs, *Introduction to Geotechnical Engineering*
Humar, *Dynamics of Structures*
Johnston, Lin, and Galambos, *Basic Steel Design, 3/e*
MacGregor, *Reinforced Concrete: Mechanics and Design, 2/e*
Mehta and Monteiro, *Concrete: Structure, Properties, and Materials, 2/e*
Melosh, *Structural Engineering Analysis by Finite Elements*
Meredith et al., *Design and Planning of Engineering Systems, 2/e*
Mindess and Young, *Concrete*
Nawy, *Prestressed Concrete*
Nawy, *Reinforced Concrete: A Fundamental Approach, 2/e*
Pfeffer, *Solid Waste Management*
Popov, *Engineering Mechanics of Solids*
Popov, *Introduction to the Mechanics of Solids*
Popov, *Mechanics of Materials, 2/e*
Schneider and Dickey, *Reinforced Masonry Design, 3/e*
Sennett, *Matrix Analysis of Structures*
Wang and Salmon, *Introductory Structural Analysis*
Weaver and Johnson, *Structural Dynamics by Finite Elements*
Wolf, *Dynamic Soil-Structure Interaction*
Wray, *Measuring Engineering Properties of Soils*
Yang, *Finite Element Structural Analysis*

Load and Resistance Factor Design of Steel Structures

Louis F. Geschwindner

Department of Architectural Engineering
The Pennsylvania State University

Robert O. Disque

Besier Gibble Norden

Reidar Bjorhovde

Department of Civil Engineering
University of Pittsburgh

Prentice Hall, Englewood Cliffs, New Jersey 07632

Library of Congress Cataloging-in-Publication Data

GESCHWINDNER, LOUIS F.
    Load and resistance factor design of steel structures / Louis F.
Geschwindner, Robert O. Disque, Reidar Bjorhovde.
        p.   cm. — (Prentice Hall international series in civil
engineering and engineering mechanics)
    Includes bibliographical references and index.
    ISBN 0-13-539156-3
    1. Building, Iron and steel. 2. Structural design. I. Disque,
Robert O. II. Bjorhovde, Reidar, III. Title. IV. Series.
TA684.G38 1994
624.1'821—dc20   93-28893
                                                        CIP

Acquisitions editor: Bill Zobrist
Editorial/production supervision and interior design: Joan Stone
Copy editor: Virginia Dunn
Cover designer: DeLuca Design
Production coordinator: Linda Behrens
Editorial assistant: Susan Handy
Cover photo: Chevron Tower, Houston, TX. Structural engineer: Walter P.
Moore and Assoc. Architect: CRS, Inc. Photo courtesy Walter P. Moore and
Assoc./Richard Payne.

PRENTICE HALL INTERNATIONAL SERIES IN CIVIL ENGINEERING AND
ENGINEERING MECHANICS
William J. Hall, Editor

The author and publisher of this book have used their best efforts in preparing this book.
These efforts include the development, research, and testing of the theories and programs
to determine their effectiveness. The author and publisher make no warranty of any kind,
expressed or implied, with regard to these programs or the documentation contained in
this book. The author and publisher shall not be liable in any event for incidental or con-
sequential damages in connection with, or arising out of, the furnishing, performance, or
use of these programs.

Printed in the United States of America

10 9 8 7 6 5 4 3 2 1

ISBN 0-13-539156-3

PRENTICE-HALL INTERNATIONAL (UK) LIMITED, *London*
PRENTICE-HALL OF AUSTRALIA PTY. LIMITED, *Sydney*
PRENTICE-HALL CANADA INC., *Toronto*
PRENTICE-HALL HISPANOAMERICANA, S.A., *Mexico*
PRENTICE-HALL OF INDIA PRIVATE LIMITED, *New Delhi*
PRENTICE-HALL OF JAPAN, INC., *Tokyo*
SIMON & SCHUSTER ASIA PTE. LTD., *Singapore*
EDITORA PRENTICE-HALL DO BRASIL, LTDA., *Rio de Janeiro*

This book is dedicated to our fathers.

Contents

Preface   **xvii**

Chapter One

Introduction   **1**

    1.1   Scope   *2*
    1.2   Principles of Structural Design   *3*
    1.3   The Parts of the Steel Structure   *5*
    1.4   Types of Steel Structures   *11*
    1.5   Design Philosophies   *16*
    1.6   Structural Safety   *19*
    1.7   Building Codes and Design Specifications   *20*
           References   *21*

Chapter Two

Load and Resistance Factor Design   **23**

    2.1   Structural Safety   *24*
    2.2   The Philosophy of LRFD   *26*
    2.3   Principles of LRFD   *28*

2.4   Structural Safety: The Reliability Index   *33*
2.5   Variability of Strength   *38*
2.6   Resistance Factors   *43*
2.7   Variability of Loads   *48*
2.8   Load Factors   *53*
2.9   Limit States   *55*
       References   *56*

Chapter Three

Loads and Load Factors   **58**

3.1   General Observations   *59*
3.2   Loads on Building Structures   *60*

    3.2.1   Dead Load   *61*
    3.2.2   Live Load   *61*
    3.2.3   Snow Load   *62*
    3.2.4   Wind Load   *63*
    3.2.5   Seismic Load   *63*
    3.2.6   Special Loads and Load Effects   *64*

3.3   Data on Building Loads   *65*

    3.3.1   Dead Load Data   *65*
    3.3.2   Live Load Data   *66*
    3.3.3   Snow Load Data   *69*
    3.3.4   Wind Load Data   *70*
    3.3.5   Seismic Load Data   *71*
    3.3.6   Data for Other Loads   *77*

3.4   Load Combinations   *78*
3.5   Analysis Factor   *80*
3.6   Design Load Factors   *81*
3.7   Computation of Factored Loads   *83*
       References   *85*

Chapter Four

Materials for Steel Structures   **87**

4.1   Scope of Chapter   *88*
4.2   Brief Historical Notes   *89*
4.3   Steel Production   *91*

4.4   Structural Steel Products   95
    4.4.1   Structural Shapes   95
    4.4.2   Tubular Products   97
    4.4.3   Plates and Bars   98
    4.4.4   Nomenclature for Structural Steel Products   103

4.5   Characteristics of Steel   105
    4.5.1   Chemical Composition   105
    4.5.2   Mechanical Properties   109

4.6   Grades of Structural Steel   115
    4.6.1   Grades of Steel for Shapes, Plates, and Bars   116
    4.6.2   Grades of Steel for Tubes and Pipe   119
    4.6.3   Grades of Steel for Sheet and Strip   121

4.7   Steels for Mechanical Fasteners   122
4.8   Steels for Welding Electrodes   125
    4.8.1   Shielded Metal Arc Welding Electrodes   126
    4.8.2   Submerged Arc Welding   128
    4.8.3   Gas Metal Arc Welding   128
    4.8.4   Flux Cored Arc Welding   129
    4.8.5   Base Metal and Weld Metal Matching   129

4.9   Steels for Stud Shear Connectors   130
4.10   Behavior of Steel at Extreme Temperatures   130
    4.10.1   High-Temperature Behavior   131
    4.10.2   Low-Temperature Behavior   134

4.11   Fatigue and Fracture   135
4.12   Corrosion Resistance and Structural Maintenance   138
4.13   Cost Factors   140
    References   141

Chapter Five

Tension Members   143

5.1   Introduction   144
5.2   Tension Members in Structures   144
5.3   Cross-Sectional Shapes for Tension Members   145
5.4   Behavior and Strength of Tension Members   148
    5.4.1   Load–Deformation Relationship   148
    5.4.2   Limit States of Strength   153

5.5   Computation of Areas   *156*
    5.5.1   Gross Area   *156*
    5.5.2   Net Area   *157*
    5.5.3   Influence of Hole Placement   *160*

5.6   Shear Lag   *166*
5.7   Block Shear   *168*
5.8   Eye-Bars and Pin-Connected Members   *178*
    Problems   *180*
    References   *182*

Chapter Six
Columns   183

6.1   The Structural Column   *184*
6.2   Factors Influencing Column Strength   *185*
    6.2.1   The Complexity of Column Behavior   *185*
    6.2.2   Residual Stress   *188*
    6.2.3   Initial Crookedness   *193*

6.3   Cross-Sectional Shapes for Columns   *196*
    6.3.1   Basic Selection Criteria   *196*
    6.3.2   Doubly Symmetric Cross Sections   *197*
    6.3.3   Singly Symmetric and Asymmetric Cross Sections   *198*
    6.3.4   Cross-Sectional Areas   *199*

6.4   Column Behavior   *200*
    6.4.1   Basic Column   *200*
    6.4.2   Elastic Behavior   *200*
    6.4.3   Columns with Initial Curvature or Eccentric Axial Load   *204*
    6.4.4   Inelastic Buckling Concepts   *206*
    6.4.5   Maximum Strength Theory   *211*
    6.4.6   Limit States Design Concepts   *212*

6.5   Column Length   *214*
    6.5.1   The Influence of Column Length   *214*
    6.5.2   Short Columns   *214*
    6.5.3   Intermediate-Length Columns   *215*
    6.5.4   Long Columns   *217*

6.6   Column Strength Approximations and Design Criteria   *218*
    6.6.1   Development of Design Criteria   *218*
    6.6.2   Limit States   *218*
    6.6.3   Variation of Column Strength   *219*

Contents   xi

6.7   LRFD Column Design   *219*
6.8   Effective Length of Columns   *223*
6.9   Columns with Slender Elements   *234*
  6.9.1   Plate Buckling   *234*
  6.9.2   Column Strength with Slender Elements   *236*
6.10   Torsional Buckling   *239*
  6.10.1   Limit States   *239*
  6.10.2   Design   *240*
6.11   Built-Up Columns   *243*
       Problems   *246*
       References   *248*

Chapter Seven

Bending Members   250

7.1   The Basic Structural Element   *251*
7.2   Behavior of Beams   *253*
  7.2.1   Local Buckling   *253*
  7.2.2   Lateral–Torsional Buckling   *254*
  7.2.3   Shear   *257*
7.3   Local Buckling   *258*
  7.3.1   Flange Local Buckling   *259*
  7.3.2   Web Local Buckling   *260*
7.4   Design Requirements for Laterally Supported Beams   *262*
7.5   Design Requirements for Laterally Unsupported Beams   *265*
7.6   Plastic Analysis   *273*
7.7   Special Requirements for Double-Angle and Tee Members   *277*
7.8   Serviceability Criteria for Beams   *279*
       Problems   *281*
       References   *283*

Chapter Eight

Plate Girders   284

8.1   Background   *285*
8.2   Homogeneous Plate Girders: Flexure   *286*
  8.2.1   Noncompact Web Plate Girders   *287*
  8.2.2   Slender Web Plate Girders   *290*

8.3   Homogeneous Plate Girders: Shear   *293*

8.3.1   Nontension Field Action: Shear   *293*
8.3.2   Tension Field Action: Shear   *294*
8.3.3   Intermediate Stiffeners: Tension and Nontension Field   *295*

Problems   *299*
References   *300*

Chapter Nine

Beam-Columns and Frame Behavior   **301**

9.1   Introduction   *302*
9.2   Second-Order Effects   *303*
9.3   Interaction Principles   *304*
9.4   Interaction Equations   *305*
9.5   Braced Frames   *307*
9.6   Unbraced Frames   *314*
9.7   Initial Beam-Column Selection   *319*
9.8   Combined Simple and Rigid Frames   *322*
9.9   Partially Restrained Frames   *329*
9.10   Bracing Design   *338*

9.10.1   Bracing: Member Support   *338*
9.10.2   Bracing: Frame Support   *340*

Problems   *342*
References   *349*

Chapter Ten

Composite Construction   **350**

10.1   Introduction   *351*
10.2   Advantages and Disadvantages of Composite Construction   *353*
10.3   Shored vs. Unshored Construction   *354*
10.4   Effective Flange   *355*
10.5   Strength of Composite Beams and Slab   *356*

10.5.1   Fully Composite Beams   *357*
10.5.2   Partially Composite Beams   *362*
10.5.3   Negative Moment Strength   *365*

10.6   Shear Stud Capacity   *366*

10.6.1   Shear Stud Placement   *367*

10.7   Composite Beams with Formed Metal Deck   *368*

    10.7.1   Deck Ribs Perpendicular to Steel Beam   *368*
    10.7.2   Deck Ribs Parallel to Steel Beam   *369*

10.8   Fully Encased Steel Beams   *374*
10.9   Selecting a Section   *374*
10.10   Serviceability Considerations   *375*

    10.10.1   Deflection During Construction   *375*
    10.10.2   Vibration Under Service Loads   *376*
    10.10.3   Live Load Deflections   *376*

10.11   Composite Columns   *377*
    Problems   *380*
    References   *381*

Chapter Eleven

Connections

382

11.1   Introduction   *383*
11.2   Basic Connection Behavior   *383*
11.3   Fully Restrained Connections   *384*
11.4   Partially Restrained Connections   *386*
11.5   Mechanical Fasteners   *388*

    11.5.1   Common Bolts   *388*
    11.5.2   High-Strength Bolts   *388*

11.6   Welding   *390*

    11.6.1   Welding Processes   *390*
    11.6.2   Types of Welds   *392*

11.7   High-Strength Bolted Shear Connections: Limit States   *394*
11.8   Weld Limit States   *398*

    11.8.1   Fillet Weld Strength   *398*
    11.8.2   Groove Weld Strength   *401*

11.9   Shear Connections: Design   *401*

    11.9.1   Behavior Requirements   *401*
    11.9.2   Double-Angle Connection: Bolted   *402*
    11.9.3   Double-Angle Connection: Welded   *408*
    11.9.4   Single-Plate Connections   *412*
    11.9.5   Seated Connection   *416*

11.10   High-Strength Bolts in Tension   *423*

    11.10.1   Behavior   *423*
    11.10.2   Design   *425*

11.11   High-Strength Bolts in Shear and Tension   *430*
        11.11.1   Behavior   *430*
11.12   Moment Connections   *433*
11.13   Column Reinforcement   *437*
        Problems   *445*
        References   *446*

Index   449

Illustrations

Chapter 1, page 1
Bell Atlantic Tower, Philadelphia, Pa.
Structural Engineer: CBM Engineers, Inc.
Architect: The Kling-Lindquist Partnership, Inc.
Photo courtesy CBM Engineers, Inc.

Chapter 2, page 23
Commerce Square, Philadelphia, Pa.
Structural Engineer: CBM Engineers, Inc.
Architect: Pei Cobb Freed & Partners
Photo courtesy CBM Engineers, Inc.

Chapter 3, page 58
Society Tower, Cleveland, Ohio
Photo ©, all rights reserved. Mort Tucker Photography Inc. and Associates.

Chapter 4, page 87
Citicorp Center, New York
Structural Engineer, LeMessurier & Assoc.
Architect: Hugh Stubbins and Associates
Photo courtesy New York Convention & Visitors Bureau

Chapter 5, page 143
Chrysler Building, New York
Architect: William Van Alen
Photo courtesy New York Convention & Visitors Bureau

Chapter 6, page 183
Empire State Building, New York
Architect: Shreve, Lamb, and Harmon
Photo courtesy New York Convention & Visitors Bureau

Chapter 7, page 250
420 Fifth Avenue, New York
Structural Engineer: Thornton-Tomasetti Engineers
Architect: Brennan Beer Gorman Architects
Photo by Roy Wright

Chapter 8, page 284
One Mellon Bank Center, Pittsburgh, Pa.
Structural Engineer: Thornton-Tomasetti Engineers
Architect: Welton Becket Associates
Photo courtesy Thornton-Tomasetti Engineers

Chapter 9, page 301
First City Tower, Houston, Texas
Structural Engineer: Walter P. Moore and Assoc.
Architect: Morris Aubry Architects
Photo courtesy Walter P. Moore and Assoc./Richard Payne

Chapter 10, page 350
Chicago's Magnificent Mile
Photo courtesy City of Chicago/Peter J. Schulz

Chapter 11, page 382
312 Walnut, Cincinnati, Ohio
Structural Engineer: Walter P. Moore and Assoc.
Architect: 3D/International and Glaser Associates
Photo courtesy Walter P. Moore and Assoc./Aker Photography

Preface

This book presents design of structural steel based on the principles of a limit states philosophy. Historically, structural design has used an allowable stress approach that has served the profession well and provided reasonably uniform techniques for design over the full range of materials used for building structures. With the publication in 1986 of the *Load and Resistance Factor Design (LRFD) Specification for Structural Steel Buildings* and the companion handbook, *Manual of Steel Construction— LRFD*, First Edition, by the American Institute of Steel Construction, the design of steel structures has been advanced to a new level. Although the procedures used in LRFD are somewhat different from those used previously in allowable stress design (ASD), the foundation upon which each philosophy has been based is the same: the actual behavior of structures composed of real structural steel elements. Following the lead set by the LRFD specification for a statistically based design, limit states specifications for other major materials used in the building industry have been forthcoming. A firm understanding of limit states design is just as important for the engineer of the future as an understanding of ASD has been in the past.

Limit states philosophies rely on an evaluation of the statistical properties of the variables affecting a design. With these as a base, design rules have been established that ensure that the ultimate capacity of the structure and its elements is greater than the applied loads, amplified by a sufficiently large load factor. Therefore, an acceptable level of safety in the final structure is established. Historically, loads and now load factors have been established by the appropriate building codes. Design specifications, such as the AISC LRFD specification, use the building code load factors as a starting point to present the material specific design rules. It is our intention to provide in this book the background material needed by designers to understand limit states design, followed by the information necessary to design steel building structures successfully according to the LRFD approach.

A discussion of the principles of structural design and the design philosophies associated with the structural design process is presented in Chapter 1. The presentation is taken further in Chapter 2 with the discussion of loads and load factors. This will give the engineer sufficient information to understand the basis of the limit states approach. It also provides an opportunity to see the foundation of the statistical methods used in the development of LRFD without the need for an extensive study of mathematical statistics. Loads on buildings and their load factors and combinations are presented in Chapter 3. These first three chapters are preliminary to the presentation of steel as a structural material found in Chapter 4.

The remainder of the book addresses the elements that make up a steel building structure and their integration into a complete structure. Tension members are treated first, in Chapter 5, along with the impact of connections on member strength. Chapter 6 then treats columns as axially loaded members. It first addresses behavior and then follows with a treatment of design. Bending members are covered in Chapter 7, where both strength and serviceability are treated, and in Chapter 8, where the special provisions for plate girders are presented. Individual elements are combined in Chapter 9 with the treatment of beam-columns and frame behavior, including partially restrained frames and combined simple and rigid frames. Composite construction is covered in Chapter 10 for both beams and columns. Connection behavior and the design of common connection types are detailed in Chapter 11.

It is assumed that the reader of this book will have access to the *Manual of Steel Construction—LRFD*, since the majority of the data needed to complete a steel structure design is found there; however, the manual is not necessary to the understanding of LRFD as presented here.

The authors acknowledge the valuable input provided by many colleagues throughout the preparation of this manuscript. They are particularly indebted to William J. Hall; Roger A. La Boube, University of Missouri—Rolla; and Charles W. Roeder, University of Washington, for their reviews of the manuscript and their insightful suggestions and recommendations. Particular thanks is expressed to Keith Grubb for his varied assistance throughout the project. In addition, our thanks are extended to those friends who assisted in providing photographs.

Louis F. Geschwindner □ Robert O. Disque □ Reidar Bjorhovde

Load and Resistance Factor Design
of Steel Structures

CHAPTER ONE

Introduction

1.1 SCOPE

A wide variety of design problems may be characterized with the label of structural steel design. Therefore, any book that attempts to cover the entire field will, by necessity, become quite voluminous. At the same time, it must be recognized that the ever-increasing degree of specialization within the profession makes such a monumental treatise unnecessary. It is becoming unusual to encounter an engineer who is equally well versed in the principles and methods of design, as applied to the specific tasks associated with bridge design, industrial structures design, and the design of the many types of building structures. Although the designer should be educated in the basic principles that would allow a professional to tackle any of these problems, the technical, legal, and political complexities of current practice are such that most engineers are required to limit their work to certain types of structures.

On the basis of these observations, this book will deal only with the design of steel structures for buildings. The design of members and connections as presented here will be applicable to all types of structures governed by the criteria of the *Specification for Load and Resistance Factor Design, Fabrication and Erection of Structural Steel for Buildings* [1]*, published by the American Institute of Steel Construction (AISC) in 1986. However, the types of problems and the areas of application that are given throughout the book have been keyed to steel building structures. The treatment of subjects generally associated with bridges and industrial structures, such as fatigue and fracture, for example, has been kept relatively brief.

The book will detail the concepts and design criteria of load and resistance factor design (LRFD) for steel structures. Although this is a relatively new approach in this country, design specifications for steel structures based on such concepts have been in use in many countries for some time. Whether referred to as LRFD, limit states design (LSD), or any of the other probability-based design standards, the basic principles are the same. This also applies to the strength design approach that has been used for reinforced concrete structures in the United States and Canada for some time.

References [2] to [11] detail the research work that led to the development of the LRFD specification [1]. References [12] to [16] represent specifications and related work that have been prepared in other countries, and Refs. [17] and [18] document the code for the design of reinforced concrete structures and some of the studies that were conducted in that connection.

---

* The italic numbers in brackets denote references, listed at the end of each chapter.

1.2 PRINCIPLES OF STRUCTURAL DESIGN

From the time an owner determines that there is a need for a building, through the development of conceptual and detailed plans, to completion and occupancy, a building project is a multifaceted task that involves many professionals. The owner and his financial analysis team evaluate the basic economic criteria for the building. The architects and engineers form the design team and prepare the initial proposals for the building, demonstrating how the users' needs will be met. This teamwork continues through the final planning and design stages, where the detailed plans, specifications, and contract documents are readied for the construction phase. During this process, input may also be provided by the individuals who will transform the plans into a real-life structure. Thus, those responsible for the construction phase of the project often assist in improving the design by taking into account the actual on-site requirements for efficient construction.

Once a project has been completed and turned over to the owner, the work of the design teams is normally over. The operation and maintenance of the building, although major activities in the life of the structure, are not usually within the province of the designers, except in those cases when significant changes in building use are anticipated. In such cases a design team should reenter the picture to verify that the proposed changes will be acceptable.

The basic goals of the design team can be summarized with the words *safety, function,* and *economy*. Thus, the building must be safe for its occupants and all others who may come in contact with it. It must fail neither locally nor overall, nor should it exhibit behavioral characteristics that test the confidence of rational human beings. To help achieve that level of safety, building codes and design specifications are published which outline the minimum criteria that must be met by any structure.

The building must also serve its owner in the best possible way to ensure that the functional criteria are met. Although structural safety and integrity are of paramount importance, a building that does not serve its intended purpose will not have met the goals of the owner.

Last, but not least, the design, construction, and long-term usage of the building should prove to be economical. The degree of financial success of any structure will depend on a wide range of factors. Some of these will have been established prior to the work of the design team, while others will be determined after the building is in operation. Nevertheless, the final design should, within all reasonable constraints, be the one that produces the lowest combined short- and long-term expenditures.

Numerous models have been proposed to describe the evolution of a building project with varying degrees of success. A recent draft proposal from the American Consulting Engineers Council (ACEC) [*19*] provides a useful breakdown for understanding the position of the structural engineer in a building project. The design phases presented in this model are (1) project definition, (2) schematic design, (3) design development, (4) contract documentation, and (5) construction administration.

The project definition phase provides an opportunity for the design team to set the bounds of the project, establishing the overall scope of the work that the structural engineer will be responsible for carrying out. The functional requirements are established and the overall constraints, such as time and budget, are set. It is this very important phase that sets the stage for the actual work that is normally thought of as structural engineering.

Schematic design is the initial planning stage, where the key is feasibility. For the structural engineer, schematic design consists of the development of preliminary framing schemes, lateral load resisting schemes, preliminary member sizes, and preliminary estimates of construction costs. These schemes are based on the constraints established during the project definition phase. The output from this phase is usually a set of documents that communicate to the other participants in the project the thinking of the structural designers. Once the schematic designs are reviewed and the overall feasibility is considered, a system is selected for further development.

The design development phase begins with the selected schematic design. It proceeds through continuous refinement of the schematic design to arrive at the final design. The refinement includes performing the iterative analyses and designs required to converge to a final solution. All materials must be selected, the members sized, and all design requirements satisfied. Perhaps the most critical aspect of this phase is the integration of the structural design with all of the other aspects of the building design, including the architectural requirements, the building environmental system needs, and the construction requirements.

The results of the work of the design development phase must be communicated to many individuals who were not directly involved in the evolution of the design. These include the owner, building code and other government officials, and a wide range of construction professionals. This communication is effected through the contract documents, which include the structural drawings and specifications. They are used by contractors to prepare estimates of costs of construction, by fabricators to prepare shop drawings, and by erectors to put the structure in place in the field. The documents must convey not only the overall intent of the structural system, but also the details of all aspects of the erected building. The contract documents are normally prepared by the members of the team responsible for the various portions of the design.

The final phase in a building project is the construction. For buildings framed by structural steel, this includes the preparation of shop drawings. These are produced by the steel fabricator to show how each individual piece of steel is to be cut and fitted for installation in the structure. The designer reviews these drawings to ascertain that the intent of the design has been adhered to.

Once the individual elements have been fabricated, the erection of these pieces into a unified structure takes place. During this stage, the design team will be involved in periodic visits to the site to review the progress of the construction, and to see that the intent of the design, as communicated through the contract documents, is carried out. The review may include such items as member sizes, material specifications, construction accuracy, and construction methods.

As can be seen from this brief review of the phases of a building project, the structural engineer is involved in all stages, from the inception of the project through the final erection. A quality finished building will result when all members of the design team participate fully in each step of the project, keeping the best interests of the owner and the public as their highest priority.

1.3 THE PARTS OF THE STEEL STRUCTURE

All structures incorporate some or all of the following basic types of structural components:

1. Tension members
2. Compression members
3. Bending members
4. Combined force members
5. Connections

The first four items of this list represent structural members. Connections (item 5) provide the contact regions between the structural members, and thus assure that all of the components work together as a unit.

Detailed evaluations of the strength, behavior, and design criteria for these members are presented in the respective chapters of this book, as follows:

Tension members: Chapter 5
Compression members: Chapter 6
Bending members: Chapters 7 and 8
Combined force members: Chapter 9
Connections: Chapter 11

The strength and behavior of frames composed of a combination of these elements are covered in Chapter 9, and the special considerations that apply to composite (steel and concrete) construction are presented in Chapter 10. The properties of the structural steel and the various shapes that are commonly used are detailed in Chapter 4. The philosophical and probabilistic bases of load and resistance factor design are detailed in Chapter 2, and the many types of loads for building structures are discussed in Chapter 3.

Tension members are most typically found as web and chord members in trusses and open-web steel joists; as diagonal members in structural bracing systems; and as hangers for balconies, mezzanine floors, and pedestrian walkways. They are also used as sag rods for purlins and girts in many building types, as well as to support platforms for mechanical equipment and pipelines. Figures 1.1 and 1.2 illustrate typical applications of tension members in actual structures.

In the idealized case, tension members transmit concentric tensile forces only. In certain structures, reversals of the overall load (wind, for example) may cause a change in the tension member force from tension to compression. Some members will be designed for this action; others will have been designed on the basis that they carry tension only.

The idealized tension member is analyzed on the assumption that its end connections are pins, which prevent any moment or shear force from being transmitted to the member. However, in an actual structure the type of connection will normally dictate that some bending may be introduced. This is also the case when the tension member is directly exposed to some form of lateral load. Moments will also be introduced if the element is not perfectly straight, or if the axial load is not applied along the centroidal axis of the member. On the whole, the primary force in the tension member is a concentric axial load, with secondary effects introduced as shear and bending.

Figure 1.1 Use of tension members in a truss. Courtesy BOVAY, Architects, Engineers and Planners.

Figure 1.2 Use of tension members as hangers.

Compression members are also known as columns, struts, and posts. They are used as web and chord members in trusses and joists, and as columns, struts, or posts (vertical members) in all types of building structures. Figure 1.3 shows a typical usage of structural compression members.

The idealized compression member carries only a concentric, compressive force. Its strength is heavily influenced by the distance between the supports, as well as by the support conditions. For these reasons the basic column is defined as an axially loaded member with pinned ends. The design rules are based on the behavior and strength of this pure compression member.

The basic column is practically nonexistent in real structures. There are many reasons for this. Realistic end supports rarely resemble perfect pins; the axial load will normally not be concentric, due to the way the surrounding structure transmits its load to the member; and beams and similar components are likely to be connected to the column in such a way that moments are introduced. All of these conditions produce bending effects in the member, making it a combined force component or beam-column. The primary force in the pinned-end column is therefore a con-

Figure 1.3   Use of columns in a
building frame.

centric axial compressive load, accompanied by the secondary effects of
bending due to a lack of straightness and nonlinear material and internal
stress properties.

Bending members are known as beams, girders, joists, spandrels or
spandrel beams, purlins, lintels, and girts. Although all are bending members, each name implies a certain structural function:

1. Beams and girders form part of common floor systems. The
   beams are most often considered as the members that frame into
   (i.e., are supported by) girders, which in turn are connected to
   and supported by the columns of the structure. In this fashion the
   girder may be considered as a higher order bending member than
   the beam. Naturally, there are variations to this basic scheme.

2. The bending members forming the perimeter of a floor plan in a
   multistory building are known as spandrels or spandrel beams.
   Their design is unique in that the load is transferred from only
   one side of the member, sometimes necessitating an analysis of
   torsional effects as well as those due to bending.

3. Bending members in roof systems are usually referred to as
   purlins.

4. Lintels are bending members that span across the top of door or window openings, carrying the load of the walls at such locations.

5. Girts are used in exterior wall systems, transferring the weight of the cladding and lateral (wind) load to the exterior columns.

Figure 1.4 shows the use of a variety of bending members in an actual structure.

The basic bending member carries transverse loads that lie in a plane that contains the longitudinal centroidal axis of the beam. The loads may be one or more concentrated forces, some form of distributed loads, or combinations of these two types. Bending action transfers the loads to the beam supports, where the reactions are taken by the elements to which the beam is connected.

Depending on the type of structure, the bending members may be simply supported or continuous over two or more spans. Naturally, this has a direct influence on the distribution of bending moments and shear forces, which normally control the size of the members. Economy, ease of construction, and the structural system usually dictate the choice of simple versus continuous beams.

This book deals only with straight, prismatic bending members. However, the LRFD specification [*1*] also provides design criteria for linearly tapered beams. The analysis of curved members is beyond the scope of this book.

The most common combined force member is known as a beam-column, implying that this is a structural element that is simultaneously subjected to bending and axial compression. Although bending and axial tension represent a realistic loading case for the combined force member,

Figure 1.4 Building structure showing bending members. Photo courtesy Greg Grieco.

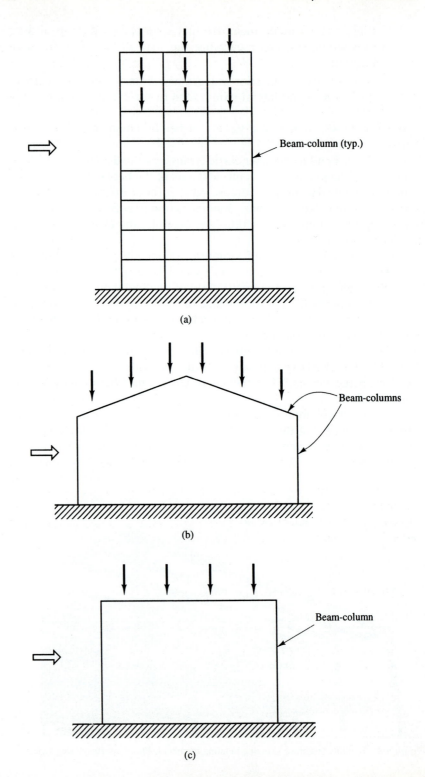

Figure 1.5 Steel frames in which the vertical members are subjected to axial loads and bending moments.

and is recognized as such through the criteria of the LRFD specification [1], the unique strength and stability problems of the beam-column warrant its separate treatment in both the LRFD specification and this text.

Figure 1.5a is a schematic illustration of a multistory steel frame, where, because of the geometric configuration of the frame as well as the types of connections and the loading pattern, the vertical members will be subjected to axial loads and bending moments. This is one typical case of practical beam-columns; other examples are the members of a gable frame (Fig. 1.5b) and the vertical components of a single story rigid frame (Fig. 1.5c).

The beam-column may be regarded as the general structural element, where axial force, shear forces, and bending moments are acting at the same time. The basic column may be thought of as a special case, representing a beam-column with no end moments or transverse loads. Similarly, the basic bending member is a beam-column with zero axial load. The considerations that have to be made in the analysis and design of columns and beams therefore also apply to beam-columns.

Because of the generalized nature of the combined force element, it is not possible to speak of primary and secondary load effects. However, it is correct to observe that column behavior will overshadow other influences when the ratio of axial load to axial load capacity becomes high. The same principle can be applied to the case of high moments, for which the beam behavior will outweigh other effects. The beam-column is therefore an element where different forces interact, and for that reason the practical design approaches are based on interaction equations. This is discussed in detail in Chapter 9.

A multitude of connections are used to join the various components of a steel structure. Whether they connect the axially loaded members in a truss or the beams and columns of a multistory frame, the connections ensure that all of the structural members will function as a unit, namely, the completed structure. The fastening elements of structural steel connections in today's construction are almost entirely limited to bolts and welds. Their behavior and strength are discussed in Chapter 11.

1.4 TYPES OF STEEL STRUCTURES

It is difficult to classify steel structures into neat categories, due to the wide variety of systems available to the designer. The elements of the structure, as defined in Section 1.3, must be combined in such a way as to form the total structure of a building. This combination of members is usually referred to as the framing system.

Steel-framed buildings come in a multitude of shapes, sizes, and combinations with other materials. A few examples are given to set the stage for the application of structural design that is presented in the subsequent chapters.

Bearing Wall Construction.   Primarily used for one- or two-story buildings, such as storage warehouses, shopping centers, office buildings, and schools, bearing wall construction normally utilizes brick or concrete block masonry walls, upon which are placed the ends of the flexural members supporting the floor or roof. The flexural members are usually hot-rolled structural steel shapes, alone or in combination with open-web steel joists or cold-formed steel shapes.

Beam-and-Column Construction.   This represents the most commonly used system for steel structures today. It is suitable for large-area buildings such as schools and shopping centers, which often do not have more than two stories, but which may have a great many spans. It is also suitable for buildings with a greater number of stories. The columns are placed according to a regular, repetitious grid, and beams, girders, and joists are used for the floor and roof systems. The regularity of the floor plan lends itself to economy in fabrication and erection, since most of the members will be of the same size. Further economy may be gained by using continuous beams or drop-in spans with cantilever beams, as illustrated schematically in Fig. 1.6. For multistory buildings with this type of construction, the use of composite (steel and concrete) flexural members affords additional savings. Further advances can be expected as designers become more conversant with the use of composite columns and other elements of mixed construction systems [25, 28].

Beam-and-column structures rely on either the connections or a separate bracing system to resist lateral loads. A frame in which all connections are moment resistant provides stability against the action of lateral loads (wind, earthquake) through the bending stiffness of the overall frame.

A frame without member end restraint needs a separate lateral load resisting system, often afforded by having the elements along one or more of the column lines act as braced frames. One of the most common types of bracing is the vertical truss, which is designed to take the loads imposed by wind and seismic action. Other bracing schemes involve end shear walls for the building, as well as interior, reinforced concrete cores. The latter type of solution is also known as a braced core system, and it can be highly efficient because of the rigidity of the box-shaped cross section of the core. The core serves a dual purpose in this case: In addition to providing the bracing system for the building, it also serves as

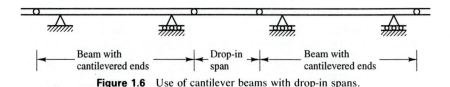

Figure 1.6   Use of cantilever beams with drop-in spans.

the vertical conduit in the completed structure for all of the necessary services (elevators, staircases, electricity and other utilities, etc.).

Combinations of types of construction are also common. For example, frames may have been designed as moment resistant in one direction of the building and as truss braced in the other. Of course, such a choice recognizes the three-dimensional nature of the structure. Figure 1.7 shows an idealized illustration of several types of beam-and-column framed structures.

Long-Span Construction.   This type of construction encompasses steel-framed structures with long spans between the vertical load-carrying elements, such as covered arenas. The long distances may be spanned by one-way trusses, a space truss, or plate and box girders. Arches or cables could also be used, although they will not be considered here.

Long-span construction is also encountered in buildings that require

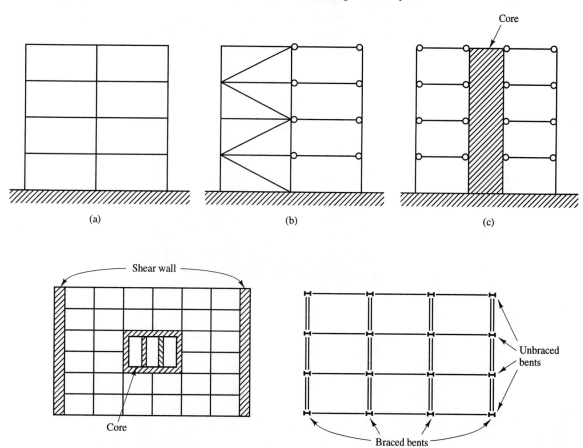

Figure 1.7   An idealized illustration of several types of beam-and-column framed structures: (a) moment-resistant frame; (b) truss-braced frame; (c) core-braced frame; (d) floor plan of shear wall and core-braced building; (e) floor plan of building with a combination of braced and unbraced bents.

large, column-free interiors. In such cases the building may be a core- or otherwise braced structure, where the long span is the distance from the exterior wall to the core.

Many designers would also characterize single-story rigid frames as part of the long-span construction systems. Depending on the geometry of the frame, such structures can span substantial distances, often with excellent economy.

High-Rise Construction.   High-rise construction refers to multi-story buildings. The large heights and unique problems that are encountered in the design of such structures warrant treating them independently from beam-and-column construction. In addition, the work of several designers over the past twenty years has resulted in a number of new concepts in multistory frame design [*23–25, 28*].

Particular care must be exercised in the choice and design of the lateral load resisting system. It is not just a matter of extrapolating from the principles used in the analysis of lower rise structures, since there are many effects that play a significant role for high-rise buildings, but have little or no impact on frames with a smaller number of levels. These effects are crucial to the proper design of the high-rise structure. They are referred to as second-order effects, since they cannot be quantified through a normal, linearly elastic analysis of the frame.

Figure 1.8   High-rise buildings: (a) the John Hancock Center; (b) the World Trade Center, courtesy New York Convention and Visitors Bureau.

An example of second-order effects is the additional moment induced in a column due to the eccentricity of the column loads that
develops when a structure is displaced laterally (the structure sways or
drifts). When added to the moments and shear forces that are produced by
the gravity and wind loads, the resulting effects may be significantly
larger than those that were computed without the second-order analysis.
A designer who does not incorporate both may be making a serious,
unconservative error.

The framing systems for high-rise buildings reflect the increased
importance of lateral load resistance. Thus, attempts at making the perimeter of a building act as a unit or tube have been proven quite successful.
This tube may be in the form of a truss, as in the John Hancock Center in
Chicago, Illinois (Fig. 1.8a) or a frame, as in the World Trade Center in
New York City (Fig. 1.8b); a solid wall tube with cutouts for windows, as
in the Amoco building in Chicago (Fig. 1.9a); or several interconnected or
bundled tubes, the most famous example being the Sears Tower in Chicago (Fig. 1.9b), which at 1450 feet (110 stories) is the tallest building in
the world today.

Single-Story Construction.   Many designers would include the
single-story frame as part of the long-span construction category. These
structures lend themselves particularly well to fully welded construction.

Figure 1.9   High-rise buildings: (a) the Amoco building; (b) the Sears Tower.

The pre-engineered building industry has capitalized on the use of this system through fine-tuned designs of frames for storage warehouses, industrial buildings, temporary and permanent office buildings, and similar types of structures.

1.5 DESIGN PHILOSOPHIES

A successful structural design is one that results in a structure that is safe for its occupants, can carry the design loads without overstressing any components, does not deform excessively, and is economical to build and operate for its intended life span. Although economy may appear to be the primary concern of an owner, it is difficult to consider price factors when developing design criteria. Costs of labor and materials will vary from one geographic location to another, making it almost impossible to devise a uniform set of provisions that will result in a structural solution that is equally economical in all locations. For these reasons, and because the foremost task of the designer is to produce a safe and serviceable structure, design criteria are usually based on strictly technical models and considerations.

To carry out a structural design, it is necessary to quantify the causes and effects of the loads that will be exerted on each element throughout the life of the structure. This is generally termed the load effect or the required resistance. At the same time, it is necessary to account for the behavior of the material and the shapes that go into making up these elements. This is normally referred to as the design strength or resistance.

Structural design in the simplest terms may be defined as the determination of member sizes and connectivity, so that the resistance of the structure is greater than the load effect. The degree to which this is accomplished is often termed the margin of safety. Numerous approaches for accomplishing this goal have been used over the years.

Although past experience might seem to indicate that the structural designer knows exactly the magnitude of the loads that are applied to the structure and exactly the strength of all of the elements, this is usually not the case. Design loads are provided by many standards and codes [32–36], and although the values that are given are specific, significant uncertainty attaches to the magnitudes. These considerations are addressed in detail in Chapter 3.

As is the case for loading, there is significant uncertainty associated with determination of the behavior and strength of the structural members. The true indication of load-carrying capacity is given by the magnitude of the load that causes the failure of a component or the structure as a whole. Failure may occur either as a physical collapse of a part of the building or it may be construed as having occurred if deflections, for instance, are larger than certain predetermined values. Whether the failure occurs as a lack of strength or stiffness, it is understood that these phenomena reflect limits of behavior or strength of the structure. For that reason it is said that the structure has reached a specific limit state. A strength failure is termed an ultimate limit state, while a failure to meet operational requirements is termed a serviceability limit state [2, 3].

Regardless of the approach to the design problem, the goal of the designer is to ensure that the load on the structure and its resulting load effect, such as bending moment, shear force, and axial force, are, in all cases, sufficiently below all of the applicable limit states. This insures that the structure meets the requisite level of safety or reliability.

There are four fundamental approaches in use today for the design of steel structures. Each represents a different way of formulating the same problem, and each has the same goal. The four approaches are:

1. Allowable stress design (ASD)
2. Plastic design (PD)
3. Load factor design (LFD)
4. Load and resistance factor design (LRFD)

Allowable Stress Design.   Also called working stress design, ASD is the historical approach to structural design. It is currently used in the design of wood structures, it is an alternative method for reinforced concrete design [17], and it is an acceptable approach to the design of steel structures [29]. ASD is based on the concept that the maximum stress in a component is not to exceed a certain permitted or allowable value under normal service conditions. The load effects (moments, etc.) are determined on the basis of an elastic analysis of the structure, while the allowable stress incorporates a factor of safety (FS). The magnitude of the FS and the resulting allowable stress depend on the particular governing limit state, against which the design must produce a certain margin of safety. This may be expressed in general terms as

$$\text{Allowable stress} = \frac{\text{Governing limit stress}}{\text{Factor of safety}} \qquad (1.1)$$

The governing limit stress depends on the structural element type and stress condition being considered. For any single element, there may be a number of limiting stresses that must be checked.

The factor of safety that is implied in each of the present allowable stress design criteria varies significantly. It is a function of both the material and the particular component being considered. It may be influenced by factors such as member length, load source, member behavior, and anticipated quality of workmanship. The FS is an empirical, experience-based factor, and although ASD-designed structures have performed adequately over the years, the actual level of safety is not known. This is the prime drawback of allowable stress design.

Plastic Design. Plastic design is an approach that has been available as an optional method for steel design since 1961, when it was introduced as Part 2 of the ASD specification. The limiting condition for the structure and its members is the attainment of the load that would cause plastic collapse, usually called the ultimate strength or the plastic collapse load. For the individual member this means that its plastic moment capacity has been reached. However, due to the ductility of the material and the member, the ultimate strength will normally not have been reached at this stage: Additional load on the structure is taken by the less highly stressed members until so many members have exhausted their individual capacities that no further redistribution is possible. This is associated with the formation of a plastic hinge mechanism [*25, 30*].

The plastic collapse load equals the service load multiplied by a certain load factor. The limit state for a structure that is designed according to the principles of plastic design is therefore the attainment of a mechanism. All of the structural members must be able to develop the full yield stress in all fibers at the most highly loaded locations.

There is a fine line of distinction between the load factor of PD and the factor of safety of ASD. The former is the ratio between the plastic collapse load and the service (specified) load for the structure, while the latter is an empirically developed, experience-based term that represents the relationship between the elastic strength of the structure and the various limiting conditions for the components. Although numerically close, the load factor of plastic design and the factor of safety of allowable stress design are not the same parameter.

Load Factor Design. Plastic design is actually a special case of load factor design (LFD). LFD has been used in European countries for years, and American bridge designers have been able to take advantage of this method as well [*31*]. Essentially, this is a method that takes into account a measure of the variability of the loads that may be imposed on the structure. It also considers the probabilities of occurrence of load combinations, so that a smaller load factor is applied if an unusual set of loads are combined. In effect, LFD may be considered a step in the development of a more comprehensive limit states design procedure. In other words, LFD is a forerunner of the present probability-based design methods.

Load and Resistance Factor Design. Also known as limit states design, LRFD incorporates explicitly the effects of the random variability of strength and load. A design may be based on either elastic or ultimate strength analyses. Ultimate limit states are checked for strength considerations, while serviceability limit states are checked for actual service conditions. Because this method covers the effects of random variations of both strength and loads and formulates the safety criteria on that basis, it is possible to arrive at a nearly uniform level of safety for the structure and all of its components. This is not possible with ASD and PD, for example, for which the actual safety levels vary within wide margins.

The concepts and principles of LRFD are discussed in detail in Chapter 2. Suffice it to observe at this stage that all design methods in effect are subsets of limit states design, using various forms of simplification and modification. The most advanced form covers the variability of loads and strengths, and is represented by what is now known as load and resistance factor design in the United States; it is termed limit states design in Canada and many other countries.

1.6 STRUCTURAL SAFETY

From the preceding discussions of design philosophies it is understood that although the basic goal of any design process is to assure that the end product will be a safe and reliable structure, the ways in which this is achieved may vary substantially.

In the past, the primary goal for safety was to ensure that there would be an adequate margin against the consequences of overload. Load factor design and its offshoots were developed to take these considerations into account. In real life, however, a great many other factors play a role as well. These include, but are not limited to, the following:

1. Variations of material strength
2. Variations of cross-sectional size and shape
3. Accuracy of method of analysis
4. Influence of workmanship in shop and field
5. Presence and variation of residual stresses
6. Lack of member straightness
7. Variations of locations of load application points

These items consider only some of the sources of variation of the strength of a structure and its components. An even greater source of variation is the loading, which is further complicated by the fact that different types of load have different variational characteristics.

It is readily seen that a method of design that does not attempt to incorporate the effects of strength and load variability on structural safety will not only be riddled with sources of uncertainty, but more importantly with sources of uncertainty that are unaccounted for. These represent the real danger to safety, because neither have they been identified nor have their effects been quantified. The realistic solution is, therefore, to deal with safety as a probabilistic concept. This is the very foundation of load and resistance factor design, where the probabilistic characteristics of load and strength are evaluated, and the resulting safety margins are determined statistically. The method recognizes that there is always a finite, albeit very small, probability that structural failure will actually take place. However, the method does not attempt to attach specific values to this probability. The important point in LRFD is that the variabilities are considered, and that uniform levels of structural reliability are obtained. No probability of failure is specified or implied.

1.7 BUILDING CODES AND DESIGN SPECIFICATIONS

The design of building structures is regulated by a number of official, legal documents that are known by their common name of building codes. These cover all aspects of the design, construction, and operation of buildings, and are not limited to just the structural design aspects.

At this time there are three model codes in use in the United States [32–34]. These have been published by private organizations and are adopted, in whole or in part, by state and local governments as the legal requirements for structures that are to be built within their area of jurisdiction.

To the structural engineer, the most important sections of a building code are those dealing with the service loads that must be used in the design, and the requirements that pertain to the design of structures of specific materials. The service loads are usually taken from the national standard published by the American Society of Civil Engineers (ASCE) [35]. (This standard was previously issued by the American National Standards Institute (ANSI), but was taken over by ASCE in the 1980s.) Additions are sometimes made to the load standard by local building authorities, to account for unique conditions such as high winds, snow loads, or seismic effects.

The AISC specifications [1, 29] are normally incorporated by reference in the building code(s). The specifications therefore become part of the code, and thus part of the legal requirements of any locality where the model code is adopted. Naturally, this does not remove the responsibility for the design from the engineer. Regardless of the specification rules, the engineer must be satisfied that the structure can carry the intended loads safely, without endangering the occupants, and that the owner's functional requirements have been met.

A final observation: The words *code* and *specification* are often used to mean the same thing. It must be understood that normally the specification will be a subset of the code, primarily because a specification deals with only one area of the concerns of a building code. Some confusion also exists because of the several codes that are available in the United States. Other countries usually have only one building code and one specification for structural steel design. The closest and most obvious example is Canada, whose code [36] and specification [12] are highly regarded by professionals around the world.

REFERENCES

[1] American Institute of Steel Construction (AISC), *Specification for Load and Resistance Factor Design, Fabrication and Erection of Structural Steel for Buildings*, AISC, Chicago, IL, 1986.

[2] Ravindra, M. K., and Galambos, T. V., Load and resistance factor design for steel, *Journal of the Structural Division, ASCE,* Vol. 104, No. ST9, September 1978 (pp. 1337–1354).

[3] Galambos, T. V., Load and resistance factor design, *Engineering Journal, AISC,* Vol. 18, No. 3, Third Quarter, 1981 (pp. 74–82).

[4] Yura, J. A., Galambos, T. V., and Ravindra, M. K., The bending resistance of steel beams, *Journal of the Structural Division, ASCE,* Vol. 104, No. ST9, September 1978 (pp. 1355–1370).

[5] Bjorhovde, Reidar, Galambos, T. V., and Ravindra, M. K., LRFD criteria for steel beam-columns, *Journal of the Structural Division, ASCE,* Vol. 104, No. ST9, September 1978 (pp. 1371–1387).

[6] Cooper, P. B., Galambos, T. V., and Ravindra, M. K., LRFD criteria for plate girders, *Journal of the Structural Division, ASCE,* Vol. 104, No. ST9, September 1978 (pp. 1389–1407).

[7] Hansell, W. C., Galambos, T. V., Ravindra, M. K., and Viest, I. M., Composite beam criteria in LRFD, *Journal of the Structural Division, ASCE,* Vol. 104, No. ST9, September 1978 (pp. 1409–1426).

[8] Fisher, J. W., Galambos, T. V., Kulak, G. L., and Ravindra, M. K., Load and resistance factor design criteria for connectors, *Journal of the Structural Division, ASCE,* Vol. 104, No. ST9, September 1978 (pp. 1427–1441).

[9] Ravindra, M. K., Cornell, C. A., and Galambos, T. V., Wind and snow load factors for use in LRFD, *Journal of the Structural Division, ASCE,* Vol. 104, No. ST9, September 1978 (pp. 1443–1457).

[10] Galambos, T. V., and Ravindra, M. K., Properties of steel for use in LRFD, *Journal of the Structural Division, ASCE,* Vol. 104, No. ST9, September 1978 (pp. 1459–1468).

[11] Ellingwood, B. R., Galambos, T. V., MacGregor, J. G., and Cornell, C. A., *Development of a Probability Based Load Criterion for American National Standard A58*, NBS Special Publication No. 577, National Bureau of Standards (now the National Institute for Standards and Technology), Washington, DC, June 1980.

[12] Canadian Standard Association (CSA), *Steel Structures for Buildings—Limit States Design*, CSA Standard No. CAN3-S16.1-89, CSA, Rexdale, Ontario, Canada, 1989.

[13] Commission of the European Communities (EC), *Eurocode No. 3: Design of Steel Structures*, Edited Draft, Issue 3, Brussels, Belgium, April 1990.

[14] Norwegian Standards Association (NSA), *Stalkonstruksjoner: Beregning og Dimensjonering* (Steel Structures: Analysis and Design), Norwegian Standard No. NS 3472, NSA, Oslo, Norway, May 1973.

[15] Schweizerischer Ingenieur und Architektenverein (SIA), *Steel Structures*, Schweizer Norm (Swiss Standard) No. SN 161, Zurich, Switzerland, April 1979.

[16] Marek, P., *Limit States Design of Steel Structures* (in Czech), R. D. Jesenik, Praha, Czechoslovakia, 1981.

[17] American Concrete Institute (ACI), *Building Code Requirements for Reinforced Concrete*, ACI Standard No. 318-86, ACI, Detroit, MI, 1986.

[18] MacGregor, J. G., Safety and limit states design for reinforced concrete, *Canadian Journal of Civil Engineering,* Vol. 3, No. 4, December 1976 (pp. 484–513).

[19] American Consulting Engineers Council (ACEC), *National Practice Guidelines for the Structural Engineer of Record*, Interim Draft, Coalition of American Structural Engineers National Guidelines Committee, ACEC, Washington, DC, October 1988.

[20] Bjorhovde, R., Brozzetti, J., and Colson, A., Eds., *Connections in Steel Structures: Behavior, Strength and Design*, Proceedings of International Workshop in Cachan, France, May 1987; Elsevier Applied Science, London, 1988.

[21] Bjorhovde, R., Colson, A., Haaijer, G., and Stark, J. W. B., Eds., *Connections in Steel Struc-

tures: Behavior, Strength and Design, Proceedings of Second International Workshop in Pittsburgh, Pennsylvania, April 1991; American Institute of Steel Construction, Chicago, IL, 1991.

[22] American Society of Civil Engineers (ASCE), *Planning and Environmental Criteria for Tall Buildings*, Monograph on Planning and Design of Tall Buildings, Vol. PC, ASCE, New York, 1978.

[23] American Society of Civil Engineers (ASCE), *Tall Building Systems and Concepts*, Monograph on Planning and Design of Tall Buildings, Vol. SC, ASCE, New York, 1980.

[24] American Society of Civil Engineers (ASCE), *Tall Building Criteria and Loading*, Monograph on Planning and Design of Tall Buildings, Vol. CL, ASCE, New York, 1980.

[25] American Society of Civil Engineers (ASCE), *Structural Design of Tall Steel Buildings*, Monograph on Planning and Design of Tall Buildings, Vol. SB, ASCE, New York, 1979.

[26] Gaylord, E. H., and Gaylord, C. N., *Structural Engineering Handbook*, 3d ed., McGraw-Hill, New York, 1989.

[27] Steel Joist Institute (SJI), *Standard Specifications for Open Web Steel Joists, Longspan Steel Joists, and Deep Longspan Steel Joists*, SJI, Richmond, VA, 1978.

[28] Iyengar, S. H., *Composite or Mixed Steel-Concrete Construction for Buildings*, Special Publication, ASCE, New York, 1977.

[29] American Institute of Steel Construction (AISC), *Specification for the Allowable Stress Design, Fabrication, and Erection of Structural Steel for Buildings*, AISC, Chicago, IL, July, 1989.

[30] Disque, R. O., *Applied Plastic Design in Steel*, Van Nostrand-Reinhold, New York, 1971.

[31] American Association of State Highway and Transportation Officials (AASHTO), *Standard Specifications for Highway Bridges*, 14th ed., AASHTO, Washington, DC, 1989.

[32] Building Officials and Code Administrators International (BOCA), *The BOCA National Building Code*, 11th ed., BOCA, Country Club Hills, IL, 1990.

[33] International Conference of Building Officials (ICBO), *Uniform Building Code*, 1988 ed., ICBO, Whittier, CA, 1988.

[34] Southern Building Code Congress International (SBCCI), *The Standard Building Code*, SBCCI, Birmingham, AL, 1988.

[35] American Society of Civil Engineers (ASCE), *Minimum Design Loads for Buildings and Other Structures*, ASCE Standard No. 7-88, ASCE, New York, 1988.

[36] National Research Council of Canada (NRC), *National Building Code of Canada*, NRC Associate Committee on the National Building Code, Ottawa, Ontario, Canada, 1990.

CHAPTER TWO

Load and Resistance Factor Design

2.1 STRUCTURAL SAFETY

As observed in Chapter 1, one of the primary goals of a structural design is to assure that the structure will perform adequately throughout its intended life, providing comfort and safety for the occupants. Provided that all of the conditions that form the basis for a design are met, there will be an acceptably small probability that the structure will fail or otherwise malfunction at some point during its intended life. It should be clearly understood that no design philosophy is founded on an absolute guarantee that failure cannot take place. Current methods differ widely in their approach to structural safety, but common to all is the realization that there is always a possibility that failure will occur.

Allowable stress design is based on deterministic principles, where there is a definite, one-to-one correlation between input and output. The design is based on "known" load values, material and member properties, and member strength and behavior characteristics while the methods of analysis are based on a direct relationship between external and internal forces and deformations. Generally, an analysis will be more accurate the more force and deformation effects that are taken into account.* Therefore, the real problem of a deterministic design is that the true uncertainty of the solution is not known. All of the built-in safety, therefore, may be wiped out in a single, random application of overload, or a single, random occurrence of an understrength member at a critical location.

The random nature of strength, load, and structural safety is consequently not recognized in deterministic design approaches. In addition, the factor of safety may give a false impression of safety, not only because it is often based on experience alone, but also because it may give contradictory results. Consider the following example: The strength of a structure is indicated by R, and the governing load effect by Q. For a safe structure it is required that

$$Q < R \tag{2.1}$$

or, in other words, that the load effect must be less than the strength. The limiting case of a safe structure, the limit state, is reached when

$$Q = R \tag{2.2}$$

Equation (2.2) expresses a factor of safety of 1.0, and is obviously not used in practice. In such instances it is required that the factor of safety is a value larger than 1.0, implying a corresponding margin of safety.

---

* An analyst's claim of having perfected the "exact" analysis of a certain problem is an engineering contradiction in terms: The analysis is only exact insofar as it satisfies the assumptions that have been made.

Consider now a structure being exposed to an overload in the amount of ΔQ, at the same time that it has a strength deficiency of ΔR. Equation (2.1) must still be satisfied, as follows:

$$Q + \Delta Q \leq R - \Delta R \qquad (2.3a)$$

and the limit state is reached when the equality sign applies. Equation (2.3a) can be reformulated to

$$Q\left(1 + \frac{\Delta Q}{Q}\right) \leq R\left(1 - \frac{\Delta R}{R}\right) \qquad (2.3b)$$

If we define the factor of safety required to satisfy the inequality as the radio of strength to load, in the limit Eq. (2.3b) becomes

$$\text{Required factor of safety (FS)} = \frac{R}{Q} = \frac{1 + \Delta Q/Q}{1 - \Delta R/R} \qquad (2.4)$$

For the following numerical examples it will be assumed that strength and load are statistically independent. This is a common and acceptable assumption for most problems involving the statistical treatment of structural engineering problems.

Case 1.   The structure has a strength deficiency of 15 percent, occuring with a relative frequency of 1/1000, as well as an overload of 40 percent, occurring with the same frequency. The data are therefore

$$\Delta R_1 = 0.15$$

$$\Delta Q_1 = 0.40$$

The joint probability of having these two values of ΔR and ΔQ occur simultaneously is therefore $(1/1000) \times (1/1000) = 1 \times 10^{-6}$. By Eq. (2.4), the required factor of safety for this case becomes

$$(\text{FS})_1 = \frac{1 + 0.40}{1 - 0.15} = \frac{1.40}{0.85} = 1.65$$

which seems like a perfectly reasonable value for this combination of overload and understrength. [1]

Case 2.   The structure has a strength deficiency of 25 percent with a relative frequency of 1/1000, and an overload of 40 percent with the same relative frequency. For this case the data are

$$\Delta R_2 = 0.25$$

$$\Delta Q_2 = 0.40$$

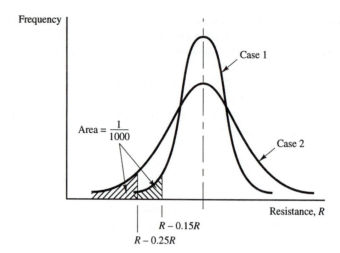

Figure 2.1 Case 1 distribution vs. case 2 distribution.

The joint probability of ΔR_2 and ΔQ_2 is 1×10^{-6}, as was also found for case 1. However, it is observed that since $\Delta R_2 > \Delta R_1$, but both occur with the same frequency, the spread or variability of the case 2 distribution is greater than that of case 1. Figure 2.1 illustrates this point.

The required factor of safety for this case becomes

$$(FS)_2 = \frac{1 + 0.40}{1 - 0.25} = \frac{1.40}{0.75} = 1.87$$

It is therefore found that the required factor of safety of case 2 is significantly higher than that of case 1, as a result of the fact that the strength variability of case 2 is much larger. Since the allowable stress design for each case would have been the same, based on R and Q, it is possible that cases with the same probability of occurring would have different factors of safety. The factor of safety approach, then, does not produce consistent safety for the structure and its components. As will be demonstrated, a more rational approach to achieve logical and consistent results is to recognize that strength and load are random variables, and to treat the safety concept on a probabilistic basis.

2.2 THE PHILOSOPHY OF LRFD

Bearing in mind the general definition of a limit state as it was used in the first part of the chapter, the philosophy of load and resistance factor design can be expressed as follows:

Applying the principles of probability theory and statistics to the analysis of load and strength (= resistance), LRFD is based on the concept that the structure will not exceed the limit states that govern its strength and behavior for any realistic load or load combination.

In essence, LRFD and limit states design (LSD) are one and the same; the former name is used in the United States and the latter in a number of other countries. In principle, however, LRFD should be regarded as a subset of LSD; in reality all methods of design are representative of some form of LSD. Current practical usage treats the two as one and the same.

The limit states considered for structural steel design are discussed in detail in Section 2.9. For now it is sufficient to observe that a distinction is typically made between ultimate limit states (ULS) and serviceability limit states (SLS). Ultimate limit states represent strength conditions for the structure, and exceeding a ULS implies a local or overall structural failure. Exceeding a serviceability limit state, on the other hand, will normally not entail any danger to the structure, since it simply means that it is not behaving or serving the way it was intended. The ULS, therefore, relates to maximum strength; the SLS considers the performance under normal operating conditions.

Variations of strength are taken into account through the factored resistance equation:

$$R = \phi R_n \qquad (2.5)$$

where R is factored resistance, ϕ is the resistance factor, and R_n is nominal resistance. The resistance factor is also called the performance factor (Canada) and capacity reduction factor (reinforced concrete design, ACI Code) and covers the random strength variations that are likely to be encountered in real life. The resistance factor is usually less than 1.0. The nominal resistance is often referred to as the "handbook" value of the component strength, since it is based on the nominal (=published) material properties, shape dimensions, and so on. For example, the nominal resistance of a compact beam is given by

$$R_n \equiv M_p = ZF_y \qquad (2.6)$$

where M_p is the fully plastic moment, Z is the plastic section modulus of the beam shape, based on the manufacturers' data for the dimensions, and F_y is the specified minimum yield stress of the steel, based on the respective material standard [2].

The variations of the loads are taken into account through the factored load or required resistance equation:

$$\text{Design load} = Q = f(D,\ L,\ W,\ T) = \sum_{i=1}^{k} \gamma_i Q_{ni} \qquad (2.7)$$

where $f(D,\ L,\ W,\ T)$ shows that the design load is a function of the superimposed dead, live, lateral (=wind or earthquake), and temperature effect loads, respectively. The load factor γ_i $(i = 1$ to $k)$ for load type i takes into account the random variation of the load, as well as the fact that the influence of load variations may vary. For example, when loads act in combination it is less likely that all of them will assume their maximum values simultaneously. The values of the load factors are generally greater than or equal to 1.0, depending on the type of load and load combination, and whether the analysis deals with strength or serviceability.

The term Q_{ni} in Eq. (2.7) represents the general nominal load effect, which may be a bending or torsional moment, a shear force, or an axial force. The superimposed loads are translated into load effects by multiplication with an influence coefficient. This is what is normally thought of as structural analysis. For instance, the maximum moment in a uniformly loaded, simply supported beam is $wl^2/8 = (D + L)sl^2/8$, where s is the spacing of adjacent beams. The influence coefficient is the term $sl^2/8$. (The magnitude of L in this case would incorporate the effect of tributary area size through the live load reduction factor. This is discussed in Chapter 3.) The nominal or service loads, D, L, W, and T, are usually determined from the governing building code as dicussed in Chapter 1.

Substituting Eqs. (2.5) and (2.7) into Eq. (2.1), we arrive at the basic LRFD design criterion:

$$\sum_{i=1}^{k} \gamma_i Q_{ni} \leq \phi R_n \tag{2.8}$$

which simply says that the factored load effect must be less than or equal to the factored resistance. This is the limit states concept expressed in a single equation, and is equally applicable to ultimate and serviceability limit states.

A final observation regarding the use of probability theory and statistics should clarify the concerns of many who are not familiar with the application of LRFD. The designer will never have to make statistical computations to make use of LRFD. It is sufficient to know that such mathematical tools were used to develop the method. Naturally, it is always an advantage to be thoroughly familiar with the background of a design specification. The better the specification is understood, the better will be the resulting design. The remainder of this chapter, therefore, addresses the probabilistic development of the LRFD specification.

2.3 PRINCIPLES OF LRFD

It is recognized that load and resistance are both random variables, since their magnitude at any given time cannot be predicted with certainty. Through measurements taken over a long period of time a certain data base can be established, which enables the user to see how the values vary, as well as whether there are any numbers that have a tendency to occur more frequently than others. In other words, the frequency distributions for loads and resistances can be established, but only after costly and time-consuming observations have been made.

Although theoretically desirable, it has been shown to be neither practical nor strictly necessary to obtain the complete distributions. Rather, data are needed on the central tendencies and variabilities of the random variables of interest. The average or mean is representative of the former, while the variance or second central moment and its derived terms (standard deviation, coefficient of variation) provide information about the latter. These parameters lend themselves to mathematical treatment easier than variables such as the median or the mode of a distribution of random variables [1].

The analytical model that was used to develop the LRFD criteria stems from the work of Cornell [3], Lind [4], and several others. The research and development work that led up to the LRFD specification was performed by Galambos and his associates.* The model is based on a first-order second-moment (FOSM) probabilistic approach, where only the first two moments of the random variables are used to develop the statistical properties of strength, load, and structural safety. Although research is continuing to produce more refined methods of analysis, the state-of-the-art in probabilistic design is properly represented by the FOSM methods.

The probabilistic concepts of LRFD are illustrated in the following. It is assumed that the strength of the structure can be represented by a single random variable, R. In the same fashion, the load effect is given as the single random variable Q. These variables are assumed to be statistically independent [1], which is satisfactory for most structural engineering applications. Their respective variation is best illustrated by showing the frequency distribution curves for R and Q, or, as they are most properly called, the probability density functions for the strength and the load effect. Since structural safety is the primary concern, the two curves are most conveniently given in the same diagram, as shown by Fig. 2.2.

For a safe structure it is required that $Q < R$ (Eq. (2.1)). Conversely, the structure will fail if $Q \geq R$, which is represented by the area where the

* These primary references were presented in Chapter 1 as [2–10] and should be drawn upon in addition to the references cited in this chapter.

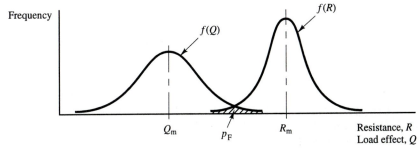

Figure 2.2   Probability density functions for strength and load effect.

two curves overlap in Fig. 2.2. Since Q and R are random variables, there is always a chance that the load effect will be greater than the strength. If the equations for the variation of Q and R are known, then the size of the overlap area can be determined. Since the size of the overlap gives the probability that $Q \geq R$, it is called the probability of failure p_f, and is expressed as

$$p_f = P(Q \geq R) \tag{2.9}$$

Instead of dealing with separate frequency distribution curves for Q and R, it is usually more convenient to define a new variable as $(R - Q)$; the safety criterion becomes

$$\text{Safe structure: } (R - Q) > 0 \tag{2.10a}$$

$$\text{Failed structure: } (R - Q) \leq 0 \tag{2.10b}$$

The distribution characteristics for $(R - Q)$ are directly derived from the data for R and Q. For example, the mean value is found as

$$(R - Q)_m = R_m - Q_m \tag{2.11}$$

where the subscript m indicates that the variable is at its mean value. Equation (2.11) holds true for all independent variables; the frequency distribution of $(R - Q)$ has properties that are derived from those of R and Q. Methods for the determination of the characteristics of the function $f(R - Q)$ are given in detail in Ref. [1] and many other books.

The failure criterion, Eq. (2.9), can now be expressed as

$$p_f = P[(R - Q) \leq 0] \tag{2.12}$$

which gives

$$p_f = P\left[\left(\frac{R}{Q} - 1\right) \leq 0\right] = P\left[\left(\frac{R}{Q}\right) \leq 1\right] \tag{2.13}$$

A further revision is now introduced by taking the natural logarithm of the expression inside the brackets:

$$p_f = P\left[\ln\left(\frac{R}{Q}\right) \leq 0\right] \tag{2.14}$$

since $\ln 1 = \exp(\ln 1) = 0$. (The expression $\exp(\cdots)$ is the same as $e^{\cdots}$.)

The distribution characteristics and the relevant parameters of $(R - Q)$ and $\ln(R/Q)$ are given in Figs. 2.3 and 2.4, respectively. The latter will be used in the following developments, because it turns out that

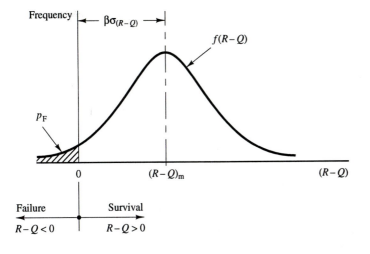

Figure 2.3   Characteristics of $(R - Q)$.

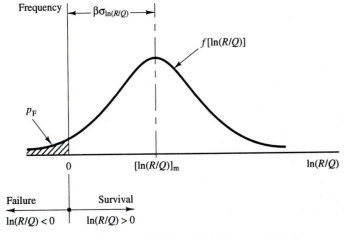

Figure 2.4   Characteristics of ln (R/Q).

$(R - Q)$ is more sensitive to variations in R than Q [5]. The natural logarithm parameter shows no such preferences.

A new variable, U, is now introduced:

$$U = \frac{ln(R/Q) - [\ln(R/Q)]_m}{\sigma_{\ln(R/Q)}} \qquad (2.15)$$

where $[\ln(R/Q)]_m$ is the mean of the logarithm of R/Q, and the symbol in the denominator represents the standard deviation of $\ln(R/Q)$. The expression for the probability of failure becomes

$$p_f = P[\ln(R/Q) \leq 0]$$

$$= P\left\{ \frac{\ln(R/Q) - [\ln(R/Q)]_m}{\sigma_{\ln(R/Q)}} \leq - \frac{[\ln(R/Q)]_m}{\sigma_{\ln(R/Q)}} \right\}$$

where the second expression is simply an expanded version of the first. When U is introduced, the expression becomes

$$p_f = P\left\{U \leq -\frac{[\ln(R/Q)]_m}{\sigma_{\ln(R/Q)}}\right\} \qquad (2.16\text{a})$$

which is written as

$$p_f = F_U\left\{-\frac{[\ln(R/Q)]_m}{\sigma_{\ln(R/Q)}}\right\} \qquad (2.16\text{b})$$

Equation (2.16b) expresses the cumulative distribution (distribution function [1]) of the normalized variable U.

The term $\ln(R/Q)$ is a normally distributed variable which defines the log-normal distribution [1]. For this reason U is a normally distributed variable as well, having a mean value of 0 and a standard deviation of 1. The frequency and cumulative distributions for U are shown in Fig. 2.5.

The term β that is indicated in Figs. 2.3–2.5 is an important variable in all probabilistic design procedures. In simplistic terms, it defines the "distance" from the mean of the strength vs. load distribution to the point where failure takes place. In other words, the farther away the mean of $\ln(R/Q)$ (for example) is from the value of zero, the smaller will be the probability of failure. Theoretically, therefore, one can make p_f become as small as is desired, although it can never be equal to zero: There is always some chance that the structure will fail.

The term β defines the distance to the dividing line between failure and survival. It is therefore a measure of the safety or reliability of the system (= structure), and is commonly known as the reliability index. It is given by the expression

$$\beta = \frac{[\ln(R/Q)]_m}{\sigma_{\ln(R/Q)}} \qquad (2.17)$$

The use of β is discussed in detail in Section 2.4. The complete distributions of R and Q are rarely known, and methods have consequently been devised to produce estimates of mean values and measures of variability. However, since β relates to the probability of failure, and because the complete R and Q distributions are not known, the probability of failure is but a concept, a relative measure of structural performance. The actual p_f cannot be computed without the complete set of data, and values that are given must therefore be treated accordingly. Indeed, there are no absolutes in the random world.

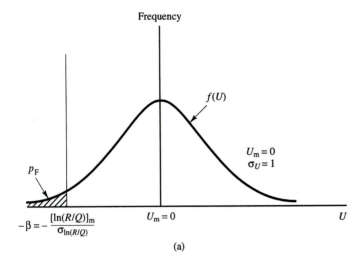

$$-\beta = -\frac{[\ln(R/Q)]_m}{\sigma_{\ln(R/Q)}}$$

$U_m = 0$

$f(U)$

$U_m = 0$
$\sigma_U = 1$

p_F

U

(a)

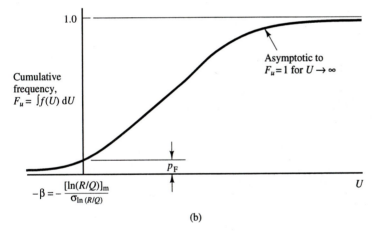

1.0

Cumulative
frequency,
$F_u = \int f(U) \, dU$

Asymptotic to
$F_u = 1$ for $U \to \infty$

p_F

$$-\beta = -\frac{[\ln(R/Q)]_m}{\sigma_{\ln (R/Q)}}$$

U

(b)

Figure 2.5 Frequency and
cumulative
distributions for U.

2.4 STRUCTURAL SAFETY: THE RELIABILITY INDEX

The reliability index β has been shown to be a measure of the reliability or safety of the structure. Whether it is related to the true probability of failure or is just a relative measure, its usefulness can be summarized by the following statement:

> With β chosen as a constant value for the structure and all its elements, the probability of failure will be the same throughout, regardless of the variability of the strengths and the loads.

This definition, in fact, represents the beauty of LRFD. Safety is no longer an empirical, judgment-based concept, but rather a rationally developed measure of structural performance.

With the log-normal distribution that is used for R/Q in the development of LRFD, and assuming that the complete distributions of R and Q are known, the probability of failure takes on the following values:

$$\beta = 5 \text{ gives } p_f = 2.9 \times 10^{-6}$$
$$\beta = 4 \text{ gives } p_f = 3.2 \times 10^{-5}$$
$$\beta = 3 \text{ gives } p_f = 1.4 \times 10^{-3}$$
$$\beta = 2 \text{ gives } p_f = 2.3 \times 10^{-2}$$

The magnitude of p_f indicates the area under the "tail" of the distribution of $\ln(R/Q)$, for which $\ln(R/Q) \leq 0$.

In the practical development of a design specification the value of β is specified such that a reasonable level of safety is achieved. There are two primary ways by which β is chosen:

1. The specification writers agree on using a certain β value, which corresponds to a certain theoretical level of reliability.
2. The reliability index is chosen such that the level of safety in the new specification will be approximately the same as in the old one.

There are advantages and disadvantages to both methods. For example, if the first approach is chosen, then the entire distributions of R and Q must be known (a disadvantage), but the safety development can proceed on absolute terms. However, we have already shown that it is impractical to expect to obtain the complete data for R and Q; thus, method 1 can be ruled out.

Method 2 expresses the principle of calibration, and has the advantages that the specification writer can build on past experience, as well as the fact that the complete load and strength distributions do not have to be known. Although it is less accurate than method 1, the calibration technique is the most realistic approach. Examples of calibration are shown later in this chapter.

As given by Eq. (2.17), the magnitude of β incorporates the variability of load and strength under the same umbrella. It is usually more convenient to be able to deal individually with the properties of R and Q, and for that reason some simplifications will be made on the basis of first-order probabilistic theory. In the strict sense it is not correct to say that the mean of a function is equal to the function of the mean; that is,

$$[f(x, y, \text{ etc.})]_m \neq f(x_m, y_m, \text{ etc.}) \tag{2.18}$$

On the other hand, if the function is well conditioned and the variables exhibit a reasonable degree of variability, first-order theory dictates that the inequality sign of Eq. (2.18) may be changed to an "approximately equal" sign. Applied to the expression for the reliability index, this gives

$$[\ln(R/Q)]_m \approx \ln[(R/Q)_m] \approx \ln(R_m/Q_m) \qquad (2.19)$$

The natural logarithm is a well-conditioned function, and Eq. (2.19) is sufficiently accurate if the coefficient of variation of R/Q is less than 25–30 percent. The latter holds true for almost all civil engineering applications [1, 5].

The standard deviation for the natural logarithm of R/Q can be simplified by making a Taylor series expansion of $\ln(R/Q)$ and neglecting the higher order terms of the series. The method of partial derivatives [1] then yields the standard deviation as the positive square root of the variance of $\ln(R/Q)$:

$$\sigma_{\ln(R/Q)}^2 \approx \left\{\frac{\partial[\ln(R/Q)]}{\partial R}\right\}_m^2 \left(\sigma_R^2\right) + \left\{\frac{\partial[\ln(R/Q)]}{\partial Q}\right\}_m^2 \left(\sigma_Q^2\right) \qquad (2.20a)$$

which is derived from the general expression for the variance of a function f; thus,

$$\sigma_f^2 \approx \left[\sum \left(\frac{\partial f}{\partial x_i}\right)_m^2 \left(\sigma_{x_i}^2\right)\right] \qquad (2.20b)$$

where $(\partial f/\partial x_i)_m^2$ is the square of the partial derivative of the function with respect to the variable x_i, evaluated at the mean, and $\sigma_{x_i}^2$ is the variance of x_i.

By carrying out the differentiations of $\ln(R/Q)$ in Eq. (2.20a), we obtain

$$\sigma_{\ln(R/Q)}^2 \approx \frac{\sigma_R^2}{R_m^2} + \frac{\sigma_Q^2}{Q_m^2} = V_R^2 + V_Q^2 \qquad (2.21)$$

because $\partial[\ln(R/Q)]/\partial R = (1/(R/Q))(1/Q) = 1/R$. This becomes $1/R_m^2$ when squared and evaluated at the mean. A similar development leads to the appearance of Q_m^2 in the denominator of the second term.

The terms V_R and V_Q represent the coefficients of variation of the strength and the load, respectively, which are defined as

$$V_R = \frac{\sigma_R}{R_m} \quad \text{and} \quad V_Q = \frac{\sigma_Q}{Q_m} \qquad (2.22)$$

The accuracy of the Taylor series expansion can be appreciated by noting that the higher order term that has been neglected is the product $\sigma_R^2 \sigma_Q^2$. Since both of these variances normally are small, the product will be negligible.

Equations (2.19) and (2.21) can now be combined to give the simplified expression for the reliability index:

$$\beta \approx \frac{\ln(R_m/Q_m)}{\sqrt{V_R^2 + V_Q^2}} \tag{2.23}$$

which is rewritten as

$$\ln\left(\frac{R_m}{Q_m}\right) = \beta\sqrt{V_R^2 + V_Q^2} \tag{2.24}$$

Taking the exponential of both sides of Eq. (2.24) gives

$$\exp\left[\ln\left(\frac{R_m}{Q_m}\right)\right] = \exp(\beta\sqrt{V_R^2 + V_Q^2})$$

which leads to

$$\frac{R_m}{Q_m} = \exp(\beta\sqrt{V_R^2 + V_Q^2})$$

or, to separate the mean strength and load variables,

$$R_m = Q_m \exp(\beta\sqrt{V_R^2 + V_Q^2}) \tag{2.25}$$

which gives the central safety factor design criterion

$$Q_m\theta \leq R_m \tag{2.26}$$

which states that the mean strength must be greater than or equal to the mean load effect times a certain factor θ.

The factor θ is given by the expression

$$\theta = \exp(\beta\sqrt{V_R^2 + V_Q^2}) \tag{2.27}$$

and is commonly referred to as the central safety factor [4], since it combines the uncertainties associated with the strength and the load effect. Although this is conceptually clear, it is better for the design process if θ can be split into factors that relate to the strength variability and the load variability independent of each other. In that fashion the strength-related or resistance factors are independent of the loads, and simply reflect the variability of the strength of the structural element. Similarly, the load-related or load factors reflect the variability of the loads on the element, which logically should be independent of the strength.

The simplest way of separating V_R and V_Q is to provide a linear approximation of the square root term in Eq. (2.27):

$$\sqrt{V_R^2 + V_Q^2} \approx a(V_R + V_Q) \tag{2.28}$$

where a is referred to as the separation function. This follows the work of Lind [4], who has shown that $a = 0.75$ is sufficiently accurate for the common range of V_R/V_Q, which is from $\frac{1}{3}$ to 3 (error $= \pm6$ percent).

Introducing Eq. (2.28) into (2.27) and (2.26) leads to

$$\theta = \exp(\beta\sqrt{V_R^2 + V_Q^2}) \approx \exp[a\beta(V_R + V_Q)] \tag{2.29a}$$

Therefore,

$$\theta = (e^{a\beta V_R})(e^{a\beta V_Q}) \tag{2.29b}$$

The design criterion, Eq. (2.26), becomes

$$Q_m(e^{a\beta V_R})(e^{a\beta V_Q}) \leq R_m \tag{2.30}$$

which is rewritten as an equation with a load "side" and a resistance "side":

$$Q_m(e^{a\beta V_Q}) \leq R_m(e^{-a\beta V_R}) \tag{2.31}$$

The nominal or service loads Q_n and resistances R_n are now introduced into Eq. (2.31) to produce the LRFD design criterion, Eq. (2.8):

$$Q_n \frac{Q_m}{Q_n} e^{\alpha\beta V_Q} \leq R_n \frac{R_m}{R_n} e^{-\alpha\beta V_R} \tag{2.32a}$$

which is better expressed as

$$Q_n\gamma \leq R_n\phi \tag{2.32b}$$

where the terms γ and ϕ are the total load factor and the resistance factor, respectively, given by the equations

$$\gamma = \frac{Q_m}{Q_n} e^{\alpha\beta V_Q} \tag{2.33a}$$

$$\phi = \frac{R_m}{R_n} e^{-\alpha\beta V_R} \tag{2.33b}$$

Note that the term α that appears in Eqs. (2.32) and (2.33) is not the same as the separation function a that was used originally. The latter assumed a value of 0.75 through linearization of $\sqrt{V_R^2 + V_Q^2}$ over the range $\frac{1}{3} \leq V_R/V_Q \leq 3$. The separation factor α has been determined such that the error in the central safety factor is minimized over the entire range of variation of all of the load and strength parameters. Lind [4] has shown that α takes on the value of 0.55 for all practical purposes.

The resistance factor therefore becomes

$$\phi = \frac{R_m}{R_n} e^{-0.55\beta V_R} \tag{2.34}$$

and V_R expresses the variation of the strength of the particular element being considered. Since the reliability index appears in the negative exponent for e, choosing a higher β value (i.e., a smaller probability of failure) will produce a smaller ϕ value. Therefore, since the resistance factor expresses the randomness of the strength, the smaller the ϕ value, the greater is the strength variation.

Sample computations of resistance factors are given in Section 2.6. Since the strength of a random member normally is less than the nominal value, the ϕ value will be a number something less than 1.0.

When incorporating the value of $\alpha = 0.55$, the total load factor becomes

$$\gamma = \frac{Q_m}{Q_n} e^{0.55\beta V_Q} \tag{2.35}$$

which combines the variability of all types of loads into a single term. This is both impractical and inaccurate, since different loads exhibit different variability characteristics. For example, it is to be expected that dead load can be determined more accurately (and hence varies less) than live load. The design criteria should take such effects into account by analyzing the load effects of individual loads independently. This has been done and is discussed in detail in Sections 2.7 and 2.8 and in Chapter 3.

2.5 VARIABILITY OF STRENGTH

We have shown that the variability in the strength of a structural component is reflected by the resistance factor. The sources of that variability can be traced to a number of random influences as indicated by the following examples:

1. Material property variations
2. Variations in cross-sectional size of members
3. Variations in internal stresses

4. Variations in beam span, column length, etc.
5. Differences between assumed and real loading points
6. Differences between theoretical and real member strength

Many other parameters can be given; the list is but a sample of some of the major sources of uncertainty. Figure 2.6 further illustrates these variations by showing (a) the frequency curve for the yield stress of 33-ksi mild structural steel [6], (b) test results for centrally loaded columns [7], and (c) column curves for the maximum strength of a large number of column types [7]. Similar results have been provided for other strength and geometrical properties of materials and structural shapes [6–9]. The data are not as plentiful as regards the accuracy of the methods of analysis, for example, or the random differences between the nominal and real structural dimensions as they are measured in a structure [*10, 11*].

On this basis, therefore, we find that the resistance factors that have been developed are based on limited sets of data. On the other hand, the model allows for continual updating of the factors, should significant additional results become available. This is the difference between the models that are founded purely on statistical analyses of the available strength data, and those that make use of a probabilistic formulation of the mechanistic strength model, as is done in LRFD. Statistical analyses give only the variability characteristics for a certain data set, and new ϕ values should be developed as soon as additional results are available. Probabilistic formulations are based on analytical models for the behavior and strength, recognizing that the input parameters are random variables. If the methods of analysis are acceptable, then this approach produces consistent safety measures that are not influenced by the sudden appearance of new, possibly extreme test data.

The analytical relationships for the resistance of the various structural components are deterministic in origin, but the variables that are used in the equations are random. The resistance, therefore, also becomes a random variable. It is usually expressed in terms of the nominal resistance of the component, which is based on the nominal or handbook values of material and geometric properties. For example, the nominal capacity of a tension member that fails through yielding on the gross cross section is given as

$$P_n = A_g F_y \qquad (2.36)$$

where A_g is the gross cross-sectional area of the member, based on the producers' catalogs for the shape that is used, and F_y is the minimum specified yield stress, as given by the material standard. This deterministic relationship does not recognize that A_g and F_y, in reality, are random variables, and that P_n, therefore, must be random as well. We can account for this variability by making use of an elementary statistical eval-

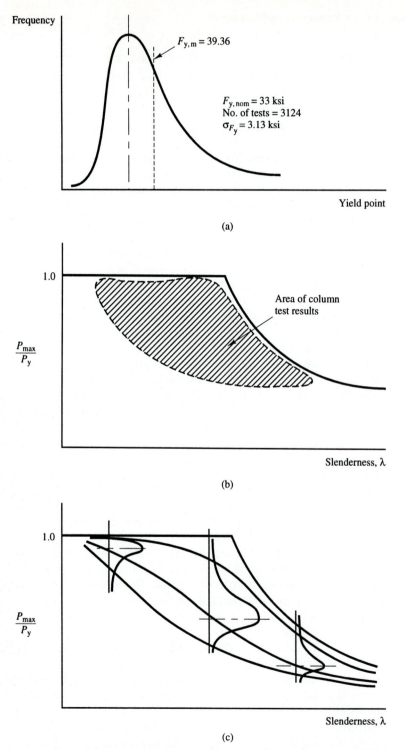

Figure 2.6 Sources of variability: (a) yield stress of 33-ksi (nominal) steel [6]; (b) column strength [7]; (c) column strength and variation as a function of slenderness [7].

uation which says that the random variable P_n is equal to the product of two random variables, A_g and F_y. If it is now assumed that A_g and F_y are statistically independent* [1], then the mean and coefficient of variation of the yield load become

$$(P_n)_m = (A_g)_m (F_y)_m \qquad (2.37a)$$

$$V_p = \sqrt{V_A^2 + V_y^2 + V_A^2 V_y^2} \qquad (2.37b)$$

where V_A and V_y are the coefficients of variation of the area and the yield stress, respectively. Since these are normally less than 0.20, the product of the squares will become a very small number, and therefore may be neglected. The coefficient of variation of the yield load is then simplified to

$$V_p = \sqrt{V_A^2 + V_y^2} \qquad (2.37c)$$

If the variability characteristics of A_g and F_y are known, then the data given by Eqs. (2.37a) and (2.37c) can now be substituted into Eq. (2.34) to produce a value for the resistance factor ϕ. In this case, $R_m = (P_n)_m$, $R_n = P_n$, and $V_R = V_p$.

An additional variability is introduced, since the explanation above assumes that the deterministic relationship is correct. In other words, if A_g and F_y are known, it is assumed that Eq. (2.36) will give the same result as a full-size test of a structural tension member. However, it is known that small changes in area throughout the member, a slight eccentric application of the axial load, and certain types of cross-sectional shapes, for example, will produce test results that differ from the strict relationship of Eq. (2.36). This effect has not been taken into account in the above procedure, but can be significant, especially for resistance models of greater complexity than that of the elementary tension member.

The random uncertainty that is associated with going from the theoretical model, as expressed by Eq. (2.36), for example, to the actual structural component should therefore be taken into account in an improved probabilistic design procedure. This is achieved in LRFD by modifying Eq. (2.5) to form the basic resistance equation:

$$R = R_n MFP \qquad (2.38)$$

where R_n is the nominal, deterministic expression for the strength of the member ($A_g F_y$ for the tension member example) and M, F, and P are

---

* Statistical independence is achieved if the outcome or value of one random variable does not depend on that of another one. Although the yield stress of steel is influenced by the thickness of the material, for practical purposes it is sufficiently accurate to proceed on the assumption that given a range of A_g, the value of F_y will vary independently of the material thickness.

random variables that take into account the variations in material strength, cross-sectional parameters, and "test vs. theory" accuracy, respectively. Therefore, R is the random resistance of the component. Since R_n and R are given in the form of limit state load effects (moments, shear forces, axial forces), the terms M, F, and P will be dimensionless quantities.

The "product" relationship of the basic resistance equation is used because a great many strength relationships for structural steel members assume that form. The tension member is one example; another is the fully plastic moment of a compact beam under uniaxial bending. Thus,

$$R_n = M_{pn} = Z_n F_y \qquad (2.39)$$

where Z_n is the nominal (handbook) value of the plastic section modulus of the shape, and F_y is the minimum specified yield stress for the steel. Equation (2.38) is now applied to obtain the random fully plastic moment of a beam as

$$R = M_p = M_{pn}MFP$$

or

$$R = (Z_n F_y)MFP \qquad (2.40)$$

The mean and coefficient of variation of R, which are needed to compute the value of ϕ, can now be found as

$$R_m = (Z_n F_y) M_m F_m P_m \qquad (2.41a)$$

and

$$V_R = [V_M^2 + V_F^2 + V_P^2 + V_M^2 V_F^2 + V_F^2 V_P^2 + V_M^2 V_P^2 + V_M^2 V_F^2 V_P^2]^{1/2} \qquad (2.41b)$$

where the properties of M, F, and P are needed. Since the coefficients of variation normally are less than 0.20, all of the product terms in Eq. (2.41b) can be neglected, and a simple form is obtained for V_R:

$$V_R = \sqrt{V_M^2 + V_F^2 + V_P^2} \qquad (2.41c)$$

Sample calculations of resistance factors are presented in Section 2.6. In general, M, F, and P cover the following random effects:

1. Also called the *material factor*, M takes into account the variations of the governing material property (normally the yield

stress). It is expressed as the ratio between the actual (random) material property value and the standardized value (e.g., minimum specified yield stress) to achieve the necessary nondimensional form. Data for M are relatively plentiful [6, 8, 9].

2. Also called the *fabrication factor*, F reflects the variations that are found in the geometrical properties of structural members. These variations arise through the dimensional tolerances of rolling and welding, from initial crookedness (camber, sweep) in members, and from erection tolerances, for example. Basically, F covers the differences that are bound to exist between the ideal and the actual member in the structure. It is also expressed as a ratio; for example, as actual gross area divided by nominal gross area, in the case of the tension member. Data for F are limited [8–11].

3. Also called the *professional factor*, P reflects the uncertainties that are associated with the relationship between the theoretical (mechanistic) strength model and the real structural member. For example, the evaluation of the behavior of the member will have been based on assumptions such as perfect material elasticity and plasticity, to mention one of the most common simplifications. Needless to say, means and coefficients of variation of P are difficult to assess on the basis of observations alone. Where test data are available, P is normally expressed as the ratio between the test value (say, the actual plastic moment of a beam) and the theoretical value that has been computed on the basis of the actual material and geometrical properties. Some data are available for various structural components [6, 7, 12, 13].

The resistance properties that are determined from Eqs. (2.38) and (2.41c) reflect the randomness of the primary strength property of a member.

2.6 RESISTANCE FACTORS

As already described, the resistance factor reflects the random variability of the strength of a structural component. The basic resistance equation, Eq. (2.38), gives the relationship between the nominal value and the mean value of the random variable, as well as the coefficient of variation of the resistance. Both of these are used in the resistance factor equation, Eq. (2.34), to determine the value of ϕ that is required to produce the necessary level of structural reliability, expressed by the reliability index β.

In the following examples, computations are made to demonstrate how the ϕ values are determined for typical structural members and strength criteria.

Resistance Factor for Tension Member

Ultimate limit state: Yielding on gross cross section

Nominal resistance: $R_n \equiv P_n = A_g F_y$
where A_g is the nominal gross area, and F_y is the minimum specified yield stress.

Basic resistance equation: $R = R_n MFP$
which becomes

$$R = (A_g F_y)\left(\frac{F_{sy}}{F_y}\right)\left(\frac{A_{ga}}{A_g}\right)\left(\frac{\text{Observed strength}}{\text{Computed strength}}\right)$$

where F_{sy} is the actual static yield stress of the material (yield stress measured at zero strain rate [*14*]), A_{ga} is the actual gross area of the member, and the last term represents the professional factor P. The observed strength is taken from a full-size tension member test; the computed strength is based on $A_g F_y$, using the actual material properties for the member in question.

Mean resistance: $R_m = R_n M_m F_m P_m$

Coefficient of variation of resistance:

$$V_R = \sqrt{V_M^2 + V_F^2 + V_P^2}$$

Properties of the material, fabrication, and professional factors
1. *Material factor:* The material factor is given as the ratio of measured static yield stress (random) to minimum specified yield stress (constant for a given steel grade and member size). It can be written as

$$M = \frac{F_{sy}}{F_y}$$

and the mean value is

$$M_m = \left(\frac{F_{sy}}{F_y}\right)_m = \frac{(F_{sy})_m}{F_y}$$

The magnitude of the yield stress depends on the thickness of the material, and for a wide flange shape, for example, the yield stress of a specimen taken from the flange will be lower than that of a specimen from the web. Much information is available about the variability of the yield stress [*8*, *15*]; here, a representative value will be used.

For flange material the mean value of M is 1.05, and for

web material it is 1.10. The corresponding coefficients of variation are 0.10 and 0.11. For the typical range of material thicknesses in a variety of tension members (angles, tees, tubes, plates), it is realistic to use the following data:

$$M_m = 1.08$$
$$V_M = 0.10$$

2. *Fabrication factor:* From the available data, as well as the fact that geometrical properties are specified with certain positive and negative tolerances, it is reasonable to assume that in the long run the mean value will actually be equal to the nominal value. This has been found in several investigations [7, 8, 11]. At the same time, good quality control will yield a coefficient of variation no larger than 0.05; in fact, measurements on typical building and bridge components have given values of V_F between 0.02 and 0.03. For the purposes of this example, the following data will be used for F:

$$F_m = 1.0$$
$$V_F = 0.05$$

3. *Professional factor:* Tension members represent one of the cases where there is very good correlation between test results and the computed ("predicted") gross section strength. For a number of reasons the test results tend to be higher, and the mean and coefficient of variation to be used here are

$$P_m = 1.02$$
$$V_P = 0.05$$

Computation of R_m and V_R: With the data that have been provided, the mean and coefficient of variation of the resistance can be computed. Hence,

$$R_m = R_n M_m F_m P_m$$
$$= (A_g F_y)(1.08)(1.0)(1.02)$$

i.e.,

$$\underline{R_m = (1.10)(A_g F_y) = (1.10)R_n}$$
$$V_R = \sqrt{V_M^2 + V_F^2 + V_P^2}$$
$$= \sqrt{0.10^2 + 0.05^2 + 0.05^2}$$

i.e.,

$$\underline{V_R = 0.123}$$

Computation of resistance factor ϕ_t: Using the reliability index value of $\beta = 3.0$, the resistance factor for a tension member that fails through yielding on the gross section will be

$$\phi = \frac{R_m}{R_n} \exp(-0.55\beta V_R)$$

which gives

$$\phi_t = \frac{1.10 R_n}{R_n} \exp[(-0.55)(3.0)(0.123)]$$

or

$$\underline{\underline{\phi_t = 0.90}}$$

This is also the value of ϕ_t that is used in the LRFD specification.

Resistance Factor for Bending Member (Uniaxial Bending)

Ultimate limit state: Plastic hinge formation; no local or overall (lateral) buckling.

Nominal resistance: $R_n \equiv M_p = Z_n F_y$
where Z_n is the nominal plastic section modulus, and F_y is the minimum specified yield stress.

Basic resistance equation:

$$R = R_n MFP$$

which becomes

$$R = (Z_n F_y)\left(\frac{F_{sy}}{F_y}\right)\left(\frac{Z}{Z_n}\right)\left(\frac{\text{Observed plastic moment}}{\text{Computed plastic moment}}\right)$$

The properties in this equation take on characteristics similar to those of the equation for tension members. The mean and coefficient of variation of the resistance are computed according to the same basic equations.

Properties of M, F, and P
   1. *Material factor:* The comments that were made relative to material properties in the discussion of the resistance of a tension member still hold true. However, since the bending resistance (i.e., moment capacity) of a beam is more significantly influenced by the flanges than the web, it is rational to

use the data for the flange material to set the values of M and V_M. Hence,

$$M_m = 1.05$$

$$V_M = 0.10$$

2. *Fabrication factor:* The fabrication factor is given as the ratio between the actual (random) plastic section modulus and its nominal (handbook) value. Since the tolerances for geometric properties are such that in the long run the mean Z value will be equal to the nominal value, F_m can be expected to equal 1.0. Also similar to the tension member case, the coefficient of variation of Z can be expected to be no more than 5 percent. Consequently, the data to be used for F_m and V_F are

$$F_m = 1.0$$

$$V_F = 0.05$$

3. *Professional factor:* Compact beams in uniaxial bending is another case where test and theory agree quite well. As a result, the following data are used for P:

$$P_m = 1.02$$

$$V_P = 0.06$$

Computation of R_m and V_R:
The mean resistance can now be found as

$$R_m = (Z_n F_y)(1.05)(1.0)(1.02)$$

i.e.,

$$\underline{R_m = 1.07 R_n}$$

The coefficient of variation becomes

$$\underline{V_R = \sqrt{0.10^2 + 0.05^2 + 0.06^2} = 0.127}$$

Computation of resistance factor ϕ_b: Using the same reliability index value, $\beta = 3.0$, the resistance factor for a compact beam that fails by reaching the fully plastic moment at a section will be

$$\phi_b = \frac{1.07 R_n}{R_n} \exp[(-0.55)(3.0)(0.127)]$$

which gives

$$\underline{\underline{\phi_b = 0.87}}$$

The value that is used for ϕ_b in the LRFD specification is 0.90. The above result is sufficiently close, considering its limited data base evaluation as compared to the specification.

The preceding examples have demonstrated the method of arriving at rational values for the resistance factors. These problems were simple, but the principles for solving them apply to all resistance criteria.

Note that the value of the reliability index β influences the load factors as well. The number 3.0 was chosen to be consistent with that used as the target value for structural members subjected to gravity loads (dead, live, and snow loads).

2.7 VARIABILITY OF LOADS

It has been shown that the variability of the loads on the structure, or more specifically, the load effects on the various structural components, is reflected through the load factors. Just as the resistance factors assume different values for different elements and limit states, the load factors also take on magnitudes that vary with the type of load. Further, the load factors also depend on the reliability index, as expressed by Eq. (2.35).

The primary types of loads, their variabilities, and how they are determined are discussed in detail in Chapter 3. The nominal or service loads generally represent the loads that are on the structure at any given time, and the structural response under these conditions reflects its serviceability. For comparison, it may be said that the service loads are similar in nature to the nominal or handbook values that are used for cross-sectional dimensions of structural shapes, to pick one example. Factored loads, on the other hand, represent the loads on the structure at insipient failure, that is, when an ultimate limit state has been reached. The consequences of this happening are normally much more serious than the problems arising from serviceability considerations, and for that reason LRFD focuses more attention on the magnitudes of loads that are likely to lead to ultimate limit states.

Structural design is also based on the fact that the different types of loads will interact or combine to produce effects that may not appear for other load criteria. At the same time it is recognized that while two loads may happen to reach their lifetime maximum or extreme values simultaneously, it is highly unlikely that additional load types also will reach their limit at this point in time. LRFD covers these phenomena by specifying different load factors for certain loads when they are used in combination with certain other loads. Examples of load combinations are given in Chapter 3.

Generally speaking, the loads on the structure must be translated into member load effects (moments, shear forces, axial forces) to permit the design to proceed. Uncertainties are introduced through this procedure, and although they can be incorporated into the overall load factors (and have been, in certain jurisdictions), it is more correct to keep these variabilities separate from those that pertain to the loads only. The LRFD approach to this problem is outlined in the following.

For the above reasons it is also preferable to determine factored loads rather than factored load effects. In other words, the nominal (service) loads are multiplied by the load factors before internal stress resultants are computed for the members. The reason for this is that some of the influence coefficients that are used to transform loads into load effects exhibit random variability, even though they are treated as deterministic variables. For example, the influence coefficient for the maximum moment in a uniformly loaded, simply supported beam (see Section 2.2) is given as $wl^2/8 = (D + L)sl^2/8$. Both the spacing of adjacent beams s and the beam span l are in effect random variables. In addition, this influence coefficient also reflects the tributary area, which in turn has an effect on the magnitude of the live load [*5, 16, 17*].

As an illustration of the treatment of loads in LRFD, the formulation that is used to develop the mean and coefficient of variation of the load effect will be demonstrated for the case of combined dead and live load. The load effect is assumed to take on the general form

$$Q = E(C_D AD + C_L BL) \qquad (2.42)$$

where D and L represent the random dead and live load intensities; A and B are random variables that express the inaccuracy that is inherent in the transformation of the load into load effects; C_D and C_L are the above-mentioned influence coefficients; and E is a random variable that reflects the error or uncertainty of the structural analysis itself. Additional load types will appear in the same form within the parentheses; to illustrate the method we will incorporate only two load types.

The dead load is assumed to be constant for the life of the structure, although variations will occur between identical structures. The live load also varies randomly in the same fashion, but may undergo large changes with time. It is therefore convenient to talk about sustained versus transient live load; the former represents the portion of the load that varies very little unless the structural occupancy is changed. At such times there may be large variations in the live load; it may even be reduced to something near zero. The transient live load, on the other hand, will vary from day to day or even within shorter time spans. The number of occupants in a room is typical of the transient live load: At times it assumes a zero or near-zero value; at other times it may be a substantial addition to the sustained live load on the structural member. Figure 2.7

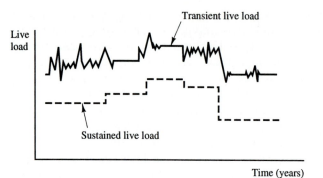

Figure 2.7 Variation of live load with time.

gives a schematic illustration of the variation of live load with time, and also points out the differences between transient and sustained live load.

The probabilistic modeling of live load is particularly complicated because it represents a stochastic phenomenon, and distribution characteristics will be different for the many types of live load. The duration or return period of the load is also important, since a designer is interested in knowing the largest possible load that the structure is likely to experience in its lifetime (lifetime maximum live load). Many studies have dealt with the concepts of live load modeling; the interested reader is referred to Refs. [*16–19*] for more information.

Further load variability is introduced because D and L incorporate not only the variation of the dead and live loads themselves, but also the uncertainties that are associated with load idealization. For example, L is assumed to be uniformly distributed; in reality it follows an entirely random pattern. A similar effect comes from the treatment of the dead load. Other uncertainties can be traced to the influence of nature, of representing localized loads (from columns and the like) as if they are truly concentrated loads, and so on. Therefore, whereas the resistance and its variability are difficult random phenomena, the complexities of load analysis make these problems even more intractable. Part of the reason for this is certainly the sparsity of load models.

The random variables A and B that appear in Eq. (2.42) reflect the differences between the actual and theoretical internal stress resultants in the structure. The reason for these differences is to be found in the methods by which idealized loads are transformed into load effects.

The random variable E in Eq. (2.42) is also known as the analysis factor. It covers the uncertainties that arise from the fact that a three-dimensional structure is being treated as a two-dimensional assembly of individual members and connections, for example. Basically, E covers the effects of using simplified models to represent complex structures and phenomena. For that reason it is to be expected that as our computational capabilities improve, the influence of the random errors indicated by E will continue to decrease. Thus, as more accurate methods of analysis are devised, and the easier it is to produce models that closely reflect the actual behavior of the structural material, the lower will be the value of E. In the extreme, when there is no error in the analysis between real and assumed behavior, E will take on the deterministic value of 1.0.

It should be noted that E could easily be applied to the resistance side of the basic LRFD design criterion, since it focuses equally on the modeling of structural behavior and strength and loads. This is a matter of choice; in the LRFD format it was decided to apply the analysis factor to the load side of the equation.

By applying the principles of probability theory to the load effect equation, Eq. (2.42), we derive the following for its mean and coefficient of variation.

Mean load effect:

$$Q = E(C_D AD + C_L BL) = ET \qquad (2.42a)$$

which gives the mean value of

$$Q_m = E_m T_m = E_m(C_D A_m D_m + C_L B_m L_m) \qquad (2.43)$$

The variance of T is obtained from

$$\sigma_T^2 = C_D^2 A_m^2 D_m^2(V_A^2 + V_D^2) + C_L^2 B_m^2 L_m^2(V_B^2 + V_L^2) \qquad (2.44)$$

which is used to develop the coefficient of variation of T as

$$V_T^2 = \frac{\sigma_T^2}{T_m^2} = \frac{C_D^2 A_m^2 D_m^2(V_A^2 + V_D^2) + C_L^2 B_m^2 L_m^2(V_B^2 + V_L^2)}{C_D A_m D_m + C_L B_m L_m} \qquad (2.45)$$

Coefficient of variation of load effect
From Eq. (2.42a) we find that

$$V_Q = \sqrt{V_E^2 + V_T^2} \qquad (2.46)$$

and this can now be completed by substituting for V_T^2 from Eq. (2.45). This is now the derived coefficient of variation of the load effect.

The values of the terms that enter into Eqs. (2.43) to (2.46) are difficult to determine under the best of circumstances. However, it can be assumed with a certain level of accuracy that

$$E_m = A_m = B_m = 1.0 \qquad (2.47)$$

and the reasoning behind these numbers reflects the probability that E, A, and B are equally likely to assume values greater than and less than 1.0. The coefficients of variation of these three variables are considerably more elusive, but it is usually assumed that

$$V_E = 0.05$$

$$V_A = 0.04 \tag{2.48}$$

$$V_B = 0.20$$

Thus, the value of V_E of 0.05 means that the shear force or bending moment in a member, for example, is accurate to within ± 10 percent of the actual load effect, and that this is accurate to the 95 percent confidence level. Similar reasoning applies to the value of V_A, whereas the higher value of V_B reflects the greater variability in the live load itself.

The variability of the dead load is relatively small, and it is therefore sufficiently accurate to use

$$D_m = \text{Code specified dead load } D_c$$

$$V_D = 0.04$$

The great live load variability is influenced by a number of factors, as already pointed out. It turns out that the size and form of the tributary area play a major role [16, 17]. Thus, for buildings the mean live load can be computed from

$$L_m = 14.9 + \frac{763}{\sqrt{A_I}} \tag{2.49}$$

and the standard deviation is

$$\sigma_L = \sqrt{11.3 + \frac{15,000}{A_I}} \tag{2.50}$$

where A_I is defined as the influence area* for the member. These formulas are discussed in detail in Chapter 3. Studies have shown that Eq. (2.49) produces good results; they also have indicated that V_L becomes almost insensitive to the size of the influence area and a realistic value of

$$V_L = 0.13$$

is reasonable for a large range of area sizes. The magnitudes of the above means and coefficients of variation will be used in the following section to compute load factors for dead and live load.

---

* $A_I = 2 \times$ (Tributary area for a beam); it is $4 \times$ (Tributary area for a column).

2.8 LOAD FACTORS

It is recalled that the central safety factor design criterion is given by Eq. (2.26) as

$$Q_m\theta \leq R_m \tag{2.26}$$

where the central safety factor is

$$\theta = \exp(\beta\sqrt{V_R^2 + V_Q^2}) \tag{2.27}$$

Lind's [4] linearization and error minimization subsequently led to a separation of the criterion into a load and a resistance side, Eq. (2.31), modified to

$$Q_m e^{\alpha\beta V_Q} \leq R_m e^{-\alpha\beta V_R} \tag{2.51}$$

The load side can be further rephrased by a linear approximation, making use of Eqs. (2.43) through (2.46) [4, 5]. The new expression becomes

$$\begin{aligned} Q_m e^{\alpha\beta V_Q} = e^{\alpha\beta V_E} [(1 &+ \alpha\beta\sqrt{V_A^2 + V_D^2})C_D D_m \\ &+ (1 + \alpha\beta\sqrt{V_B^2 + V_L^2})C_L L_m] \end{aligned} \tag{2.52}$$

where the terms are all of those that were used in the preceding section of this chapter. Equation (2.52) can now be rewritten by introducing the analysis, dead, and live load factors, γ_E, γ_D, and γ_L, respectively. Hence,

$$Q_m e^{\alpha\beta V_Q} = \gamma_E(\gamma_D C_D D_m + \gamma_L C_L L_m) \tag{2.53}$$

where the load factors are

Analysis: $\gamma_E = \exp(\alpha\beta V_E)$

Dead load: $\gamma_D = 1 + \alpha\beta\sqrt{V_A^2 + V_D^2}$ (2.54)

Live load: $\gamma_L = 1 + \alpha\beta\sqrt{V_B^2 + V_L^2}$

The basic load and resistance factor design criterion, Eq. (2.8), can now be phrased as follows for the case of combined dead and live load:

$$\gamma_E(\gamma_D C_D D_m + \gamma_L C_L L_m) \leq \phi R_n \tag{2.55}$$

Additional load types and combinations can be added inside the brackets on the left side of Eq. (2.55).

Computation Examples:

Using a separation factor α of 0.55 and a reliability index of 3.0, the following load factors are determined on the basis of the data provided in Section 2.7:

Analysis factor:
$$\gamma_E = \exp(\alpha\beta V_E)$$
$$= \exp[(0.55)(3.0)(0.05)]$$
$$\text{i.e., } \underline{\underline{\gamma_E = 1.086}}$$

Dead load factor:
$$\gamma_D = 1 + \alpha\beta\sqrt{V_A^2 + V_D^2}$$
$$= 1 + (0.55)3.0\sqrt{0.04^2 + 0.04^2}$$
$$\text{i.e., } \underline{\underline{\gamma_D = 1.093}}$$

Live load factor:
$$\gamma_L = 1 + \alpha\beta\sqrt{V_B^2 + V_L^2}$$
$$= 1 + (0.55)3.0\sqrt{0.20^2 + 0.13^2}$$
$$\text{i.e., } \underline{\underline{\gamma_L = 1.394}}$$

Note that a designer will not have to deal with the analysis factor independently from the load factors of dead and live load. Rather, the analysis factor has been incorporated into the terms within the parentheses in Eq. (2.55), such that the criterion reads

$$\gamma_E\gamma_D C_D D_\mathrm{m} + \gamma_E\gamma_L D_L L_\mathrm{n} \le \phi R_\mathrm{n} \tag{2.56}$$

and the apparent dead and live load factors therefore are

Dead load:
$$\gamma_{DD} = \gamma_E\gamma_D = (1.086)(1.093)$$
i.e.,
$$\underline{\underline{\gamma_{DD} = 1.19}}$$

Live load:
$$\gamma_{LL} = \gamma_E\gamma_L = (1.086)(1.394)$$
i.e.,
$$\underline{\underline{\gamma_{LL} = 1.51}}$$

These load factors can now be compared with those that have been proposed for design. It will be seen that the dead load factors are quite close: 1.9 vs. 1.2 in the American standard; 1.19 vs. 1.25 in the Canadian limit states design standard. The live load factors also compare favorably: 1.51 vs. 1.6 in the American standard, and 1.51 vs. 1.5 in the Canadian standard. The correlation with the current reinforced concrete design code is not as good (1.19 vs. 1.4; 1.51 vs. 1.7), but the development backgrounds are quite different. Further analyses are expected to provide good agreement.

2.9 LIMIT STATES

Some discussion has already been given of the various types of limit states that provide the focal points for the modes of behavior of the structure. Basically, the LRFD criterion, Eq. (2.8), states that the combined effects of the factored loads must be less than or equal to the factored resistance of the component of the structure. If not, the component is said to have reached (or exceeded) the governing limit state, and by definition it has failed. It does not matter whether the "failure" is an actual physical collapse or simply the component reaching a deformation limit: By not satisfying the basic design requirement the limit has been reached.

There are basically two categories of limit states that are pertinent to the structural design process:

1. Ultimate limit states (ULS)
2. Serviceability limit states (SLS)

In some cases a third category, damage limit states, is added between the ULS and SLS. Damage limit states have not received wide acceptance, possibly because they are more applicable to brittle type materials, where certain types of cracking behavior would be more serious than reaching a mere serviceability limit state. For steel structures, use is made only of the ultimate limit states and the serviceability limit states.

Ultimate Limit States.   Violation of a ULS will normally involve the failure or loss of all or parts of a structure. For example, a compact steel beam has reached the ultimate limit state when a plastic hinge mechanism forms. Some other typical ultimate limit states are as follows:

1. Frame instability
2. Column buckling
3. Lateral torsional beam buckling
4. Connection tear-out
5. Yielding on gross section of tension member
6. Incremental structural collapse
7. Progressive structural collapse
8. Overturning
9. Fatigue failure ⎫
10. Brittle fracture ⎬ These are special cases; see below.

The special conditions that apply to fatigue failure and brittle fracture are rooted in the fact that these limit states are true ULS; however, they are reached under the action of service loads.

Servicability Limit States.   Violation of an SLS does not necessarily involve the actual failure of a structure or any of its components. However, it has reached a limit of acceptable behavior, and for that reason may be classified as having failed. The following are some typical serviceability limit states:

1. Excessive elastic deflection of a member
2. Excessive permanent deformation of a member
3. Excessive rotations in a connection
4. Excessive vibration in a floor system
5. Long-term deformations

For all practical purposes items 1, 4, and 5 may be of the greatest concern to a structural steel designer, although item 5 is more prevalent for composite construction. However, it is necessary to recognize that design for serviceability is as important as design for strength, and both should form the basis for any structural design.

REFERENCES

[1] Benjamin, J. R., and Cornell C. A., *Probability, Statistics and Decision for Civil Engineers,* McGraw-Hill, New York, 1970.

[2] American Society for Testing and Materials (ASTM), *1989 Annual Book of ASTM Standards,* Part 4, ASTM, Philadelphia, PA, 1989.

[3] Cornell, C. A., A probability-based structural code, *Journal of the American Concrete Institute,* Vol. 66, No. 12, December 1969.

[4] Lind, N. C., Consistent partial safety factors, *Journal of the Structural Division, ASCE,* Vol. 97, No. ST6, June 1971 (pp. 1651–1669).

[5] Galambos, T. V., and Ravindra, M. K., *Tentative Load and Resistance Factor Design Criteria for Steel Buildings,* Research Report No. 18, Department of Civil Engineering, Washington University, St. Louis, MO, September 1973.

[6] Beedle, L. S., and Tall, L., Basic column strength, *Journal of the Structural Division, ASCE,* Vol. 86, No. ST7, July 1960.

[7] Bjorhovde, Reidar, *Deterministic and Probabilistic Approaches to the Strength of Steel Columns,* Thesis, Lehigh University, Bethlehem, PA, May 1972.

[8] Alpsten, G. A., *Variations in Mechanical and Cross-Sectional Properties of Steel,* Proceedings, International Conference on Planning and Design of Tall Buildings, Vol. Ib: Criteria and Loading, Lehigh University, Bethlehem, PA, August 1972.

[9] Tall, L., and Alpsten, G. A., *On the Scatter in Yield Strength and Residual Stresses in Steel Members,* Final Report, IABSE Symposium on Concepts of Safety of Structures and Methods of Design, London, UK, 1969.

[10] Tomonaga, K., *Actually Measured Errors in Fabrication of Kasumigaseki Building,* Proceedings, 3rd Regional Conference of the Joint ASCE-IABSE Committee on the Planning and Design of Tall Buildings, Tokyo, Japan, September 1971.

[11] Beaulieu, D., and Adams, P. F., *The Destabiliz-*

ing Forces Caused by Gravity Loads Acting on Initially Out-of-Plumb Members in Structures, Structural Engineering Report No. 59, University of Alberta, Edmonton, Alberta, Canada, March 1977.

[*12*] Bjorhovde, Reidar, The safety of steel columns, *Journal of the Structural Division, ASCE,* Vol. 104, No. ST3, March 1978 (pp. 463–477).

[*13*] Bjorhovde, Reidar, and Birkemoe, P. C., Limit states design of HSS columns, *Canadian Journal of Civil Engineering,* Vol. 6, No. 2, June 1979 (pp. 276–291).

[*14*] Rao, N. R. N., Lohrmann, M., and Tall, L., *Effect of Strain Rate on the Yield Stress of Structural Steel,* Fritz Engineering Laboratory Report No. 249.23, Lehigh University, Bethlehem, PA, September 1964.

[*15*] Baker, M. J., *Variations in the Mechanical Prop-erties of Structural Steel,* Final Report, IABSE Symposium on Concepts of Safety of Structures and Methods of Design, London, UK, 1969.

[*16*] Allen, D. E., Limit states design: A probabilistic study, *Canadian Journal of Civil Engineering,* Vol. 2, No. 1, 1976 (pp. 36–49).

[*17*] McGuire, R., and Cornell, C. A., Live load effect in office buildings, *Journal of the Structural Division, ASCE,* Vol. 100, No. ST7, July 1974 (pp. 1351–1366).

[*18*] Corotis, R. B., and Doshi, V. A., Probability models for live load survey results, *Journal of the Structural Division, ASCE,* Vol. 103, No. ST6, June 1977 (pp. 1257–1274).

[*19*] Ellingwood, B., and Culver, C. G., Analysis of live loads in office buildings, *Journal of the Structural Division, ASCE,* Vol. 103, No. ST8, August 1977 (pp. 1551–1560).

CHAPTER THREE

Loads
and Load Factors

3.1 GENERAL OBSERVATIONS

Normally, a design specification does not prescribe the magnitudes of the loads that are to be used as the basic input to the structural analysis, with the exception of special cases such as crane design specifications. It is the role of the specification to detail the methods and criteria to be used in arriving at satisfactory member and connection sizes for the structural material in question, given the magnitudes of the loads and their effects. The specification therefore reflects the requirements that must be satisfied by the structure in order that it will have a response that allows it to achieve the performance that is needed. Loads, on the other hand, are governed by the usage or type of occupancy of the building, which in turn is dictated by the applicable local, regional, and national laws that are more commonly known as building codes.

The building code loads have traditionally been given as nominal values, determined on the basis of material properties (e.g., dead load) or load surveys (e.g., live load and snow load). To be reasonably certain that the loads are not exceeded in a given structure, the code values have tended to be higher than the loads on a random structure at an arbitrary point in time. This may, in fact, be one of the reasons why excessive gravity loads are rarely the obvious cause of structural failures. Be that as it may, the fact of the matter is that all of the various types of structural loads exhibit random variations that are functions of time, and the manner of variation also depends on the type of load. Rather than dealing with nominal loads that appear to be deterministic in nature, a realistic design procedure should take load variability into account along with that of the strength, in order that adequate structural safety can be achieved through rational means.

Since the random variation of the loads is a function of time as well as a number of other factors, the modeling, strictly speaking, should take this into account by using stochastic analyses to reflect the time and space interdependence. Many studies have dealt with this highly complex phenomenon, especially as it pertains to live load in buildings. References [1–4], along with those cited in earlier chapters,* illustrate a number of the avenues that have been investigated. In practice, however, the use of time-dependent loads is cumbersome at best, although the relationship must be accounted for in certain cases (i.e., seismic loads). For most design situations the code will specify the magnitude of the loads as if they were static. Their time and space variation are covered through the use of the maximum load occurring over a certain reference (return) period, and its statistics. For example, American live load criteria [1, 2] are based on a reference period of 50 years, while Canadian criteria use a 30-year interval.

---

* Of particular interest will be reference [11] from Chapter 1; it should be considered an integral reference for this chapter.

The geographical location of the structure plays an important role for certain loads. It is particularly applicable to snow, wind, and seismic loads, the first being of special importance in north-central and north-eastern areas of the United States, the second in high wind coastal and mountain areas, and the last in areas having earthquake fault lines. Further details are given in Section 3.3.

Design for wind effects is complicated by a number of phenomena. Like snow and earthquake loads, wind loads are given more attention in certain parts of the country. At the same time wind loads are neither static nor uniformly varying, and are heavily influenced by the geometry of the structure as well as the surrounding structures and the landscape. To a certain degree this also applies to the magnitude of the snow load. Building codes treat these effects as static phenomena and relate them to the actual conditions through semiempirical equations. This gives the designer a better handle on a difficult problem, but can lead to difficulties when the real structure departs significantly from the bases of the code. For that reason wind loads, and sometimes earthquake and snow loads, are determined on the basis of model tests. In particular, wind tunnel testing has become a useful and practical tool in these endeavors.

The loads on the structure are normally assumed to be independent of the type of structure and structural material, with the exception of dead loads. The response of a building, however, will be different for different materials, depending on the type of load. For example, the behavior of a moment-resistant steel frame will be quite unlike that of a braced frame, when subjected to lateral loads, especially those due to an earthquake. On the other hand, the response of these two frames to gravity loads will not be all that different.

The size of a structure (height, floor area) has a significant impact on the magnitudes of most loads. All loads are influenced by the increasing height of a multistory building, for example. Similarly, the greater the floor area that is to be supported by a single member, the smaller will be the probability that the code live load will appear with its full intensity over the entire area. In such cases a live load reduction method is used to arrive at more realistic design data.

3.2 LOADS ON BUILDING STRUCTURES

There are many types of loads that may act on a building structure at one time or other, and this section provides a general description of the characteristics of the most important ones. Detailed data for each kind are given in Section 3.3.

The following loads are of primary concern to a building designer:

Gravity loads
1. Dead load

2. Live load
3. Snow load

Lateral loads
4. Wind load
5. Seismic load
6. Special loads and load effects

Special loads and load effects include the influence of temperature variations, structural foundation settlements, impact, and blast. They are given only a brief description here; the reader interested in the details is advised to seek out the specialized publications that deal with this type of loading.

Each of the primary types of loads can be divided into several subtypes. The difference between sustained and transient live load was discussed in Chapter 2.

3.2.1 Dead Load

In theory, at least, the dead load on a structure is supposed to remain constant. In reality, the word *constant* is a relative measure, because the dead load includes not only the self-weight of the structure, but also the weight of permanent construction materials (e.g., stay-in-place formwork), partitions, floor and ceiling materials, machinery and other equipment, and so on. Also included in this category are the load effects of prestressing forces.

The weight of all of these elements can be determined exactly only by actually weighing and/or measuring the pieces. This is almost always an impractical solution, and the designer will therefore rely on published material data to arrive at the dead loads. Some variation consequently will occur in the real structure, accounting for some of the deviation from constancy. Similarly, there are bound to be differences between otherwise identical structures, representing the major source of dead load variability. However, compared to the other structural loads, the dead load variations are relatively small, and the actual mean values are quite close to the code-prescribed data.

3.2.2. Live Load

Live load, or more accurately the gravity live load, is the name that is commonly used for the loads on the structure that are not part of the permanent installations. To that end it includes the weight of the occupants of the building, furniture and movable equipment, and so on. The fluctuations in this load are bound to be substantial. From being essentially zero immediately before the tenants take possession to a maximum value that may be several times as high as the dead load, the magnitude of the live load at any given time may be quite different from that specified by the building code. This is one of the reasons why numerous attempts have been made to model the live load and its variations, and why live load measurements in actual buildings continue to be made.

The live load on the structure at any given time is also called the arbitrary point-in-time live load. As shown by Fig. 2.7 and repeated here as Fig. 3.1, this is the load that is determined in a live load survey. Part of the total load may now be directly attributable to the occupants of the structure: As soon as the room is emptied, the transient live load (TLL) is reduced to zero or near zero. The load that remains, say, due to furniture and the like, is a sustained live load (SLL) that changes very little as long as the same tenant occupies the premises. Significant variations in the SLL may come about when the occupancy changes: As one company moves out, the SLL may drop to near zero. It will remain at this level until the next tenant moves in, at which time it will increase to a level that may be quite different from the SLL of the earlier occupant.

As demonstrated in Section 3.3, live loads in general are a function of the size of the floor area under consideration. The larger the area, the smaller is the chance that the full code load will appear over the entire area. This affects the magnitude of the arbitrary point-in-time loads as well as the maximum lifetime live load, which represents the maximum live load that the structure may experience in its lifetime. In American practice the life of a structure is expected to be 50 years; hence, a 50-year reference period forms the basis for many of the live load models that have been developed.

The live load on the structure at any time is normally well below the code value; the maximum lifetime (50-year) live load may be a certain amount larger. Current load statistics and code recommendations take these phenomena into account.

3.2.3 Snow Load

Although snow is a form of live load, unique conditions govern its magnitude and distribution. It is the primary roof load in many geographical areas, and it is heavily dependent on structural geometry, exposure, and local climate.

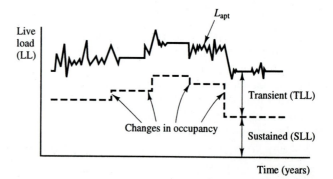

Figure 3.1 Variation of live load with time.

Snow load data are normally based on surveys that have produced isoline maps, showing areas of equal depth of ground snowfall (actually, the maps show the water-equivalent of the amount of snow). On this basis annual extreme snowfalls have been determined for a period of several years, and further analyzed through statistical models to find the lifetime maximum snow loads. The reference period is again the 50-year anticipated life of a structure.

A major difficulty is encountered in the method of translating the ground snow load into roof snow load. This is effected through a semiempirical relationship, whereby the ground snow load is multiplied by a factor that takes factors such as roof geometry into account (see Section 3.3). Much work continues to be done to improve the method of snow load computation; Refs. [5–9] give a broad illustration of some of these studies.

3.2.4 Wind Load

By its very nature, wind is a highly dynamic natural phenomenon. For this reason it is also a complex problem from a structural point of view. Not only do the wind forces fluctuate significantly, they also are influenced by the geometry of the structure (height, width, depth, plan, and elevation shape), as well as the surrounding landscape. It has already been observed that in some of the more difficult cases a designer must resort to wind tunnel tests to determine the loads and behavior of the building under high winds, and even that may not be sufficient in some cases. Nevertheless, the basic approach to wind load analysis is to treat the phenomenon as a static load problem, using the Bernoulli equation to translate wind speed into wind pressure. In a treatment akin to what is used for snow loads, a semiempirical equation then gives the wind load at a certain level as a function of a number of constants, each representing effects such as wind gusts, landscape, and structural geometry.

The data that have been provided for wind loads are all based on measured wind speeds, since these are relatively easy to determine. Meteorological data from many locations throughout the country yield the local wind speed data for daily and annual maxima; these are then used to model long-term characteristics. In particular, the maximum wind speed is needed for the 50-year reference period. In some cases this number is available; in other cases it must be extrapolated using a statistical model.

Wind loads are highly dependent on local conditions, in the same way that snow loads are. The designer must consult local requirements; there are no general, national specifications, other than a compilation of meteorological data. Reference [10] gives an extensive set of such data for the United States.

3.2.5 Seismic Load

The treatment of seismic load effects is extremely complicated, because of the stochastic character of this natural phenomenon and the many factors that influence the impact of an earthquake on a structure. In addition, since the ground rather than the building moves, the inertia effects cannot be overlooked.

For most buildings it is sufficient to deal with the seismic effects as if they represent a static load, provided the magnitude of this equivalent load reflects the dynamic background. Thus, the magnitude of the peak ground acceleration is needed, as is knowledge of the type of structure (frequency and damping properties). The structural material is also important, especially in view of the need for structural ductility. After all, the load effects produced by a major earthquake are expected to be well in excess of normal service loads, producing significant permanent deformations.

The primary load that is developed under seismic conditions is the base shear, and building codes contain detailed provisions for its computation. Newly developed proposals [*11*] give better ways of representing the influence of structural ductility, as well as the higher modes of dynamic behavior of a structure. These are discussed in some detail in Section 3.3.

3.2.6 Special Loads and Load Effects

Impact.　Most building loads are static or essentially so, meaning that their rate of application is so slow that the kinetic energy associated with their motion is insignificant. For example, a person entering a room is actually exerting a dynamic load on the structure by virtue of his motion. However, because of the small mass and slow movement of the individual, his kinetic energy is essentially zero.

When loads are large and/or their rate of application is very high, the influence of the energy that is brought to bear on the structure as the movement of the loads suddenly is restrained must be taken into account. This is the phenomenon of impact. The kinetic energy of the moving mass is translated into a load on the structure. Depending on the rate of application, the effect of the impact is that the structure "feels" a load that may be as large as two times its static value.

Impact is of particular importance for structures where machinery and the like are operating. For example, elevators, cranes, and equipment such as printing presses and weaving looms all produce impact effects. The LRFD specification has detailed rules for the computation of these influences.

Blast.　Blast effects, whether due to gas or other explosions, have influences similar to impact. They do not occur as often as impact in normal structures, but under certain circumstances they must be considered. This is particularly true for certain industrial installations, where the product that is manufactured may be of a volatile nature. Analysis and design data for blast effects are limited; certain defense specifications will have rules.

Thermal Effects.   Steel expands or contracts under changing temperatures, and in doing so may exert considerable forces if the members are restrained from moving. The use of expansion joints, properly located throughout a structure, can be important. However, for most building structures the thermal effects are less significant than others for structural strength and behavior. The movement (or restraint effect) results from the total temperature differential, and is directly proportional to the length of the structural member. Naturally, an uneven temperature distribution will cause some of the movement to be translated into a bending effect; this may have to be combined with bending from other sources in certain cases.

Foundation Settlements.   For a statically determinate structure a foundation settlement will do nothing but make the structure unsightly and possibly create some cracked partitions and the like. The structural effects are minimal, unless, of course, the settlement is very large. For a statically indeterminate structure, however, any such movement will produce additional loads on members and connections which must be taken into account. Of course, a settlement rarely occurs until after the structure has been completed; the designer should reanalyze the building to ascertain that the added stress resultants can be carried by all of the affected components.

3.3 DATA ON BUILDING LOADS

The statistical characteristics of the various load types are presented as a demonstration of the development of the different load factors. Building codes and other load standards prescribe nominal (service) loads in great detail. What is useful here is to show the relationships between the actual and the nominal loads, as well as the variability properties of the actual loads.

3.3.1 Dead Load Data

It is generally accepted that the mean dead load D_m is close to or slightly higher than the nominal value D_n. Measurements also indicate that the frequency distribution for D is close to a normal one, as expected. Similarly, because of the relative accuracy of computing structural dead loads, their variability has been found to be comparatively small.

The dead load factor that has been accepted for use in the ASCE load standard (ASCE 7-88) is based on a normal distribution with the following properties:

$$\text{Mean actual dead load: } D_m = 1.05D_n$$

$$\text{Coefficient of variation: } V_D = 0.10$$

This is further illustrated in Fig. 3.2.

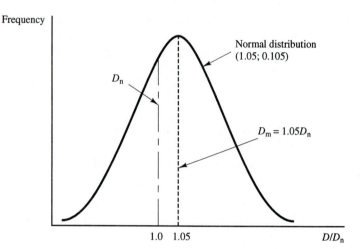

Figure 3.2 Distribution of dead
load.

The proposed dead load factor (and all other load factors as well) is intended to apply to all structural materials, which is a significant improvement over the situation that existed in the past. However, researchers have assumed that the dead load variability is independent of the type of construction material, on the premise that much of the variability can be attributed to items such as partition. This is true to some extent, but the weight of the structural members themselves is not the same. Consequently, it is reasonable to assume that the variability of the nonstructural items would be significant for the lesser weight structural materials (timber, aluminum, light-gage cold-formed steel structures), of some consequence for medium-weight frames (such as those of fabricated structural steel), and of little significance for heavy frames of reinforced concrete. Since the weight of steel members is much more tightly controlled than that of concrete, it is natural to expect concrete to exhibit greater variability.

3.3.2 Live Load Data

As already pointed out, live loads may vary widely, depending on whether arbitrary point-in-time, annual maximum, or lifetime maximum values are under consideration. The size of the floor area also plays a significant role, to the effect that both the mean and the coefficient of variation of the actual live load will decrease as the floor area increases. However, the data can be confusing and contradictory; for example, the arbitrary point-in-time live load appears almost constant for all areas.

In design, the area used to compute the total load that must be resisted by a structural member is the tributary area A_T. Simplified tributary areas for some members are illustrated schematically by Fig. 3.3. For use in the statistical evaluation of live loads, however, it was found that the influence area A_I produced more uniform reliability for the

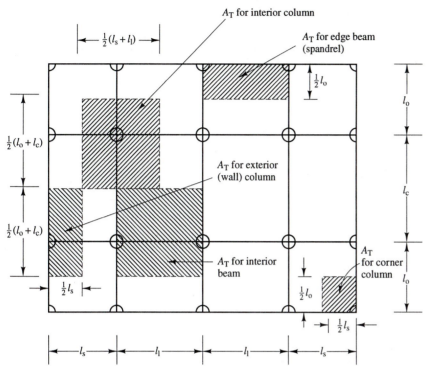

Figure 3.3   Simplified tributary areas for some structural members.

different types of loads, and hence this characteristic was used to develop the properties of the live load. Briefly, the relationship between A_I and A_T is as follows:

$$
\begin{aligned}
\text{Columns:} & \quad A_I = 4A_T \\
\text{Beams:} & \quad A_I = 2A_T \\
\text{Slabs:} & \quad A_I = A_T
\end{aligned}
\tag{3.1}
$$

The tributary area is a mathematical tool for determining the magnitude of load on a particular member; it has no physical significance. The influence area, however, is significant in that it is the actual area over which any applied load would have its effect felt by the member under consideration. No portion of a load applied outside of the influence area would be carried by the member being designed.

The basic characteristics of the arbitrary point-in-time live load and the maximum lifetime live load have already been described. Their relationship to each other, and, more importantly, to the service or nominal loads prescribed by the building codes must also be understood if the subsequent load factor development is to be appreciated. To put this in the proper perspective, Table 3.1 presents a sample of the uniformly distributed live loads that are recommended by ASCE 7-88.

TABLE 3.1   SOME UNIFORMLY DISTRIBUTED
UNREDUCED NOMINAL LIVE LOADS[a]

	Live Load[b]	
Type of Occupancy	psf	kg/m²
Apartments, classroom, etc.	40	195
Offices	50	244
Assembly halls (fixed seating)	60	293
Corridors	80	390
Retail stores, restaurants, etc.	100	488
Light storage	125	610
Library stacking areas	150	732

[a] Chapter 1, Ref. [35].

[b] SI conversion is given in terms of mass. To determine the corresponding force units, multiply kg/m² by the acceleration of gravity (approx. 9.81 m/s²) to obtain N/m² (= pascals).

When the values given in Table 3.1 are compared with live load survey results (i.e., arbitrary point-in-time data), the data yield ratios of mean actual live load to nominal live load of approximately 0.23 and coefficients of variation of 0.70 and 0.90. This means, for example, that a *typical* live load on an office floor at any given time will be approximately 12 psf.

On the other hand, the relationship between the maximum lifetime live load and the nominal loads is based on a 50-year reference period, and for that reason must be extrapolated from a live load model. The ratio of the mean maximum to the nominal load varies between 1.11 and 1.38 for an influence area of 200 ft²; the coefficient of variation for the same data ranges from 0.14 to 0.19. Both exhibit a tendency to drop as A_I increases. If the uncertainties associated with the live load model are incorporated into the overall live load coefficient of variation, the latter would assume a value of approximately 0.25. This is now independent of the size of the influence area.

The data in the load code, as sampled by Table 3.1, are the basic unreduced live loads; to obtain the load magnitude for a specific member, the tributary area must be taken into account. In the case of the 1972 ANSI standard* the live load reduction factor was expressed as

$$\text{RF} = 1 - \text{minimum of} \begin{cases} 0.0008 A_T \\ 0.60 \\ 0.23\left(1 + \dfrac{D_n}{L_n}\right) \end{cases} \tag{3.2}$$

* The American National Standards Institute (ANSI) supervised the preparation of the load codes before ASCE took over in 1983. The ANSI load code was known as ANSI A58.1.

where all of the terms have been defined previously. For the revised ANSI load standard, which now has become ASCE 7-88, the authors recognized that the 1972 version had a tendency to underestimate the maximum lifetime live loads. A new reduction factor was therefore proposed, based on the work of McGuire and Cornell and Ellingwood and Culver,† and using the influence area:

$$RF = 0.25 + \frac{15}{\sqrt{A_I}} \tag{3.3}$$

No reduction is allowed for A_I less than 400 ft², which corresponds to beam and column tributary areas of 200 and 100 ft², respectively. Equations (3.2) and (3.3) are shown in Fig. 3.4, further demonstrating the importance of being able to incorporate the effect of area. Code limits on the maximum permitted reduction must also be checked.

3.3.3 Snow Load Data

The roof snow load is determined by the following expression:

$$S = C_s q \tag{3.4}$$

where S is the roof snow load, q is the ground snow load, and C_s is the snow load coefficient. The snow load coefficient depends on factors such as the geometry (slope, primarily) of the roof, its exposure to wind, the presence of adjacent roofs, and so on. Details for determining C_s are not given here; the reader is referred to any of the applicable building codes or

† See Chapter 2, Ref. [*17, 19*].

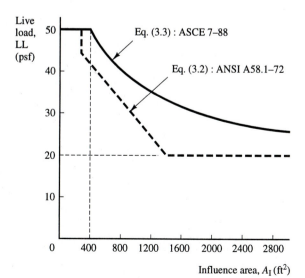

Figure 3.4 Live load reductions as a function of influence area.

ASCE 7-88. Suffice it to observe that C_s is equal to 0.8 in a great many cases, representing a flat roof with normal insulation and reasonable exposure to wind. Statistically, this type of roof would have a mean C_s of 0.5 and a coefficient of variation of 0.23.

As noted earlier, the ground snow load has been determined on the basis of a number of surveys that were used to develop isoline maps. These maps, along with continual meteorological observations, have provided statistics for the annual extreme snowfalls, and data to predict the 50-year reference period snowfall.

Specifics on the snow loads to be used in a locality are provided in ASCE 7-88 and the local building code. The statistical properties of the 50-year maximum and the annual maximum S values have been established and given as fractions of the nominal roof snow loads S_n. Thus, the mean annual maximum snow load equals $0.20S_n$, with a coefficient of variation of 0.73. Similarly, the mean maximum roof snow load is $0.82S_n$, and its coefficient of variation is 0.26. The large values of the coefficient of variation are worth observing; it is to be expected that significant deviations from the mean may occur with regularity. Although not the sole, and possibly not even the most important, cause of some roof collapses over the years, extreme snow loads are not uncommon. In addition, if they appear just prior to a major rainfall, the true snow load may be substantial. For these reasons there are snow load design requirements even within some geographical areas where such precipitation is relatively unlikely.

3.3.4 Wind Load Data

The basic wind pressure against a flat surface perpendicular to the direction of the wind is determined by Bernoulli's equation:

$$q = \tfrac{1}{2}\rho v^2 \tag{3.5}$$

where ρ is the density of the air and v is the speed of the wind. This is dynamic pressure; it is set equal to a static force in most wind load analyses. The influence of the geometry of the building, the topography of the surrounding landscape, and the turbulent character of the wind (i.e., wind gusts) are further covered through the empirical wind pressure equation:

$$p = C_e C_g C_p q \tag{3.6}$$

In this expression, q is the pressure defined by Eq. (3.5), C_e is the exposure factor, C_g is the wind gust factor, and C_p is the structural pressure coefficient or shape factor. The various building and load codes give detailed rules for the determination of the factors that enter into Eq. (3.6).

The three C factors that are part of the wind pressure equation are discussed in the following.

Exposure Factor (C_e).   The basic C_e value reflects the location of the building, the surrounding landscape, and its height. Height predominates, to the effect that C_e takes on values ranging from 1.0 (building height less than 40 ft) to 2.0 (heights between 750 and 1200 ft).

Gust Factor (C_g).   The magnitude of the gust factor is a reflection of the wind turbulence as well as the interaction between the wind and the structure and its elements. For example, a C_g value of 2.0 is used for normal structural members, and a value of 2.5 is used for cladding. These data indicate that the size of the exposed surface is also important.

Pressure Coefficient (C_p).   Of the three C factors, the pressure coefficient is the most difficult to understand and to compute. It may assume positive (i.e., pressure) or negative (i.e., suction) values, and it depends on the shape of the structure (plan and elevation) relative to the wind direction, as well as on whether the building is open or closed (airtight). In brief, the pressure coefficient relates to the flow of the wind around the structure, and a great many full-scale and wind tunnel model tests have been run to determine the value of C_p for different geometries.

For complex building shapes it is almost always advisable to run wind tunnel tests to determine the wind loads and their distribution. The boundary layer wind tunnel facilities at the University of Western Ontario in Canada and at Colorado State University in Fort Collins are well known for their efforts in this field.

The statistical characteristics of the wind loads will incorporate those coefficients that enter into Eq. (3.6), in addition to the variability of the dynamic wind pressure, Eq. (3.5). Certain assumptions have been made about the properties of the C values to the effect that they are distributed normally with means equal to the magnitudes for C_e, C_g, and C_P that are prescribed by ASCE 7-88. Their respective coefficients of variation are taken as 0.16, 0.11, and 0.12. On this basis the 50-year reference period wind load, i.e., the maximum lifetime wind load, has a mean-to-code value of 0.78, and a coefficient of variation of 0.37. Similar data have been developed for the annual and daily maximum wind loads.

The large value of the coefficient of variation is again emphasized. It is not unusual to obtain wind loads significantly in excess of those specified by the codes.

3.3.5 Seismic Load Data

The primary load influence of an earthquake is expressed as the shear at the base of the structure. Base shear is a horizontal force which is applied at the ground–structure interface and resisted by the inertia of the structure above. In elementary terms, the force is directly proportional to the mass of the structure and the acceleration of the ground:

$$F = ma \qquad (3.7)$$

In actual buildings, the conditions are considerably more complex. The mathematical model must consider both the total quantity of the mass and its distribution in plan and vertically throughout the building. In addition, the type of structural system, the materials, ground–structure interaction, and other less readily defined aspects must be included. Furthermore, the earthquake motion is not a single application of a constant acceleration; rather, the ground acceleration will vary significantly over the duration of the quake and the duration will vary from event to event. The variability of acceleration can be seen in Fig. 3.5 which shows the relationship between ground acceleration and time for one of the locations in the San Francisco area during the 17 October 1989, Loma Prieta earthquake.

Because of the complexity of the problem, building codes prescribe simplified, empirical equations for determining base shear and distributing the forces throughout the building. The static force approach permitted by the 1988 Uniform Building Code (UBC) will be outlined here. The design base shear is given by

$$V = \frac{ZIC}{R_{\mathrm{w}}} W \qquad (3.8)$$

where

$$C = \frac{1.25S}{T^{2/3}}$$

and

$$Z = \text{seismic zone factor}$$
$$I = \text{structural importance factor}$$
$$W = \text{total seismic dead load}$$
$$R_{\mathrm{w}} = \text{structural system coefficient}$$
$$S = \text{site coefficient for soil characteristics}$$
$$T = \text{period of the structure}$$

These terms are discussed briefly here. The reader interested in the detailed development of these factors is referred to Ref. [11] and its many references.

Seismic Zone Factor (Z). The United States is divided into four primary seismic zones, according to the probability of the occurrence of a major earthquake. The seismic zoning map of the United States shown in Fig. 3.6 reflects the magnitudes of peak ground acceleration that could be

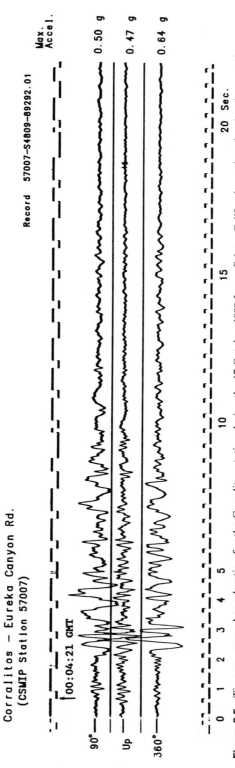

Figure 3.5 Time vs. ground acceleration for the Corralitos station during the 17 October 1989 Loma Prieta, California earthquake showing a peak horizontal acceleration of 0.64 g.

73

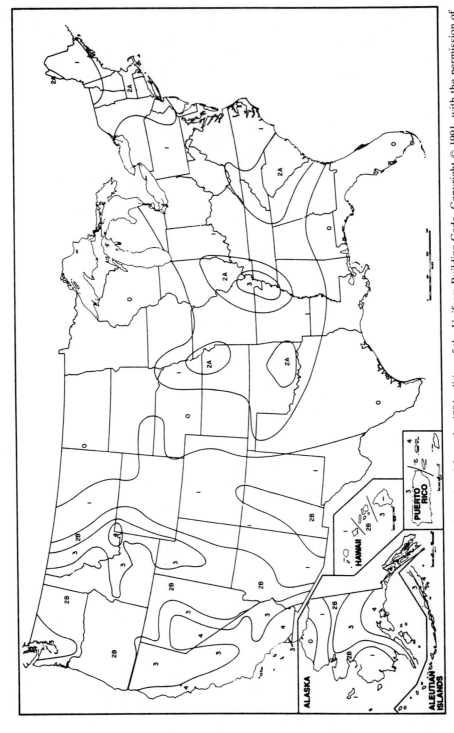

Figure 3.6 U.S. seismic zoning map. Reproduced from the 1991 edition of the Uniform Building Code, Copyright © 1991, with the permission of the publisher, the International Conference of Building Officials.

expected to be exceeded in a 50-year interval with a probability of 10 percent. Each zone is assigned a value for the zone factor, as given in Table 3.2. The higher the value of Z, the more severe the seismic action that can be anticipated. Zone 4 is the most seriously affected of the four U.S. zones. For example, the areas along the San Andreas Fault in California are all classified as Zone 4.

TABLE 3.2 SEISMIC ZONE FACTOR VALUES

Zone	Z
1	0.075
2A	0.15
2B	0.20
3	0.30
4	0.40

Structural Importance Factor (*I*). This is essentially a judgment-based factor that has evolved from the philosophy that there are certain structures that must remain serviceable even after the most severe earthquake. The UBC provides an extensive description for their four occupancy categories. For the two most critical categories, a factor of 1.25 is used, while the remaining two categories use a factor of 1.0.

Total Seismic Dead Weight (*W*). The magnitude of the dead load used in a seismic analysis is determined by combining the actual dead load with a specified portion of the partition load and snow load, all permanent equipment, and, for storage and warehouse occupancy, at least 25 percent of the floor live load.

Structural System Coefficient (R_w). This factor is also referred to as the ductility coefficient, since it reflects the ability of the structural frame to absorb energy. Comparing the force vs. displacement diagrams for the idealized ductile and brittle structures shown in Fig. 3.7, the relative amount of energy that may be absorbed during a displacement can be seen. For the two structures to absorb the same amount of energy, the areas under the curves must be equal. For the structural systems defined in the UBC, the coefficient ranges from 4 for braced heavy timber frames to 12 for special moment-resisting steel space frames.

Base Shear Coefficient (*C*). Also called the seismic coefficient, C is directly related to the flexibility of the structure. For dynamic loads such as those from earthquakes, the flexibility is best represented by the magnitude of the first natural frequency of the structure. This can be determined by any number of analytical methods, among which, Teal's approach [*13*] is a simplified and rational procedure. The expression given by the code is based on a Rayleigh analysis of the structure. Generally, the more flexible the frame, the larger will be the natural frequency and the less the inertia effect. Therefore, a flexible frame will have a lower base shear than an otherwise identical, stiff frame.

Site Coefficient (*S*). The magnitude of the site coefficient represents the interaction between soil and structure. It incorporates the flexibility and damping of the ground and ranges from 1.0 to 2.0. Some typical values are given in Table 3.3.

Period of Structure (*T*). Two methods are available to determine the fundamental period of the building for use in the static analysis

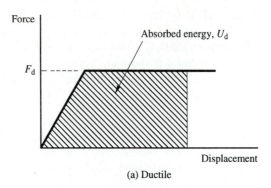

(a) Ductile

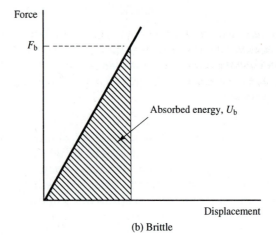

Figure 3.7 Force vs. displacement for idealized (a) ductile and (b) brittle structures.

(b) Brittle

procedure: dynamic analysis of a properly formulated mathematical model or a simplified approach using

$$T = C_t \, (h_n)^{3/4} \tag{3.9}$$

where h_n is the total height of the structure, and C_t is 0.035 for steel moment-resistant frames.

As indicated earlier, the base shear must be resisted throughout the structure by the inertia forces. Thus, the base shear is usually distributed

TABLE 3.3   SITE COEFFICIENTS FOR SOME TYPICAL SOILS

Soil Condition	S
Rock-like material	1.0
Dense stiff soil, >200 ft deep	1.2
Soft soil, 20–40 ft deep	1.5
Soft soil, >40 ft deep	2.0

linearly over the height of the building. This is illustrated schematically in Fig. 3.8. It should be observed that the linear variation reflects the first mode of vibration only. The influence of the higher modes is made by applying a single force, F_t, at the roof level. The force F_t will have a maximum value of 25 percent of the base shear and may be ignored for very stiff frames, defined as frames with a period of less than or equal to 0.7 s.

Overturning and torsional effects of seismic action must also be considered. Codes give general guidelines for the evaluation of such influences. The magnitude of the overturning moment is almost always governed by the first mode of the frame and will be lower for more flexible frames. Torsional effects normally arise because of differences between the locations of the structural center of mass and the center of resistance, although occasionally the ground motion itself will be of a twisting nature. Figure 3.9 shows an example of the locations of the center of mass and resistance for a typical building floor plan.

The statistical data for earthquake effects are overwhelmingly dominated by those of the peak ground acceleration. The load factors that have been developed are based on a number of assumptions that have been correlated with the seismic risk maps.

3.3.6 Data for Other Loads

Statistical data for the other types of loads and load effects are limited. However, the treatment of impact loads, temperature effects, and settle-

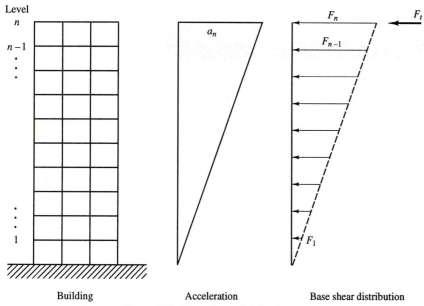

Building   Acceleration   Base shear distribution

Figure 3.8 Base shear distribution.

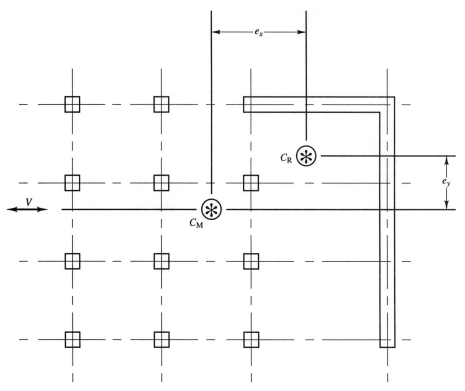

Figure 3.9 Location of the center of mass C_M and center of resistance C_R for a typical building floor plan. V = shear; $M_T = Ve_y$.

ment-induced loads can be deterministic, since it is a matter of translating energy into load effects. Construction loads from equipment and storage of materials may be more complicated, and the designer essentially has to consult with the contractor to get assessments of such influences. Ponding effects, whether due to the accumulation of rainfall on a roof or the deflection of a steel beam while concrete is being placed, should be incorporated into the design procedure. Rainfall accumulation is a meteorological phenomenon that is of importance only for structures with flat or near-flat roofs where water may be prevented from draining. The deflection of steel beams relates to the method of construction of certain composite (steel–concrete) floor systems. It is of importance only when unshored construction is used. Techniques are available to analyze for the effects of rainwater ponding [16]; similar approaches should be used for unshored beams to ascertain that they will not deflect too much or be stressed too highly.

3.4 LOAD COMBINATIONS

In the analysis and design of a structure, the governing load effect controls the sizes of members and connections. The governing load effect, in turn, is a function of the load or loads that can be expected to act on the member throughout its lifetime. In all likelihood, therefore, the stress resultant that controls the member size will represent the fact that several load types together produce the governing effects.

Under normal operating conditions, two or more load types will act on a structure at any given time. In other words, the load types combine to produce more severe conditions than if only single loads were to act. When this is considered, together with the different stochastic characteristics of the various loads, it is not reasonable to expect that all loads will exert their maximum lifetime values simultaneously on the structure. For example, the probability of having the full dead, live, and wind loads acting together is very small. In general, the more load types that act together, the less is the likelihood that they will all reach their lifetime maxima at the same time.

Strictly speaking, the governing load effect should be determined on the basis of a reliability analysis of the loads that are acting together, including their different stochastic properties. This is both complex and impractical; instead, a realistic procedure can be based on *Turkstra's rule* [*17, 18*], which effectively says that

> The governing load effect due to a certain combination of load types is found when one of the loads attains its lifetime maximum value, and all of the other loads take on their arbitrary point-in-time values.

For example, if a roof is to be designed for the combination of dead, live, and snow loads, the governing load effects would arise from either of the two following combinations:

$$D + L + S_i$$
$$D + L_i + S$$

where L and S represent lifetime maxima, and L_i and S_i are arbitrary point-in-time values. The dead load D does not vary in the same sense as L and S.

In theory, with the relatively large number of load types that may act on a structure, the number of potential load combinations will be very large. For realistic design situations, however, a great many of these combinations have been ruled out, with the result that a manageable number remains, as follows:

DESIGN LOAD COMBINATIONS

1. Dead load (alone)
2. Dead and live load
3. Dead, snow, *and* live *or* wind load
4. Dead, live, and wind load

5. Dead, seismic, *and* live *or* snow load
6. Dead *and* seismic *or* wind load

Each of the combinations represents realistic loading cases; the respective load factors and reliability indices reflect that the more loads that are combined, the less the likelihood that any of them will reach a maximum. Thus, the target reliability indices that were sought in load factor development are

(a) Combinations 1 and 2: $\beta = 3.0$
(b) Combinations 3 and 4: $\beta = 2.5$
(c) Combinations 5 and 6: $\beta = 1.75$

These target β values lead to load factors that depend on the load combination under consideration, as shown in Section 3.6.

Different methods can be used to account for the reduced probability of having two or more loads appear simultaneously at their maximum lifetime values. In the case of the LRFD specification, the effects are covered through the use of different load factors for the same load type, depending on the particular combination.

An earlier form of load combination is the one-third increase in the allowable stress that has been used in the ASD specification when gravity loads are combined with wind and other lateral loads. Although a time-honored approach, this procedure has not been verified scientifically.

The rationale behind the choice of relevant design load combinations will be better understood when the respective load factors are also given. Briefly, combinations 1 through 5 are natural choices for the joint effects of gravity and lateral loads. Combination 1 considers the bare structure before occupancy; combination 2 gives the loading case that will govern most floor systems and interior columns. Combination 3 is actually two combinations, but both describe loads that pertain to roof structures. Combination 4 is the most common loading system for structures where lateral load is taken into account; it may govern the loads in certain connections and exterior columns in high-rise frames. It plays the primary role when P-Δ effects are considered. Load combination 5 deals with the behavior of the frame under seismic action, and combination 6 reflects the particular criteria that must be considered to prevent an overturning failure. In combination 6, the seismic or wind loads are subtracted from the dead load to deal with the most critical case.

3.5 ANALYSIS FACTOR

As discussed in detail in Sections 2.7 and 2.8, the analysis factor that is incorporated into the LRFD format reflects the uncertainties associated with the analytical models that have been used, for example, by treating a three-dimensional frame as a two-dimensional one. This factor has been combined with the various load factors and does not have to be considered separately by the designer.

3.6 DESIGN LOAD FACTORS

The values of the load factors discussed in this section are those of ASCE 7-88. They are independent of the structural material and should be used only for the specific load combinations 1 through 6. The load factors should be applied to the loads only, and not to the load effects. The subscript n refers to nominal (code) value.

Load case 1: Dead load (only)

$$\text{Factored load} = 1.4D_n$$
$$\text{Dead load factor: } \gamma_{D1} = 1.4$$

Load case 2: Dead + Live load

$$\text{Factored load} = 1.2D_n + 1.6L_n$$
$$\text{Dead load factor: } \gamma_{D2} = 1.2$$
$$\text{Live load factor: } \gamma_{L2} = 1.6$$

The live load that is used should incorporate the live load reduction factor.

Load case 3: Dead + Snow + (Live or Wind) load

3(a): Dead + Snow + Live load
$$\text{Factored load} = 1.2D_n + 1.6S_n + 0.5L_n$$
$$\text{Dead load factor: } \gamma_{D3} = 1.2$$
$$\text{Live load factor: } \gamma_{L3} = 0.5$$
$$\text{Snow load factor: } \gamma_{S3} = 1.6$$

3(b): Dead + Snow + Wind load
$$\text{Factored load} = 1.2D_n + 1.6S_n + 0.8W_n$$
$$\text{Dead load factor: } \gamma_{D3} = 1.2$$
$$\text{Snow load factor: } \gamma_{S3} = 1.6$$
$$\text{Wind load factor: } \gamma_{W3} = 0.8$$

Load case 4: Dead + Live + Wind load

$$\text{Factored load} = 1.2D_n + 0.5L_n + 1.3W_n$$
$$\text{Dead load factor: } \gamma_{D4} = 1.2$$
$$\text{Live load factor: } \gamma_{L4} = 0.5$$
$$\text{Wind load factor: } \gamma_{W4} = 1.3$$

Load case 5: Dead + Seismic + (Live or Snow) load

5(a): Dead + Seismic + Live load
$$\text{Factored load} = 1.2D_n + 1.5E_n + 0.5L_n$$

Dead load factor: $\gamma_{D5} = 1.2$
Seismic load factor: $\gamma_{E5} = 1.5$
Live load factor: $\gamma_{L5} = 0.5$

5(b): Dead + Seismic + Snow load
Factored load $= 1.2D_n + 1.5E_n + 0.2S_n$
Dead load factor: $\gamma_{D5} = 1.2$
Seismic load factor: $\gamma_{E5} = 1.5$
Snow load factor: $\gamma_{S5} = 0.2$

Load case 6: Dead − (Wind or Seismic) load
This load case represents the check for overturning of the structure. Live (gravity) loads are assumed to be entirely absent.

6(a): Dead − Wind load
Factored load $= 0.9D_n - 1.3W_n$
Dead load factor: $\gamma_{D6} = 0.9$
Wind load factor: $\gamma_{W6} = 1.3$

6(b): Dead − Seismic load
Factored load $= 0.9D_n - 1.5E_n$
Dead load factor: $\gamma_{D6} = 0.9$
Seismic load factor: $\gamma_{E6} = 1.5$

The load factors and load cases are tabulated in Table 3.4, which is a handy reference table for design usage. The table also indicates the load combinations that are the most likely to govern common building design. However, the designer must ascertain which combination dominates each particular structure.

TABLE 3.4   LOAD FACTOR VALUES FOR DESIGN LOAD COMBINATIONS

	Load Case	Dead	Live	Snow	Wind	Seismic
1	D	1.4	—	—	—	—
2*	$D + L$	1.2	1.6	—	—	—
3 (a)*	$D + S + L$	1.2	0.5	1.6	—	—
3 (b)	$D + S + W$	1.2	—	1.6	0.8	—
4*	$D + L + W$	1.2	0.5	—	1.3	—
5 (a)*	$D + E + L$	1.2	0.5	—	—	1.5
5 (b)	$D + E + S$	1.2	—	0.2	—	1.5
6 (a)*	$D - W$	0.9	—	—	1.3	—
6 (b)*	$D - E$	0.9	—	—	—	1.5

Note. The most important load combinations are indicated by an asterisk. Live load must be reduced by the appropriate live load reduction factor.

3.7 COMPUTATION OF FACTORED LOADS

As an extension of the preceding discussion, the following computations illustrate a typical factored load determination. Figure 3.10 shows the floor plan of a high-rise office building; the calculations are made to find the factored loads for load case 2: Dead + Live load for some of the structural members.

> *Unreduced live load*: L_n = 50 psf
>
> > *Dead load* (all inclusive): D_n = 75 psf
>
> *Live load reduction factor*: RF = $0.25 + \dfrac{15}{\sqrt{A_I}}$
>
> > where A_I = influence area (ft²)

Service live load data are included for the bending members, since these must be checked for deflection criteria.

1. *Girder AB on line 2–2*:
    > Tributary area: A_T = (40)(20) = 800 ft²
    > Influence area: A_I = $2A_T$ = 1600 ft²

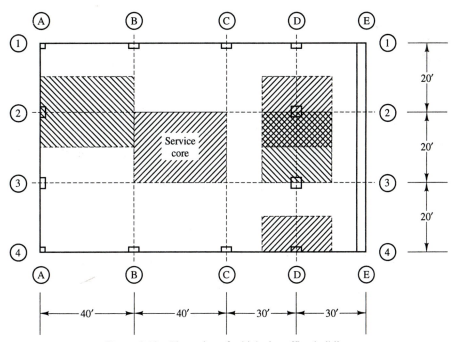

Figure 3.10   Floor plan of a high-rise office building.

$(A_I = 2A_T$ for beams and girders)

Live load reduction factor: $RF = 0.25 + \dfrac{15}{\sqrt{1600}}$

i.e.,

$$RF = 0.625$$

Factored loads per lineal foot:

Dead load $= (1.2)(75)(20)$ $= 1.8$ k/ft

Live load $= (1.6) \, [(0.625)(50)(20)] = 1.0$ k/ft

$$\text{Total} = 2.8 \text{ k/ft}$$

Service live load per lineal foot $= 0.6$ k/ft

2. *Floor beam 2–3 on line D–D:*

Tributary area: $A_T = (20)(30) = 600$ ft^2

Influence area: $A_I = 2A_T = 1200$ ft^2

Live load reduction factor:

$$RF = 0.25 + \frac{15}{\sqrt{1200}} = 0.683$$

Factored loads per lineal foot:

Dead load $= (1.2)(75)(30)$ $= 2.7$ k/ft

Live load $= (1.6) \, [(0.683)(50)(30)] = 1.6$ k/ft

$$\text{Total} = 4.3 \text{ k/ft}$$

Service live load per lineal foot $= 1.0$ k/ft

3. *Interior column D–2:*

Tributary area: $A_T = (30)(20) = 600$ ft^2

Influence area: $A_I = 4A_T = 2400$ ft^2

$(A_I = 4A_T$ for columns)

Live load reduction factor:

$$RF = 0.25 + \frac{15}{\sqrt{2400}} = 0.556$$

Factored load entering column D–2 at this level:

Dead load $= (1.2)(75)(600)$ $= 54.0$ kips

Live load $= (1.6) \, [(0.556)(50)(600)] = 26.7$ kips

$$\text{Total} = 80.7 \text{ kips}$$

4. *Exterior column D–4:*

Tributary area: $A_T = (30)(10) = 300$ ft^2

Influence area: $A_I = 4A_T = 1200$ ft^2

Live load reduction factor:

$$RF = 0.25 + \frac{15}{\sqrt{1200}} = 0.683$$

Factored load entering column D–4 at this level:
Dead load = (1.2)(75)(300)   = 27.0 kips
Live load = (1.6) [(0.683)(50)(300)] = 16.4 kips

                                       Total = 43.4 kips

Since A_I is larger than 400 ft² in all cases, the value of RF is less than 1.0
(see Section 3.3.2).

REFERENCES

[1] Galambos, T. V., Ellingwood, B., MacGregor, J. G., and Cornell, C. A., Probability based load criteria: Assessment of current design practice, *Journal of the Structural Division, ASCE,* Vol. 108, No. ST5, May 1982 (pp. 959–977).

[2] Ellingwood, B., MacGregor, J. G., Galambos, T. V., and Cornell, C. A., Probability based load criteria: Load factors and load combinations, *Journal of the Structural Division, ASCE,* Vol. 108, No. ST5, May 1982 (pp. 978–997).

[3] Corotis, R. B., and Chalk, P. L., Probability models for design of live loads, *Journal of the Structural Division, ASCE,* Vol. 106, No. ST10, October 1980 (pp. 2017–2033).

[4] Peir, J. C., and Cornell, C. A., Spatial and temporal variability of live loads, *Journal of the Structural Division, ASCE,* Vol. 99, No. ST5, May 1973 (pp. 903–922).

[5] Isyumov, N., and Davenport, A. G., A probabilistic approach to the prediction of snow loads, *Canadian Journal of Civil Engineering,* Vol. 1, No. 1, March 1974 (pp. 28–49).

[6] Isyumov, N., and Mikitiuk, M., Climatology of snowfall and related meteorological variables with application to roof snow load specifications, *Canadian Journal of Civil Engineering,* Vol. 4, No. 2, June 1977 (pp. 240–256).

[7] Taylor, D. A., A survey of snow loads on the roofs of arena-type buildings in Canada, *Canadian Journal of Civil Engineering,* Vol. 6, No. 1, March 1979 (pp. 85–96).

[8] Taylor, D. A., Snow loads for the design of cylin-drical curved roofs in Canada 1953–1980, *Canadian Journal of Civil Engineering,* Vol. 8, No. 1, March 1981 (pp. 63–76).

[9] Theakston, F., *Snow Loads on Low-Profile Roofs,* Proceedings, 1980 Canadian Structural Engineering Conference, Montreal, Quebec, Canada, February 1980.

[10] Simiu, E., *Extreme Wind Speeds at 129 Stations in the Contiguous United States,* Building Science Series Report No. 118, National Bureau of Standards, Washington, DC, March 1979.

[11] Applied Technology Council (ATC), *Tentative Provisions for the Development of Seismic Regulations for Buildings,* ATC 3-06, Applied Technology Council, Palo Alto, CA, 1988.

[12] Degenkolb, H. J., *Practical Design (Aseismic) of Steel Structures,* Proceedings, 1978 Canadian Structural Engineering Conference, Toronto, Ontario, March 1978.

[13] Teal, E. J., Seismic design practice for steel buildings, *AISC Engineering Journal,* Vol. 12, No. 4, 1975.

[14] Clough, R. W., and Penzien, J., *Dynamics of Structures,* McGraw-Hill, New York, 1975.

[15] Hurty, W. C., and Rubinstein, M. F., *Dynamics of Structures,* Prentice-Hall, Englewood Cliffs, NJ, 1964.

[16] Marino, F. J., Ponding of two-way roof systems, *AISC Engineering Journal,* Vol. 3, No. 3, July 1966.

[17] Turkstra, C. J., and Madsen, H. O., Load combi-

nations in codified structural design, *Journal of the Structural Division, ASCE,* Vol. 106, No. ST12, December 1980 (pp. 2527–2543).

[*18*] Turkstra, C. J., *Theory of Structural Safety,* Solid Mechanics Study No. 2, Solid Mechanics Division, University of Waterloo, Waterloo, Ontario, Canada, 1970.

[*19*] Galambos, T. V. (Ed.), *The Structural Stability Research Council Guide to Stability Design Criteria for Metal Structures,* 4th ed., Wiley-Interscience, New York, 1988.

CHAPTER FOUR

Materials
for Steel Structures

4.1 SCOPE OF CHAPTER

The subject of this chapter is of great importance to the structural steel designer, yet it is too often considered but briefly in the design process. Whether it is a question of selecting the appropriate steel grade for the framing members or of how to evaluate the complexities of connection design, fabrication, and long-term service, it is crucial that the engineer understands all aspects of the behavior of the materials that will be used. In essence, each material has its advantages and limitations, and many of the decisions that have to be made are only marginally of a structural nature.

The chapter details the characteristics of the various steel materials that are used for steel structures today, essentially in accordance with the listings of the LRFD specification. These include the grades of steel for shapes and plates, as well as the materials that are utilized for mechanical connectors (i.e., bolts, shear connectors) and for welded joints. Emphasis will be placed on the properties that are of primary importance to all of its users (architect, engineer, fabricator, owner):

1. Mechanical properties (strength, ductility)
2. Chemical/metallurgical properties
3. Influence of service conditions (short vs. long-term behavior, operating temperatures and environment, types of loads)
4. Influence of methods of fabrication and erection

A brief discussion is also given of some of the criteria that govern material costs, but only insofar as steel mill deliveries are concerned. The details of fabrication cost estimating are considered to be beyond the scope of this book.

A great many materials standards provide the detailed performance and composition requirements for structural and other grades of steel. References [1–4] contain all of these, in addition to the delivery and property determination (i.e., testing) procedures. In general, the American Society for Testing and Materials (ASTM) develops all of these documents in the United States as consensus standards, meaning that they represent the material requirements that have been formulated by broadly based technical committees. The members of these groups come from producers as well as users. In Canada the role of ASTM is played by the Canadian Standards Association (CSA); its methods of working and the material requirements are very similar to those of ASTM.

The primary exceptions to the ASTM-produced specifications are those developed by the American Welding Society (AWS) for welding electrodes or filler metals. However, the basic system is the same, and the user has only to recognize the needs for weld and base metal matching, as well as the particular demands of the various welding processes. In other words, the material specification for an electrode also depends on the method of welding. These criteria are discussed in Section 4.8 and in the part of Chapter 11 that deals with the design of welded connections.

4.2 BRIEF HISTORICAL NOTES

The earliest man-made iron-based materials were used for tools and weapons, dating as far back as 1350 B.C. [5]. The metals were obtained directly from iron-bearing ore by heating the ore strongly over coal fires, combining the effects of heat, carbon, and oxygen to reduce the iron from the ore (i.e., separating the iron from the oxides to which it is most commonly bound, namely, FeO, Fe_2O_3, and Fe_3O_4) through the process of smelting. Depending on the length of time the ore was kept in contact with the hot coals, the final product contained varying degrees of carbon. The early ironmasters quickly learned that a longer exposure was needed for the iron to become what is now called hardenable through rapid cooling or quenching. They also discovered that if the metal were too hard, it would be brittle, and that a subsequent heating to a low temperature would alleviate most of the unfavorable effects of the quenching. This reheating technique is now known as tempering.

The early iron had carbon contents of up to 2 percent and the tools and weapons were made from bars that had been wrought from this metal; hence the name *wrought iron*. Structural usage of this material was limited, for a number of reasons; most important was probably the fact that the methods of construction of the time were not amenable to the use of metals. Other than that, the needs for weapons and tools overshadowed all others.

The ironmasters also discovered that if the carbon content became too high, that is, by leaving the iron ore exposed to the hot coals for a long time, a very porous, low melting point material would result. This was the first appearance of cast iron, and it was generally regarded as useless because of its brittleness and temperature sensitivity. However, once it was determined how this form of iron could be cast into almost any shape that was needed, its practical applications soon multiplied.

The high compressive strength of cast iron made it one of the earliest structural metals. For example, cast iron arch bridges were used in Europe in the late 18th century, and columns of this kind were used in buildings in the United States in the first part of the 19th century [6]. By this time wrought iron also had found structural applications; of note is the Coalbrookdale (England) bridge of 1779, and some of the iron trusses and multistory frames that were built in the United States between 1840 and 1860.

Widespread use of metals for structural purposes did not come about until the advent of a material that had excellent strength and sitffness properties regardless of the direction of the applied load. The old ironmaking techniques were too limited in terms of product volume and diversity, and it was only when Kelly and Bessemer independently devised the method of blowing air through the bath of molten iron that large-scale production and significant improvements could be made to the end product [5]. This also allowed the producer to control the chemical composition of the metal, and thus steel was born.

The work of Kelly and Bessemer came to fruition around 1850. Since then a great many improvements and innovations have been made to the basic steelmaking techniques, in particular, the work of Karl Siemens in developing the open-hearth steelmaking process and of William Siemens in devising the electric furnace process. The open-hearth method has played the most important role in the American steel industry; it has only recently been supplemented by the basic oxygen process to an increasing extent, largely because of the cost of energy. Reference [5] gives extensive data on the history and current techniques of steelmaking; it is probably the single most comprehensive source of information in this field.

The number of notable structures that have been built with structural steel are legion. The first major structure in the United States was the Eads Bridge in St. Louis, Missouri [7], a milestone not only because of its use of steel, but also because it represented the first of its kind to utilize steel shapes. Other significant achievements are the Eiffel Tower in Paris, France (1889), the Brooklyn Bridge in New York (1883), and the Firth of Forth railroad bridge near Glasgow, Scotland (1890). In this century bridges such as the George Washington, the Golden Gate, the Verrazano Narrows (all in the United States), and the Humber River (England) are spectacular achievements in suspension bridge construction; the Sydney Harbor (Australia) and the New River Gorge (West Virginia) are outstanding arch bridges, to mention but a few. Steel-framed building structures abound as well; the very tall multistory building is an American invention of the Chicago and New York groups of architects. Among the best known of these structures are the Woolworth, Chrysler, Empire State, and World Trade Center buildings in New York, and the John Hancock Center, Standard Oil, and Sears Tower buildings in Chicago. The Sears Tower is presently the tallest building in the world (1450 ft).

Many of the advances in the use of steel for structures have been facilitated by ever-improving steel grades and connection methods. For example, the chemical additives that have made possible the development of high-strength steels and weathering steels have allowed designers to go to unsurpassed heights and spans, as well as to reduce overall maintenance costs. This is discussed in some detail in Section 4.5. Methods of connection have evolved from rivets to high-strength bolts, and the multifaceted techniques of welding have played a crucial role. In brief, the state-of-the-art of steel construction reflects a long-term development in all facets of steelmaking, including metallurgical as well as structural and mechanical criteria.

4.3 STEEL PRODUCTION

It is important for the user of structural steel to have more than just a rudimentary knowledge of the production of steel. A knowledge of steel production will impart an understanding not only of the various metallurgical and chemical processes that influence the final product, but also of the many heating, cooling, forming, and straightening operations that are part of the manufacturing procedure. All of these affect the performance of the material in the completed structure, and an intelligent choice of steel grades along with the proper connection methods can then be made. Reference [5] contains detailed data about all of these operations, and Ref. [8] gives a succinct explanation of the most important parts.

Figure 4.1 gives a schematic illustration of the primary steps in a modern steelmaking process. In brief, the blast furnace is charged (i.e., filled from the top) with the basic components of the first process: iron ore, limestone, and coke. Coke is a product of coal, specifically made for steel production. Hot air is then blown into the blast furnace at the bottom, and the ore and other materials are gradually reduced to iron as they descend through the furnace. The end product is tapped at the bottom and separated from the waste products (slag); the iron that has thus been made is often referred to as pig iron [8].

Visitors to a steel mill rarely see the part of the operation that has just been described. The first activity to be seen is usually the end of the second major step: the pouring or teeming of the molten steel from the open hearth or other furnace. Before this occurs, the pig iron that comes from the blast furnace is charged to the primary steelmaking furnace, which may be either of the open hearth, oxygen, or electric kind. Very often scrap steel will be added at this stage, as indicated in Fig. 4.1.

It is in these furnaces that most of the chemical or alloying elements are provided, to develop a steel with the desired properties. Although more than 95 percent of the material is made up of iron, the remaining percentage of other components is important, since they will largely control strength, ductility, and other characteristics of the finished product. The material specifications dictate the chemical composition of the grades of steel, and the producer must show proof of this, normally through the data given by the mill test certificate. In addition to showing the mechanical properties of the steel, this certificate indicates the quantities of carbon and alloying elements that can be found in the particular batch or heat of steel that has come from a single pour from the furnace.

The words *basic* or *acid* are sometimes used to describe some of the steelmaking processes, as, for example, in the basic oxygen furnace (BOF). These terms describe the type of brick lining that is used in the furnaces [5].

Oxygen is crucial to the production of steel, because it reacts with carbon to remove excess amounts of this chemical. The carbon monoxide

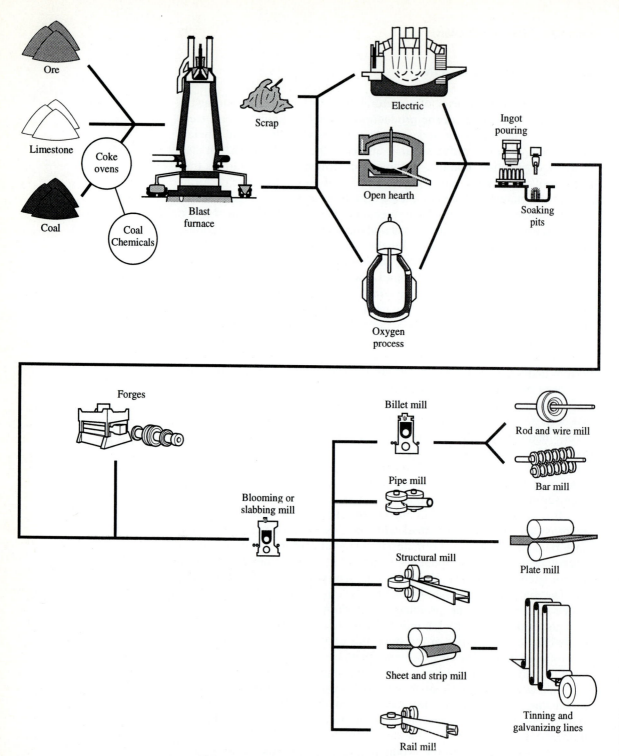

Figure 4.1 Steps in the steelmaking process.

gas that develops in this reaction has to escape from the molten metal; otherwise, large pockets of gas may be trapped in the metal as it cools. The formation of the gas occurs while the steel bubbles violently; the technical term for this phenomenon is rimming, and hence certain grades of steel are referred to as rimmed steel [1].

It is readily understood that the potential for gas pockets in a rimmed steel may make it a less desirable grade for many applications. The degree of rimming is therefore often reduced by adding so-called deoxidizing elements to the molten metal, the most common of which are silicon and aluminum. The amount that is added may stop the rimming entirely, or simply reduce it to a more acceptable level. Steel grades that have been produced with all rimming suppressed are referred to as killed steels; grades with a lesser degree of deoxidation are called semi-killed.

Most steel for structural purposes is supposed to be semi- or fully killed. One notable exception is ASTM A36 steel for plates in thicknesses less than $\frac{1}{2}$ in. and for structural shapes of group 1 [1] (shape grouping is explained in Section 4.4). The specification simply observes that A36 steel for thicker plates and other shape groups cannot be of rimmed or capped quality. (Capped steel is similar to rimmed; the details of the manufacturing process are somewhat different [5].) Another exception is ASTM A514 steel; this quenched and tempered grade must be fully killed [1].

Silicon is the most commonly used agent for deoxidation. Aluminum has the same effect, but is mostly used to produce steel with a finer grained crystalline structure. For that reason many material specifications provide optional, supplementary requirements that dictate not only the killing practice, but also the crystalline structure. Steels of this kind are usually referred to as produced to silicon-killed, fine-grained practice [1], and are generally preferred for applications where material toughness and more uniform properties are needed. A fully welded structure with severe dynamic loads and environmental service conditions may require such a material. Standards for testing for such properties have been developed by ASTM [9]. Among other grain-refining methods, the techniques of electroslag remelting, vacuum-arc remelting, and vacuum degassing are currently used for certain steel grades [1, 5].

Referring again to Fig. 4.1, once the steelmaking has been completed in the open hearth furnace (for example), the molten metal is then poured into molds that will yield large slabs or ingots when it has cooled. A typical ingot weighs about ten tons. As soon as it has solidified sufficiently, the mold is stripped and the ingot is placed in a soaking pit. This is essentially an oven that is designed to maintain a uniform temperature throughout the ingot, to facilitate the subsequent rolling operations. These take place in the many kinds of rolling mills that are found in the plant, and basically "knead" the steel ingot into the requisite shapes. As can be seen in Fig. 4.1, the various types of mills break the steel down to successively smaller sizes. For example, an ingot that is to become a wide-flange shape (see Section 4.4) is taken from the soaking pit to several passes through the rolls of the bloom mill, then through the breakdown mill, and finally through the series of vertical and horizontal rolls that make up the wide-flange mill. Figure 4.2 is a photograph from one of these operations. Similar techniques are followed in the production of tubular products and rails; the number of rolling stages for plates and plate-like products are normally fewer and less complicated.

Production of steel by casting into ingots and subsequent rolling is by far the most common method in North America. However, certain plants use the technique of continuous casting, where the molten steel is poured directly and continuously into the form of long slabs or plates. Cutting and straightening devices allow the producer to ready the material for rolling into wide-flange shapes, for example. The requirements of this procedure are unique, and some problems are difficult to avoid. In particular, the method is susceptible to the formation of large interior separations or craters [5]. USX Corporation, Bethlehem Steel Corporation, and the NUCOR Corporation currently operate continuous casting facilities in North America.

In the final stages of steel production the finished products are left on the cooling beds to cool to room temperature. In some cases controlled cooling is required; in most cases simple air exposure is used. Due to the size and shape of the members the cooling is far from even throughout the material, and as a result internal or residual stresses and crookedness develop. The residual stresses can be more or less removed by varying degrees of reheating and controlled cooling, the most extensive of which is referred to as stress-relief annealing. Straightening of excessively

Figure 4.2   A wide-flange shape being rolled in a steel mill. Courtesy Bethlehem Steel Corporation.

curved members is routinely used in the steel mills to meet the delivery requirements of ASTM A6 [*1*] or CSA G40.20 [*2*]; it may be done at room temperature or some higher level [*5*]. Straightening is normally achieved by sending the crooked member through a series of rolls, called rotorizing. In the case of large shapes the straightening is done by loading with large hydraulic jacks at certain points along the length (gag straightening).

As will be seen in Section 4.5, all of the rolling, heating, cooling, and straightening operations influence the material and its metallurgical and mechanical properties. To a structural engineer metallurgical properties are especially important when the design and fabrication of connections are considered; the mechanical properties influence all aspects of member and connection behavior.

4.4 STRUCTURAL STEEL PRODUCTS

The products of the steel mill that are of particular interest to the structural engineer can be loosely classified as plate-like products and shapes. Some of the plate-like products are further used to make tubular elements and very thin plates, called sheet and strip, which form the starting point for the production of a great variety of light-gage cold-formed shapes and plates. The most common of these include the many types of corrugated or formed steel deck, frequently used for walls and roofs of some structures, as well as for composite (steel-concrete) bending members and slabs. The design of cold-formed steel structures is beyond the scope of this book and the specifications issued by AISC; instead the criteria developed by the American Iron and Steel Institute (AISI) [*10, 11*] and their Canadian counterpart [*12*] should be used. Additional procedures that apply to the design of steel deck diaphragms have been published by the Steel Deck Institute (SDI) [*13*]. The use of steel deck in composite construction is covered in Chapter 10.

The terminology that is used to describe the various steel products in the following is the same as that used by ASTM [*1, 14*]. The standardized sizes are given by ASTM; for the structural engineer they are most conveniently available through the *LRFD Manual of Steel Construction* [*14*]. This manual should be part of every designer's library.

4.4.1 Structural Shapes

Figure 4.3a shows schematically the types of hot-rolled structural steel shapes that are used in construction today under the general name of wide-flange shapes. They are also referred to as I-beams or H-shapes, denoting the geometry of the cross section, but these are nontechnical names that should be avoided. The figure identifies the various parts of the shape and their names, specifically, the flanges and the web, and the designations that are used for depth, width, and thicknesses. The wide-flange shapes have a depth that is greater than or equal to the flange width, and the flange thickness is usually larger than that of the web.

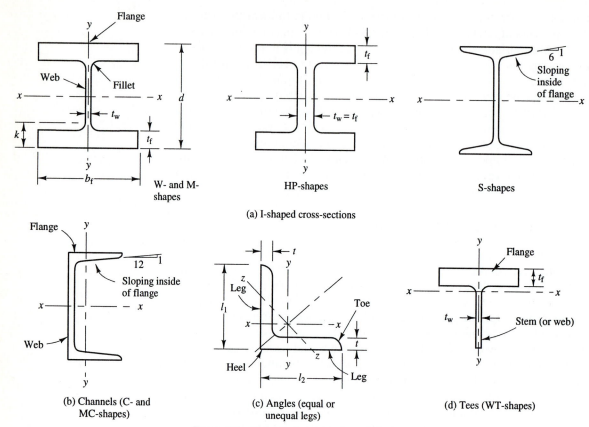

Figure 4.3   Hot-rolled structural steel shapes.

In actual usage there are a number of categories of I-shapes:

W-shapes
M-shapes
HP-shapes
S-shapes

and the characteristics of each are noted in the figure. The W-, M-, and HP-shapes have outsides and insides of the flanges that are parallel, as shown; the differences between the W and M designations are largely by ASTM definition [1]. The HP-shapes are unique in that the flange and web thicknesses are equal. The outside measurements of S-shapes may be very similar to those of the W- and M-shapes, but the inside of the flanges are sloped, as shown. (The S-shapes were previously called American Standard I-shapes.)

The W-, M-, and S-shapes are primarily used for beams and columns, whereas the HP-shapes are intended for piles. Referring to the *x* and y axes (see Fig. 4.3) as the strong (major) and weak (minor) centroidal axes of the cross section, the preferred usage of the W-, M-, and S-shapes is governed by the ratio of the strong and weak axis moments of inertis, I_x and I_y, as follows:

1. Use the shape as a column if $I_x/I_y \leq 5$
2. Use the shape as a beam if $I_x/I_y \geq 5$

The *LRFD Manual* [*14*] gives the cross-sectional data for all structural steel shapes. The above numbers should be regarded only as guidelines; in many cases unique structural conditions require a shape for a column whose moment of inertia ratio is substantially larger than 5 (for example). These concerns are addressed in Chapters 6 to 9.

Figure 4.3b–d shows the other common structural shapes. The channel is given in Fig. 4.3b; it is a singly symmetric shape with sloping insides of the flanges. Products of this kind are designated as C- or MC-shapes; the differences between these shapes are ASTM-defined. It is also noted that all M- and MC-shapes may not be readily available from local or even large area suppliers. Although preferable for many applications, the single-axis symmetry of the channel warrants special considerations when it is being used as a bending member.

One of the most common shapes is shown in Fig. 4.3c. This is the angle (or L-shape); it may have legs of the same or unequal length. In the former case it is called an equal-legged angle; in the latter it is an unequal-legged angle. Apart from the areas in the vicinity of the angle toes and heel, the leg thickness is constant. The cross-sectional axes that are given as x and y are centroidal but not principal; this causes some difficulties when single angles are used as columns, for example.

Figure 4.3d shows a structural tee; each shape of this type is made by cutting a wide-flange shape in half along the length. Called WT-shapes, the tees have unique applications, for example, as members in trusses.

Other hot-rolled cross-sectional shapes that may be encountered are the rail and the zee. Both have particular uses, but are limited for structural purposes. Data for rail sections are given in the *LRFD Manual*; the zee is used primarily as purlins in pre-engineered buildings, and is therefore not included here.

4.4.2 Tubular Products

The cross sections of the commonly used tubular products are shown in Fig. 4.4. They are all referred to as tubes, pipes, or structural tubing, with a further qualifier as to whether they are circular, square, or rectangular. The circular tubes are also known as pipe, although strictly that is the name of a certain product defined by ASTM Standard A53. All of these products are either manufactured from flat plates that are gradually bent into a circle and electric resistance welded, formed by piercing a solid billet called seamless [*5*], or spirally welded from strip steel. The first two

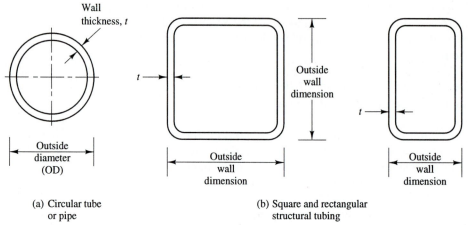

(a) Circular tube
or pipe

(b) Square and rectangular
structural tubing

Figure 4.4   Tubular products.

techniques are by far the most commonly used in the United States and Canada. The square and rectangular shapes are formed from the circular products, by running the latter through sets of rolls that gradually change the shape into the required type. This is also the reason why the corners of these types of tubes are rounded, as shown in Fig. 4.4b. The wall thickness is constant throughout the shape, with the exception of the corner areas (due to local bending effects in the forming process).

Tubular products are either completed as hot-formed (HF) or cold-formed (CF), and the implications of both names are obvious. The advantage of the HF type is that the final heat treatment or processing under an elevated temperature will relieve some of the residual stresses that otherwise would develop. This gives the HF members a somewhat improved performance, especially under compression [15]. The CF tubes lose some of this advantage, although cold-forming of the material has the effect of raising its apparent strength [15]. Cold-formed tubes are also less expensive than hot-formed ones.

In the United States the distinction between HF and CF tubular products is achieved by having separate material specifications for the two, as discussed in Section 4.6. In Canada this is obtained by specifying whether the tube is Class H or Class C, respectively [2].

4.4.3 Plates and Bars

The appearance and distinction between the various plate and bar products are illustrated schematically in Fig. 4.5. The differences are essentially industry-defined [1], to the effect that plates have a width of 8 in. or greater, with thicknesses as small as 0.18 in. The wide face is always rolled; the edges may also have been rolled, or the plate may have been produced by cutting (shearing or gas cutting) from a larger unit [1, 14]. The all-rolled product is called a universal mill (UM) plate; the other is referred to as a sheared plate. The residual stresses are significantly

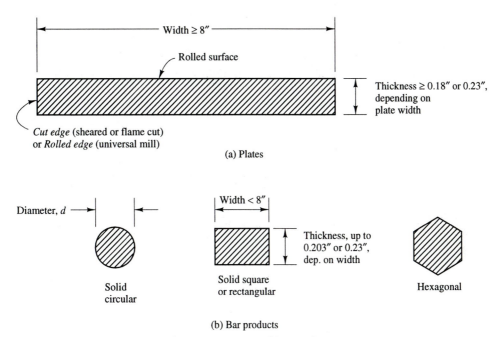

Figure 4.5   Plate and bar products.

different in the two categories. Thus, welded built-up wide-flange shapes with UM plates are not as strong in compression as those that utilize sheared plates.

Bars are solid cross-section elements that may be square, rectangular, circular, or hexagonal. The structural engineer uses mostly the first three types.

Designs involving plate-like products could theoretically specify any thickness and width, but this would be impractical, so certain preferred dimensions are used, as given by Table 4.1. The data in Table 4.1 represent preferred width, thickness, and diameter increments. Use of these promotes economy of construction.

TABLE 4.1   PREFERRED DIMENSIONS FOR PLATES AND BARS [14]

Product	Range of Thicknesses/Diameters				Width
	$t \leq \frac{1}{2}$ in.	$\frac{1}{2} < t \leq 1$ in.	$1 < t \leq 3$ in.	$t \geq 3$ in.	
Plates	$\frac{1}{32}$	$\frac{1}{16}$	$\frac{1}{8}$	$\frac{1}{4}$	Consult with steel mill
Square and rectangular bars	$\frac{1}{8}$	$\frac{1}{8}$	$\frac{1}{8}$	$\frac{1}{8}$	$\frac{1}{4}$
Circular bars	$\frac{1}{8}$	$\frac{1}{8}$	$\frac{1}{8}$	$\frac{1}{8}$	—

Note. Table gives increments in thickness or diameter.

The number of cross-sectional shapes that can be made by welding together plates and/or shapes of different sizes is almost unlimited. Figure 4.6 shows some of the most common types. None of these are standardized in the United States; in Canada the Algoma Steel Company produces a range of welded wide-flange shapes (WWF) that are available on the same basis as rolled shapes [*16*].

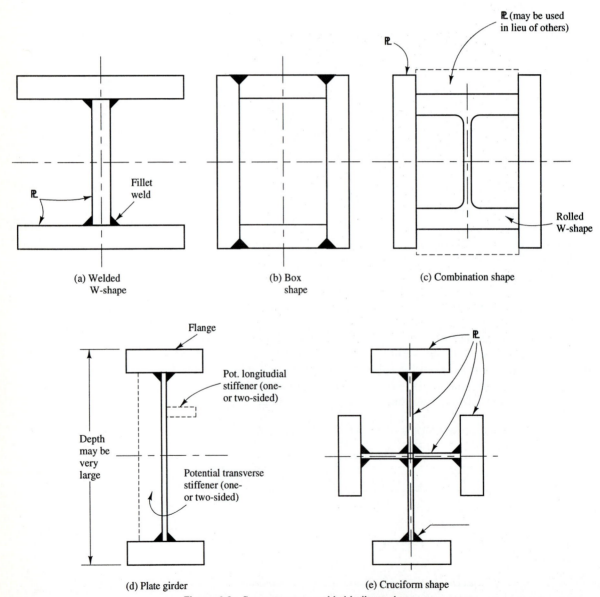

Figure 4.6   Some common welded built-up shapes.

Welded built-up shapes are particularly useful when large forces must be resisted, as for the lower level columns in multistory buildings. Some of these shapes can be extremely heavy, with plate thicknesses of 8 in. or more [*16, 17*]. An example of the use of box shapes is given by Fig. 4.7, which shows part of the facade of the 64-story headquarters building of the USX Corporation in Pittsburgh, Pennsylvania. Note that columns as well as spandrel beams are box shapes. All of these are filled with a mixture of water and antifreeze as part of the fire protection system for the structure.

In Chapter 2 we mentioned that the mechanical properties of steel, in particular, the yield stress, are a function of the thickness or mass of the material. Thus, two pieces of steel of the same nominal grade may have different strengths, depending on their origin. This is readily taken care of for plates and bars, for which the ASTM and CSA standards specify the properties as a function of the material thickness. For rolled shapes the process is a little more complicated, and for that reason the ASTM has divided the shapes into a number of structural shape size groups [*1*]. They are essentially dependent on the web thickness of the shape (leg thickness for angles, stem thickness for tee). Table 4.2 shows the details of the five shape groups. Note that the heavier the shape, the

Figure 4.7 Part of the facade of the USX building.

TABLE 4.2 STRUCTURAL SHAPE SIZE GROUPS FOR TENSILE PROPERTY CLASSIFICATION

Structural Shape	Group 1	Group 2	Group 3	Group 4	Group 5
W-shapes	W24 × 55, 62	W36 × 135 to 210 incl	W36 × 230 to 300 incl	W14 × 233 to 550 incl	W14 × 605 to 730 incl
	W21 × 44 to 57 incl	W33 × 118 to 152 incl	W33 × 201 to 241 incl	W12 × 210 to 336 incl	
	W18 × 35 to 71 incl	W30 × 99 to 211 incl	W14 × 145 to 211 incl		
	W16 × 26 to 57 incl	W27 × 84 to 178 incl	W12 × 120 to 190 incl		
	W14 × 22 to 53 incl	W24 × 68 to 162 incl			
	W12 × 14 to 58 incl	W21 × 62 to 147 incl			
	W10 × 12 to 45 incl	W18 × 76 to 119 incl			
	W8 × 10 to 48 incl	W16 × 67 to 100 incl			
	W6 × 9 to 25 incl	W14 × 61 to 132 incl			
	W5 × 16, 19	W12 × 65 to 106 incl			
	W4 × 13	W10 × 49 to 112 incl			
		W8 × 58, 67			
M-shapes	To 20 lb/ft incl				
S-shapes	To 35 lb/ft incl	Over 35 lb/ft			
HP-shapes		To 102 lb/ft incl	Over 102 lb/ft		
American standard channels (C)	To 20.7 lb/ft incl	Over 20.7 lb/ft			
Miscellaneous channels (MC)	To 28.5 lb/ft incl	Over 28.5 lb/ft			
Angles (L), structural & bar-size	To $\frac{1}{2}$ in. incl	Over $\frac{1}{2}$ to $\frac{3}{4}$ in. incl	Over $\frac{3}{4}$ in.		

Notes. Structural tees from W-, M-, and S-shapes fall in the same group as the structural shape from which they are cut. Group 4 and group 5 shapes are generally contemplated for application as compression members. When used in other applications or when subject to welding or thermal cutting, the material specification should be reviewed to determine if it adequately covers the properties and quality appropriate for the particular application. Where warranted, the use of killed steel or special metallurgical requirements should be considered.

Source: Reprinted by permission of the American Institute of Steel Construction, Inc., from the 1st edition of the *AISC-LRFD Manual* (pp. 1–9).

larger is the group number. The correlation between the various steel grades and the properties of the plates and shapes is discussed in Section 4.6.

4.4.4 Nomenclature for Structural Steel Products

To facilitate ordering procedures, material specifications, and the like, a certain nomenclature is used for all structural shapes and plate-like products. In brief, the following system is used for shapes:

$$(\text{Shape letter designation(s)})[\text{Nominal depth (in.)}] \times \begin{bmatrix} \text{Nominal weight} \\ \text{per foot (lb)} \end{bmatrix}$$

For example, the designation W21 × 57 indicates a wide-flange shape of 21 in. nominal depth and a weight of 57 lb per lineal foot. The cross-sectional and other data are given in the *LRFD Manual*.

Angles use a different system:

(Letter L (for angle))[Largest leg length] × [Smallest leg length]
    × [Leg thickness]

The above allows for equal- and unequal-legged angles alike. All measurements are given in inches.

For tubular products distinction must be made between circular and square/rectangular shapes. The following designations are to be used:

For circular tubes

(Outside diameter (in.) OD) × [Wall thickness (in.)]

For square and rectangular tubes

(Largest outside dimension (in.))
    × [Smallest outside dimension (in.)] × [Wall thickness (in.)]

For plates and bars the nomenclature is as follows:

For plates

(Letters PL or ℙ) [Plate thickness (in.)] × [Plate width (in.)]

For circular bars

(Bar)[Diameter (in.)] ∅

For square and rectangular bars

(Bar)[Smallest dimension (in.)] × [Largest dimension (in.)] ⌀

Table 4.3 gives a list of designations as typical examples of the steel product nomenclature. The reader should consult the *LRFD Manual* while going over the listing.

It is also observed that the cross-sectional data given by the *LRFD Manual* come as decimal numbers and as fractions. Decimal numbers should be used by designers; fractions are intended for use by the structural detailer in the fabricating company in making up shop drawings of the structural components and their connections.

The cross-sectional sizes and other properties of structural shapes change somewhat over the years. The designer should therefore always make use of the most recent edition of the *LRFD Manual* or the steel producers' catalogs to ensure that the products chosen are actually available. For example, a number of important changes in the sizes of wide-flange shapes were implemented by the American and Canadian steel mills in 1978; thus, many of the shapes that were listed in the earlier manuals are not produced any more.

The preceding can present a serious problem to the structural engineer that is assigned the task of evaluating an old steel-framed building. However, Ref. [*19*] contains a detailed listing of the shapes (and their properties) that were produced between 1873 and 1952, thus effectively covering the entire period of American steel shape production. The changes that were made after 1952 are indicated by the data given in the 6th (1963), 7th (1970), 8th (1980), and 9th (1989) editions of the *ASD Manual* and in the 1st (1986) edition of the *LRFD Manual*.

TABLE 4.3   EXAMPLES OF PRODUCT NOMENCLATURE

W shapes	$W21 \times 57$
M-shapes	$M10 \times 9$
S-shapes	$S18 \times 70$
HP-shapes	$HP13 \times 60$
Standard channels	$C9 \times 20$
Miscellaneous channels	$MC12 \times 37$
Tees, cut from W-shapes	$WT10.5 \times 28.5$ $(= \frac{1}{2}$ of $W21 \times 57)$
Tees, cut from M-shapes	$MT5 \times 4.5$ $(= \frac{1}{2}$ of $M10 \times 9)$
Equal-legged angles	$L3 \times 3 \times \frac{3}{8}$
Unequal-legged angles	$L5 \times 3 \times \frac{3}{8}$
Circular tubes*	$5 \times OD \times 0.258$ ASTM A. . .
Square tubes*	$6 \times 6 \times \frac{1}{2}$ ASTM A. . .
Rectangular tubes*	$16 \times 12 \times \frac{3}{8}$ ASTM A. . .
Plates	$\text{Pl} 1 \times 8$
Circular bars	Bar $\frac{3}{4} \emptyset$
Square bars	Bar 1 ▱
Rectangular bars	Bar $\frac{3}{8} \times \frac{3}{4}$ ▱

Note. Structural tubing (*) must include the ASTM specification number.

4.5 CHARACTERISTICS OF STEEL

The many property characteristics of steel can be generally grouped under the following headings:

1. Chemical composition
2. Mechanical properties

Mechanical properties include data on ductility and toughness. All of the criteria that fall into these major groups are important to consider in the choice of steel for a structure, although most structural engineers have a tendency to confine their evaluation to the mechanical properties.

The descriptions that are given in this section detail the conditions at room temperature, around 70°F. Only small changes occur within the temperature range 30–110°F; it is only when significantly lower or higher temperatures are encountered that their effect on the material properties should be considered. This topic is discussed in Section 4.10.

4.5.1 Chemical Composition

The primary types of structural steel are usually classified according to the following chemical composition categories:

1. Carbon-manganese steels
2. High-strength, low-alloy (HSLA) steels
3. High-strength quenched and tempered alloy steels

By their chemistry, the steels that are used for bolts, stud shear connectors, and welding electrodes also fall into the above general classes.

The carbon–manganese steels are also referred to as carbon steels and mild structural steels. The primary chemical elements (in addition to iron, of course) are carbon and manganese, although the materials standards place restrictions on certain other elements as well. Restrictions apply in particular to sulfur and phosphorus, both of which have a significant detrimental effect on the ductility and weldability of the steel. The materials of this type are generally the least expensive; they have quite adequate strength and ductility characteristics, and are therefore by far the most used grades. All sizes of shapes and plates are commonly available in these grades, and their weldability is usually good.

One of the most prominent of the carbon–manganese steels is the ASTM grade A36, with a specified minimum yield stress (see Section 4.5.2) of 36 ksi. It is currently used for well over half of the structural steel tonnage for framing members and detail elements.

In the relatively recent past a type of steel known as high-strength carbon steel saw some usage. The higher strength was primarily achieved by raising the carbon contents, but this had the effect of making the steel poorly suited for welding. Also known as general construction steels, their limited weldability essentially forced these grades off the market. A typical grade was ASTM A440, which was dropped as a recognized grade by ASTM in 1979. Some existing structures were built with that grade, however, and a designer involved in the evaluation of such a building for expansion or other refurbishing would be well advised to make use of bolted connections wherever possible. This comment applies even more importantly if some of the earlier (pre-1930, approximately) steel-framed structures are being studied. The chemical compositions of the steel grades of that period are very different from those of today. The designer given such an assignment should consult with a metallurgical engineer and the representatives of a steel producer to obtain the chemical composition for the material at hand.

The high-strength, low-alloy (HSLA) steels represent a relatively recent development in steelmaking. The higher strength (42- to 65-ksi yield stress, as opposed to 30 to 36 ksi for the C-Mn steel types) is achieved by adding small amounts of a number of chemical elements, such as columbium (Cb), molybdenum (Mo), and vanadium (V), to the steel matrix, instead of raising the carbon contents. Adding these elements increases the strength, but at a cost of lowered ductility. All of the HSLA steels are weldable, some of them have improved corrosion resistance, and most are available in several grades. Two of the most common ASTM grades of HSLA steel are A572 and A588, both of which are available as shapes and plates in most size categories (see Table 4.2) for the 42- to 50-ksi yield stress classes.

The high-strength quenched and tempered (Q&T) alloy steels that are used for structural purposes are essentially available only as grade ASTM A514 today. With a yield stress level of 90 to 100 ksi, the increase in strength over that of the HSLA variety is achieved through heat treatment. Thus, the rapid cooling or quenching produces a material with a very fine-grained and hard crystalline structure, called martensite [5]. The quenched material has very high strength, but low ductility, and for that reason it is subsequently sent through a tempering or reheating process. This causes some loss of strength, but also a significant increase in ductility.

Proper use of Q&T steels requires more than the usual attention to chemistry, metallurgy, and strength. For example, unique welding procedures must be used, and the designer should pay particular attention to stability effects, especially for local buckling (see Chapters 6 and 7). A514 is available only in plate form, up to 6 in. thick.

The materials that are used for the two grades of high-strength bolts (see Section 4.7) are also of the Q&T variety. This is an important consideration, especially for the higher of the two grades (ASTM A490). This and related subjects are detailed in Chapter 11.

We now discuss briefly the most important chemical elements and their influence on the properties of steel. Complete data can be found in Ref. [5]. The required chemical compositions of the various steel grades are given by the respective ASTM standards [1]. Iron, the single most important element in steel, is not discussed. As already observed, iron constitutes roughly 95 percent of the structural steel matrix, and materials with a lower percentage are not normally classified as "structural." In other words, steel without iron would not be steel.

Carbon (C).   Next to iron, carbon is by far the most important chemical element in steel. Increasing the carbon content produces a material with higher strength and lower ductility. Structural steels, therefore, have carbon contents between 0.15 and 0.30 percent; if the carbon content goes much higher, the ductility will be too low, and for magnitudes less than 0.15 percent the strength will not be satisfactory.

Manganese (Mn).   Manganese appears in structural steel grades in amounts ranging from about 0.50 to 1.70 percent. It has effects similar to those of carbon, and the steel producer uses these two elements in combination to obtain a material with the desired properties. Manganese is a necessity for the process of hot rolling of steel by its combination with oxygen and sulfur [5].

Aluminum (Al).   As observed in the discussion of the production of steel, aluminum is one of the most important deoxidizers in the material, and also helps form a more fine-grained crystalline structure. It is usually used in combination with silicon to obtain a semi- or fully killed steel.

Chromium (Cr).   Chromium is present in certain structural steels in small amounts (for example, between 0.10 and 0.90 percent in ASTM A588). It is primarily used to increase the corrosion resistance of the material, and for that reason often occurs in combination with nickel and copper. Stainless steels will typically have significant amounts of chromium. Thus, the well-known "18-8" stainless steel contains 18 percent nickel and 8 percent chromium.

Columbium (Cb).   Columbium is a strength-enhancing element, and is one of the important components in some of the HSLA steels. Its effects are similar to those of manganese and vanadium (see below); it also has some corrosion-resistance influence. Cb appears in types 1 and 3 of ASTM A572; it is restricted to between 0.005 and 0.05 percent in type 1, and between 0.02 and 0.15 percent in type 3. It must be used in combination with vanadium for type 3.

Copper (Cu).   Copper is another of the primary corrosion-resistance elements. It is typically found in amounts not less than 0.20 percent, and is the primary anticorrosion component in steel grades like A242 and A441.

Molybdenum (Mo).   Molybdenum has effects similar to manganese and vanadium, and is often used in combination with one or the other. It particularly increases the strength of the steel at higher temperatures [5] and also improves corrosion resistance. Typical amounts of molybdenum are 0.08 to 0.25 percent for certain grades of A588 steel, and 0.15 to 0.65 percent for various types of A514 [1].

Nickel (Ni).   In addition to its favorable effect on the corrosion resistance of steel, nickel enhances the low-temperature behavior of the material by improving the fracture toughness (see Section 4.10). It is used in structural steels in varying amounts; for example, certain grades of ASTM A514 have Ni contents between 0.30 and 1.50 percent; some types of A588 have nickel contents from 0.25 to 1.25 percent.

Phosphorus (P) and Sulfur (S).   Both of these elements are generally undesirable in structural steel. Sulfur, in particular, promotes internal segregation in the steel matrix. Both act to reduce the ductility of the material. All steel grade specifications, therefore, place severe restrictions on the amount of P and S that are allowed, basically holding them to less than about 0.04 to 0.05 percent. Their detrimental effect on weldability is significant.

Silicon (Si).   Along with aluminum, silicon is one of the principal deoxidizers for structural steel. It is the element that is most commonly used to produce semi- and fully killed steel, and normally appears in amounts less than approximately 0.40 percent.

Vanadium (V).   The effects of this chemical element are similar to those of Mn, Mo, and Cb. It helps the material develop a finer crystalline structure and gives increased fracture toughness. Vanadium contents of 0.02 to 0.15 percent are used in ASTM grades A572 and A588, and in amounts of 0.03 to 0.08 percent in A514.

Other Chemical Elements.   Certain steel grades utilize small amounts of other alloying elements, such as boron, nitrogen, and titanium. These elements normally work in conjunction with some of the major components to enhance certain aspects of the material performance. Detailed data are given in Ref. [5].

Weldability of Structural Steel.   Although the earliest form of welding was introduced in 1881 [6], it was only after 1930 that this method of joining began to achieve ever-increasing importance to steel-framed structures. A great many manual, semiautomatic and automatic welding processes are now commercially available, and the most important are discussed in Chapter 11. However, for the commonly used methods to be applicable without major modifications, it is imperative that the structural steel has a chemical composition that allows the fusion of metals to take place without serious flaws. This is called the weldability of structural steel.

All currently available structural steel grades are weldable, although certain processes must be used for certain steel types. For example, the welding requirements of ASTM A514 are considerably more restrictive than those of A36. Details of these and the many other aspects of welding are given in the *Welding Handbook* of the American Welding Society (AWS), whose section on "Metals and Their Weldability" is given as Ref. [20] (the other sections will be referenced as needed).

It is important that the structural engineer be able to assess whether a given grade of steel is weldable or not. This is crucial in the event that all records of the original delivery have been lost, and the chemical composition has been established through testing. Although a number of methods are available, the simplest determines the carbon equivalent (CE) of the steel, and compares it to a certain limit:

$$CE = C + \frac{Mn + Si}{6} + \frac{Cr + Mo + V}{5} + \frac{Ni + Cu}{15} \lesssim 0.50 \quad (4.1)$$

where *C, Mn,* etc., are the percentages of the respective elements in the grade at hand. Good weldability is virtually assured if Eq. (4.1) is satisfied.*

A final comment regarding welding may be appropriate at this point. Although certain steels are easier to weld than others, it is not correct to state that any particular type cannot be welded. It is always possible to develop custom procedures for even the "dirtiest" materials, but it must be recognized at the same time that such procedures may require large amounts of time and money. If the steel is of a type that needs a new procedure, it is rarely economical to specify the use of welding. Rather, bolted or other kinds of mechanical fastening will generally be preferable. These conclusions are normally reached when the material in question is one of the "old" grades, generally produced before 1930 or so.

4.5.2 Mechanical Properties

In addition to giving the details of the chemical composition of a heat of steel, the mill test certificate will also include data on the mechanical and other relevant properties of the material. It is the prerogative and responsibility of the engineer of record to make certain that the material that has been delivered meets all of the necessary requirements. The designer should therefore insist upon seeing this document, to compare it with the relevant ASTM standard(s). It is usually supplied to the structural steel fabricator at the time of delivery of the material, and the fabricator will maintain a file of all such certificates. Without this documentation, the engineer has the following options for action (LRFD specification, Section A3.1.2, "Other Metals"):

---

* The carbon equivalent given by Eq. (4.1) is but one of many such formulas, but is generally regarded as the one most suitable to assess the weldability of structural steel.

1. Reject the material and request a new shipment.
2. Request complete metallurgical and mechanical tests.
3. Accept the material as is, but allow it for use only for parts of minor or no importance.

Option 1 may be conservatively prudent, but the delivery time for replacement material may be as long as 3–6 months. The economical consequences of such a delay may be so severe as to eliminate this line of action. A similar observation applies to Option 3, where de facto rejection is the result, since the material cannot be used for any loadbearing or other important structural components. The only realistic alternative is therefore Option 2, the costs of which must be borne by the group responsible for procuring the certificate. (The assignment of ''blame'' may not always be obvious, and many such cases will end up in the courts.)

Table 4.4 gives an illustration of a typical mill test certificate. Readers who are particularly interested in this subject may want to contact a local fabricator (or steel mill, if one is nearby), to be shown actual samples of such documents. The meaning of the chemical composition data has already been discussed; we now detail the various mechanical properties, how they are determined, and what they mean.

The single most useful device to describe the behavior and mechanical properties of the material itself is the stress–strain curve. Typical examples of this are shown in Fig. 4.8, which gives the curves for three of

TABLE 4.4   THE MILL TEST CERTIFICATE

Name of producer and location of steel mill	Order Data	Certification of Correct Measurements
Purchaser Data	Destination of shipment and mode of transportation	Signature: _____ Date: _____
Product and Steel Grade Description		
Material Data (size and number of pieces, weight, heat number, test I.D. numbers, etc.)	Mechanical Property Data (yield stress, ultimate tensile stress, elongation at fracture and gage length, reduction of area at fracture)	
Chemical Composition Data (percentages of carbon and all alloying elements)	Other Test Types and Results (toughness, grain size, etc.)	

Note. The data in this table are representative of those found in common mill test certificates. The exact appearance may vary from company to company. Some producers will attach extra sheets for documentation of additional data, e.g., when special chemical, metallurgical, or mechanical tests have been made.

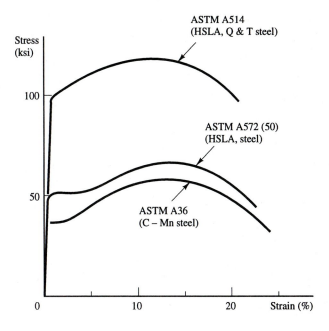

Figure 4.8 Examples of the stress–strain curve.

the common grades of structural steel. For a metallic material like steel the stress–strain curve is obtained through the tension test, where a specimen of standardized dimensions is placed in a testing machine and loaded until failure, following a given procedure. In the United States the required testing method is given by ASTM Standard A370 [*1*], which details such items as specimen size and shape, where the specimen is to be taken from, the loading speed of the testing machine, and gage length. For example, the test specimen of a wide-flange shape is to be taken from the web of the shape, and the gage length must be either 8 or 2 in., as detailed by the representative material standard.

The curves shown in Fig. 4.8 illustrate the overall stress–strain characteristics of the steels. To that end they pinpoint the magnitudes of the highest stress reached during the test, the ultimate tensile stress (or simply tensile strength, for short) F_u, and the strain at which the specimen failed, referred to as the elongation at fracture ε_u. Note that ε_u is not the strain at which F_u is reached. The cross-sectional area of the test specimen is also measured after failure to determine the amount of reduction of area, another indicator of the overall deformability or ductility of the material. The magnitude of the stress at the instant of fracture is not important; it is a highly variable property that happens to be paired with ε_u, and therefore does not indicate the maximum strength.

To a designer the most important part of the stress–strain curve is the initial portion, which demonstrates the immediate response of the material, as well as the levels of load at which the deformations start to become significant. After all, the normal (service) loads on the structure are expected to produce strains well below the ultimate capacity of the material if the structure is to continue to serve its original purpose without large, permanent deformations. As an illustration of these important properties, Figs. 4.9 and 4.10 show the initial portions of the stress–strain curves of a carbon–manganese (mild) structural steel and a high-strength steel, respectively. The diagrams emphasize the characteristics of the two types of steel, as well as the differences between them. For instance, the mild structural steel has a pronounced yield plateau, at which the strain increases without a corresponding increase in stress. The high-strength steel, on the other hand, exhibits a continuously increasing level of stress, although the stress gradient decreases rapidly as the material approaches its tensile strength.

As shown in Figs. 4.9 and 4.10, the initial portion of the stress–strain curve is a straight line, indicating that in this region the material is linearly elastic. The slope of the straight line is the modulus of elasticity E; it varies relatively little from test specimen to test specimen. In fact, E is not influenced very much by the grade of steel either, and is therefore regarded as a constant for all steels. It has been assigned a value of $E = 29,000$ ksi.*

After the initial wavering of the stress–strain curve for the C-Mn steel as it reaches the yield plateau, the curve settles into a horizontal line. The level of stress at this stage is defined as the yield point of the material, indicating the stress at which the material deforms at a constant load. The term yield point applies only to materials with this kind of plateau in the stress–strain curve; it is also referred to by the general name yield stress F_y. Yield stress is used throughout this book to designate yield point,

---

* E is not normally determined by a tension test. Rather, a simply supported bending test is used to record deflections, and elastic beam theory is then utilized to calculate the magnitude of E for the material in question.

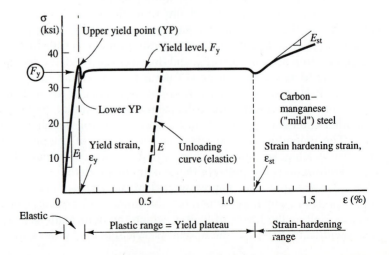

Figure 4.9 Initial portion of a stress–strain curve for a carbon–manganese (mild) structural steel.

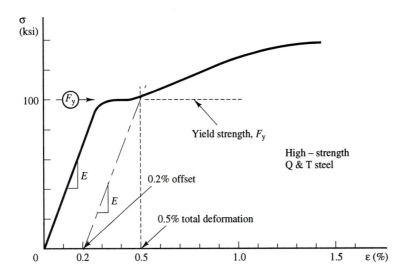

Figure 4.10 Initial portion of a stress–strain curve for a high-strength steel.

along with yield strength, which is the deformation-defined property of the steels that do not have the plateau. This is shown in Fig. 4.10, where F_y is defined as the stress at which the steel exhibits a permanent or plastic deformation of 0.2 percent. The technique is referred to as the offset method. Another approach defines the yield strength as the stress at which the tension specimen has a total elongation of 0.5 percent.

Stress–strain curves are usually obtained from specimens that have been tested at a certain speed of loading, as defined by ASTM A370. In other words, the yield stress is associated with a certain strain rate. However, the loads that are imposed on a building structure are mostly static; i.e., the material experiences a constant, zero-strain-rate load. Studies have shown that the strength of the steel increases with the strain rate (see Chapter 2, Ref. [*14*]); for that reason it would be more accurate to record the yield stress (for example) at a zero strain rate. This is the distinction between dynamic and static yield stress: Dynamic yield stress is obtained from the standardized tests, and static yield stress represents the zero-rate value. Depending on the type of steel, the static yield stress may be 5 to 10 percent lower than the dynamic yield stress. The static values were used in the development of the LRFD criteria. The use of dynamic values for design purposes is acceptable because the F_y mandated by the ASTM standards is a specified minimum value, and the mill test data are normally well above this magnitude.

Figure 4.9 also illustrates the various strain ranges of the material characteristic. The intersection of the yield plateau line and the linearly elastic line defines the yield strain ε_y, and the intersection with the strain-hardening portion of the curve gives the strain-hardening strain ε_{st}. The slope of the latter portion of the curve is defined as the strain-hardening modulus of elasticity E_{st}; it is not a well-defined property for most steels. It is commonly given a value between 600 and 1000 ksi.

Typical strains in a structure at service load will amount to one-half to two-thirds of the yield strain, although local conditions (around holes, for example) may produce yielding. This is neither dangerous nor undesirable. By its very nature, steel is a ductile material, and any such kind of minor "overstressing" is accepted by the structure through the redistribution of forces. That is, the less highly strained regions are able to absorb the loads that cannot be taken by the most severely loaded ones.

The relative size of the three principal regions of the strain are typically as follows:

$$\begin{bmatrix} \text{Elastic} \\ \text{region} \end{bmatrix} : \begin{bmatrix} \text{Plastic} \\ \text{region} \end{bmatrix} : \begin{bmatrix} \text{Strain-hardening} \\ \text{region} \end{bmatrix} = 1:15:50$$

This gives some perspective to the material behavior in the structure, especially as it relates to the problems of "overstressing." Rather than thinking of stress, the designer should consider the strains that will develop. With the above ratios in mind, it is easily understood that even a fair degree of overstraining will not impair the structure.

Other mechanical properties of interest are the shear yield stress F_{vy}, the shear modulus of elasticity G, and Poisson's ratio, ν. The magnitude of F_{vy} depends on the material failure model that is used. For a ductile material like steel the yield criterion of von Mises is considered the most suitable [21]; it gives a shear yield stress as

$$F_{vy} = \frac{F_y}{\sqrt{3}} \approx 0.57 F_y \qquad (4.2)$$

The value of G is usually determined by the theory of elasticity formula:

$$G = \frac{E}{2(1 + \nu)} \qquad (4.3)$$

and becomes approximately 11,200 ksi for an E value of 29,000 and a Poisson's ratio of 0.3. The value of G in the strain-hardening range is neither well understood nor of much use.

It is tacitly implied that the properties of steel loaded in tension are the same as those for compression. For most practical purposes this is true, although repeated (cyclic) loads past the yield point will cause a gradual lowering of the subsequent yield value through the phenomonen known as the Bauschinger effect [21]. This effect is of no significance for statically loaded structures.

It has already been mentioned that the elongation at fracture is one of the measures of the ductility of the steel. When considered together with the full stress–strain curve, the area under the curve represents the total energy that is absorbed by the material by the time it fails. This area is occasionally referred to as the modulus of toughness. However, since the conditions in a real structure are normally bi- or triaxial states of strain and stress, the energy absorption of the uniaxially loaded test specimen is not representative of realistic conditions. Other testing techniques have been devised to cover this aspect of structural behavior, the most common of which are the Charpy impact test (also called the Charpy V-notch test) [1] and the crack tip opening displacement (CTOD) test [22]. Both measure the resistance of the steel to the start and propagation of a crack, and hence reflect the material toughness in a better way. These tests can also be used to cover the influence of temperature on the toughness of the material, specifically through the temperature transition (TT) curve. These aspects of the behavior of steel are covered in some detail in Section 4.10. The low-temperature characteristics are of particular importance, since steel becomes less ductile as the temperature drops.

A final material property that is of interest to the structural engineer is the thermal coefficient of linear expansion, which is of use in assessing the movements that will take place in the structure as the temperature changes. This is important when locations of expansion joints are to be determined, as well as the forces that will develop if the movements are restricted. The coefficient of expansion is essentially a constant within the range of operating temperatures for most building structures, with a value of 0.00065 in./in. per 100°F [14].

4.6 GRADES OF STRUCTURAL STEEL

The grades of structural steel are standardized by the ASTM into certain categories, and the individual standards give the detailed requirements for each material. This has already been discussed to some degree in this chapter. The grade designations follow a certain system, and the structural designer should adhere to this in specifying the steel that is to be used. Specifically, the grade is identified as

ASTM A[Standard number]

For example, ASTM A572 refers to the material that is governed by standard number A572; the user must refer to that particular document for details of the requirements.

Most of the ASTM standards are jointly approved by the American National Standards Institute (ANSI) as ANSI standards as well; other organizations also have endorsed many of these documents. The user should also be aware that each standard is reviewed periodically; an additional number is normally attached to the grade designation, showing the most recent year of reapproval [1]. For instance, the grade ASTM A572-79 shows that this standard was reapproved in 1979. A designer will not usually include the year in the specification data, but rather refer to the current issue of the standard.

The structural steel grades that are briefly described in the following have all been approved for design in accordance with the LRFD specification. The engineer must take special care that only such grades are to be used; the development and approval of new grades are published as they appear. The engineer should also be aware when standards are withdrawn, as was the case for ASTM A440 in 1979 [1].

4.6.1 Grades of Steel for Shapes, Plates, and Bars

The grades of steel that are accepted for the structural usage of hot-rolled shapes and plates are ASTM A36, A242, A441, A514, A529, A572, and A588. Table 4.5 details the primary strength characteristics and types of steel of these grades, along with their availability in the various shape groups (see Table 4.2) and plate and bar thicknesses. These grades are the ones that are currently approved under the LRFD specification. The only unlisted ASTM grade that would be acceptable in addition to the ones given in the table is ASTM A709, which incorporates a range of steel grades that are specially designed for use for bridge structures. As such each of the grades under A709 has a counterpart in A36, etc., with the additional requirement that the A709 grades also must satisfy certain fracture toughness criteria. For that reason the A709 tends to be more expensive than the other grades.

The following gives a brief description of the grades given in Table 4.5 and their primary structural use.

ASTM A36.   The mainstay of the steel grades that are currently used for steel-framed structures, A36 is suitable for most structural purposes. It is used for bolted and welded construction alike, and all detail elements (connection clip angles and seats, for example) are made from A36, regardless of the grade of the primary components themselves. A36 is also the only grade that can be obtained in thicknesses greater than 8 in., albeit with a lower specified minimum yield stress (see Table 4.5). All shape and plate sizes are produced in A36. The steel has excellent ductility but limited fracture toughness; if the latter characteristic is a needed property, the 36 grade of ASTM A709 should be used. A36 was introduced in 1960, replacing the now discontinued grade ASTM A7, which had a yield stress of 33 ksi.

ASTM A242.   One of the first of the atmospheric corrosion-resisting (weathering) steels, A242 is a high-strength, low-alloy steel that is available in three grades, depending on the size of the shape or plate. The corrosion resistance is approximately four times that of normal A36 (twice that of A36 with a specified copper contents) [1]. A242 is suitable for all bridge and building structures, but should be specified only if the higher corrosion resistance can be utilized. It is now largely superseded by A588 and the 50-ksi yield stress grades of A709.

TABLE 4.5 ASTM STEEL GRADES FOR HOT-ROLLED SHAPES, PLATES, AND BARS

| Steel Type | ASTM Designation | F_y Minimum Yield Stress (ksi) | F_u Tensile Stress[a] (ksi) | Shapes[b] Group per ASTM A6 | | | | | Plates and Bars | | | | | | | | | | |
				1[b]	2	3	4	5	To ½ Incl.	Over ½ to ¾ Incl.	Over ¾ to 1¼ Incl.	Over 1¼ to 1½ Incl.	Over 1½ to 2 Incl.	Over 2 to 2½ Incl.	Over 2½ to 4 Incl.	Over 4 to 5 Incl.	Over 5 to 6 Incl.	Over 6 to 8 Incl.	Over 8
Carbon	A36	32	58–80																■
	A36	36	58–80[c]	■	■	■	■	■	■	■	■	■	■	■	■	■	■	■	
	A529	42	60–85	■					■										
High-strength, low-alloy	A441	40	60													■	■	■	
		42	63			■							■	■	■				
		46	67		■						■	■							
		50	70	■					■	■									
	A572– Grade 42	42	60	■	■	■	■	■	■	■	■	■	■	■	■	■			
	A572– Grade 50	50	65	■	■	■	■	■	■	■	■	■	■						
	A572– Grade 60	60	75	■	■	■			■	■	■								
	A572– Grade 65	65	80	■	■				■	■									
Corrosion resistant, High-strength, low-alloy	A242	42	63			■								■	■				
		46	67		■						■	■							
		50	70	■					■	■									
	A588	42	63														■	■	
		46	67													■			
		50	70	■	■	■	■	■	■	■	■	■	■	■	■	▨	▨		
Quenched & tempered alloy	A514[d]	90	100–130												■	■	■		
		100	110–130						■	■	■	■	■	■					

[a] Minimum unless a range is shown.
[b] Includes bar-size shapes.
[c] For shapes over 426 lb/ft, minimum of 58 ksi only applies.
[d] Plates only.

■ Available; □ not available; ▨ available only by special order.

ASTM A441. A441 is one of the early HSLA steel grades (approved by ASTM in 1960). It has excellent weldability in all four strength categories, and can be obtained in all shape and plate sizes (although with lower values of F_y and F_u for the larger sizes, as is normally the case). A441 has now been replaced by A572 for buildings (and A709 for bridges) in most instances.

ASTM A514. A514 is the only American high-strength quenched and tempered steel grade, and as such it is the material with the highest yield stress. It was first approved by ASTM in 1964 [1]. It is available only in the form of plates, and can be obtained in thicknesses up to 6 in. The production procedure makes it the most expensive structural steel, and the chemical composition and crystalline structure require special consideration as regards welding and other heat-generating fabrication techniques. However, it is an excellent steel that can be used to good advantage in numerous applications. For example, it is not uncommon to find A514 flanges in certain high-stressed regions of bridge plate girders. The bridge applications of A514 are now essentially replaced by the Q&T 100-ksi grade of A709.

ASTM A529. A529 is a steel grade that is almost exclusively used by the metal building industry in its production of so-called pre-engineered frames. It is a carbon-manganese steel with $F_y = 42$ ksi; it can be obtained only for shapes in group 1 (see Table 4.2) and for plates and bars with thicknesses less than $\frac{1}{2}$ in. Its chemical and mechanical properties are similar to those of A36.

ASTM A572. Originally approved by ASTM in 1966 [1], the A572 standard includes HSLA steels in four strength grades. It is now undoubtedly the most versatile of the higher strength steels, and is largely replacing A441. It has excellent weldability, and can be obtained for all shape groups and plate thicknesses up to 6 in. (depending on the strength category). For example, both the 42- and 50-ksi-yield stress grades can be specified for all shape sizes. (The A572 (50) is the only higher strength grade of steel for which this is possible, along with A588 (50).)

ASTM A588. A588 is the most recent of the atmospheric corrosion-resisting steels, having been approved by ASTM in 1968 [1]. It has corrosion characteristics that are similar to those of A242, but A242 can be obtained only for a limited series of shapes and plate products. As a result, A588 is now the predominant weathering steel. All shape sizes are produced in A588 (50); plates up to 4 in. thick can also be obtained in this strength category.*

---

* Table 4.5 notes that it is possible (by special order) to obtain A588 (50) for plates up to 6 in. thick. This is beyond ASTM practice, however, and is arranged with the individual producer. The purchaser is well advised to require that such materials be of UT (ultrasonic testing) quality.

Canadian Steel Grades.   The Canadian structural steel grades are specified by the CSA standard for structural quality steel [*3*]. A range of different grades are available, and their properties are closely matching the many American grades. The most common Canadian grade is given as CSA G40.21-44W, indicating a weldable (W) steel with a specified minimum yield stress of 44 ksi. There is no direct counterpart to ASTM A36, but all of the other ASTM grades have Canadian equivalents. Thus, the CSA G40.21-50A grade is similar to A588 (50), and the CSA G40.21-100Q steel is identical to A514 for all practical purposes.

Other Steel Grades.   European and Japanese steel grades are used on occasion for structural purposes in the United States. The user should compare the given characteristics of the grade with the ASTM categories, and ascertain that it will be acceptable to the building authorities. However, under normal circumstances the project specifications will state explicitly that any steel that is to be used must have been made to ASTM requirements.

4.6.2 Grades of Steel for Tubes and Pipe

The grades of steel that are used for hollow structural sections are ASTM A53, A500, A501, and A618. Table 4.6 details the primary characteristics of these grades, including the unique criteria that apply in the form of methods of manufacturing. These materials are also the ones that are currently approved for use with the LRFD specification.

The following gives a brief description of the steel grades listed in Table 4.6, their production method characteristics, and primary structural applications.

ASTM A53.   A53 designates circular tubes or pipe according to the method of production as well as the material strength. Thus, A53 pipe products come as Type E, for electric resistance welded, and Type S, for seamless pipe. The meaning of these manufacturing methods was explained briefly earlier; details can be obtained from Ref. [*5*]. A53 materials come in two strength categories, A and B, and only B is acceptable for structural usage. This grade is a carbon–manganese steel with a yield stress of 35 ksi, with other properties similar to those of A36.

Steel pipes delivered in accordance with A53 come as standard weight, extra strong, and double-extra strong [*14*]. These designations signify increasing wall thicknesses, such that, for example, a pipe with a nominal diameter of 4 in. will have wall thicknesses of 0.237, 0.337, and 0.674 in., respectively, in the three categories. All are normally available through local distributors.

ASTM A500.   Products manufactured to this specification are either round or shaped (square or rectangular) hollow structural sections that are cold-formed. Both types are made in three strength categories, as shown in Table 4.6, but only the shaped HSS in grade B can be counted

TABLE 4.6 ASTM STEEL GRADES FOR TUBES AND PIPE

Steel	ASTM Specification	Grade	F_y Minimum Yield Stress (ksi)	F_u Minimum Tensile Stress (ksi)	Shape Round	Square and Rectangular	Availability
Electric-resistance welded	A53 Type E	B	35	60	■		Note 3
Seamless	Type S	B	35	60	■		Note 3
Cold-formed	A500	A	33	45	■		Note 1
		B	42	58	■		Note 1
		C	46	62	■		Note 1
		A	39	45		■	Note 1
		B	46	58		■	Note 2
		C	50	62		■	Note 1
Hot-formed	A501	–	36	58	■	■	Note 1
High-strength, low-alloy	A618	I	50	70	■	■	Note 1
		II	50	70	■	■	Note 1
		III	50	65	■	■	Note 1

Note: 1, Available in mill quantities only; consult with producers; 2, normally stocked in local steel service centers; 3, normally stocked by local pipe distributors.
■ Available; □ not available.

Source: Reprinted by permission of the American Institute of Steel Construction, Inc., from the 1st edition of the *AISC-LRFD Manual* (pp. 1–14).

on to be available from a local distributor. This grade has a yield stress of 46 ksi, but is otherwise similar to A36 in properties and structural usage. It is strictly a carbon steel, since only grade C has a specified contents of manganese [*1*].

ASTM A501.   A501 is, for all practical purposes, the A36 for tubular products. It is used for hot-formed round and shaped tubing, and comes in one strength grade only. As for A53, the round tubes are made as standard weight, extra strong, and double-extra strong. Since tubes of this grade are available only in mill quantities (i.e., in larger quantities, ordered from the steel producer), it is generally preferable to specify the use of A53 or A500, as applicable. The differences in production techniques also make these somewhat less expensive, as a general rule.

ASTM A618.   A618 specifies the material properties of hot-formed round and shaped tubes in an HSLA steel grade. The standard divides the material into three categories that differ primarily in chemical composition; the yield stress is equal to 50 ksi for all. Grades II or III are generally preferable, especially since grade II has enhanced corrosion resistance, and both grades II and III have specified contents of silicon (see Section

5.1). Copper contents may be specified for grades I and III, improving their corrosion resistance as well.

A618 tubes are usually available only in mill quantities. Unless there are specific reasons for using such members, the designer is probably better off specifying tubes according to A500. On the other hand, round tubes in high-strength steel can be obtained only in larger amounts. Both of these observations should be taken into account in the selection process.

Hot-Formed vs. Cold-Formed Tubes. The differences in the production methods for hot-formed (HF) and cold-formed (CF) tubes tend to make hot-formed tubes somewhat more expensive, and also less readily available. On the other hand, in certain applications the designer may want to use HF tubes because of their improved strength. The production technique results in smaller residual stresses in such tubes. For that reason they may prove economical for columns and other compression members with high axial loads [*15*]. The final choice must be made by the designer in consultation with the steel mill and the fabricator, but it is believed that the ready availability of many of the CF tubes makes them the preferred selection.

4.6.3 Grades of Steel for Sheet and Strip

Sheet and strip steel products* are used primarily for roof and wall cladding in steel-framed structures, and now also to a significant extent in the form of corrugated steel deck for floor slabs. Prefabricated metal buildings utilize a number of other cold-formed products, notably for purlins, girts, and the like. In general, therefore, the structural usage of the light-gage, cold-formed elements that are made from sheet and strip is limited, and most are governed by the design specifications of the American Iron and Steel Institute (AISI) [*10, 11*] or the Canadian Standards Association [*12*]. For completeness, however, it is important that the designer is aware of the common grades of steel that are used for such products.

The grades of steel for sheet and strip that are approved for use under the LRFD specification are ASTM A570, A606, and A607. Their primary characteristics are given in Table 4.7.

Briefly, A570 is the carbon–manganese steel for sheet and strip, and is used for a variety of cold-formed shapes. The larger thickness and lower grades have excellent ductility, but ductility drops off significantly as the thickness is reduced toward the minimum value. The same tendency applies to A606 and A607; the thicknesses that are used for sheet and strip lead to elongations at fracture as low as 10 percent (grade 50, A570).

A606 defines a steel with improved corrosion resistance whose properties are similar to those of A242 and A588. Both this grade and

---

* References [*5*] and [*8*] describe the production methods of these products. The respective material standards define the geometric differences between sheet and strip.

TABLE 4.7 SOME PROPERTIES OF ASTM GRADES OF STEEL FOR SHEET AND STRIP

ASTM Standard	Grade (strength or other)	Yield Stress (ksi)	Ultimate Stress (ksi)	Range of Thickness (in.)
A570	30	30	49	
	33	33	52	
	36	36	53	0.0255
	40	40	55	to
	45	45	60	0.2299
	50	50	65	
A606[a]	Hot-rolled	45–50	65–70	To
	Cold-rolled	45	65	0.2299
A607[b]	45	45	60	
	50	50	65	
	55	55	70	To
	60	60	75	0.2299
	65	65	80	
	70	70	85	

Note. Refer to the individual ASTM standards for detailed data.

[a] A606 is supplied in two chemical composition types (Types 2 and 4).

[b] A607 is supplied as hot-rolled or cold-rolled.

A607 are HSLA steels, having improved strength characteristics while maintaining good ductility. A606 and A607 are used for cold-formed shapes and deck; A606 is also utilized for corrosion-resistant cladding for buildings.

4.7 STEELS FOR MECHANICAL FASTENERS

The behavior and properties of the various types of mechanical fasteners that are used for the connections of steel-framed structures are covered in detail in Chapter 11. Here only the respective ASTM standards are discussed, primarily to focus on the characteristics of the materials that are used for the fasteners. This forms a necessary introduction to the study of connections in general.

The ASTM grades that are covered in the following are A307, A325, A354, A449, and A490. All of these refer to threaded fasteners or bolts, and are approved for use under the LRFD specification. The only exception is A354; the reason for its inclusion here will be explained.

It is specifically noted that rivets, which were long the main fastener for steel structures, are not included in the LRFD specification. This method of connection is for all practical purposes not used any more for steel construction, and most fabricating shops do not even have the necessary equipment. Rivets have essentially been made obsolete by the advent of high-strength bolts. The strength, installation, and cost of bolts make them superior fasteners.

As will be seen in later chapters of the book, the three-dimensional character of the states of stress and strain in the bolts and the surrounding material suppress the importance of having a yield plateau for the bolt material. Rather, the behavior and strength of a bolted connection are functions of the ultimate tensile stress of the materials [23], and for that reason only this property is prescribed by the standards.

ASTM A307.   Commonly referred to as black bolts, the A307 standard deals with bolts of mild steel. The specified minimum ultimate tensile stress for the material is 60 ksi, essentially the same as for A36. The bolts have adequate strength and ductility when they are used for appropriate purposes, but will twist off relatively easily if they are tightened too much. It is therefore imperative that the A307 and the high-strength bolts can be identified and separated. This is usually achieved by the markings that are mandated for the HS bolts.

A307 bolts have come to be used mostly for temporary tasks, such as to hold the components of the structure together during erection. However, these bolts are good fasteners when used properly, and A307 bolts are now gaining wider acceptance as structural connectors.

ASTM A325.   A325 bolts are the primary grade of high-strength fasteners. The steel is a quenched and tempered type that is quite similar to A514. The ultimate tensile stress is thus 120 ksi for bolts with diameters $\frac{1}{2}$ to 1 in., and 105 ksi for diameters from $1\frac{1}{8}$ to $1\frac{1}{2}$ in. It should be noted that A325 bolts are not made with diameters smaller than $\frac{1}{2}$ in. or larger than $1\frac{1}{2}$ in.; for such sizes ASTM A449 must be used.

A325 fasteners are produced as Type 1, 2, or 3, depending on the chemical composition. Type 3 is a corrosion-resistant bolt, required for use in structures where the primary steel grade is either A242, A588, or A709. Types 1 and 2 differ in carbon contents [0.30 vs. 0.23 percent (maximum)] and manganese contents (0.50 vs. 0.70 percent), and Type 2 also has a specified amount of boron. In general, Type 2 is therefore a better bolt for normal structural applications. However, the user must be aware that a high-temperature service environment for the structure requires the use of Type 1 bolts, and also that unless clearly specified, bolt manufacturers may supply Type 1 or 2 as they see fit. These concerns are normally noted by the fabricator, but it is necessary for the designer to be aware of the differences and the pertinent ASTM requirements.

Compared to the A307 bolts, A325 fasteners have a significantly higher strength, less material ductility, and a finer-grained crystalline structure. The quality control criteria for the latter are much more stringent, giving the A325 less variability in strength and dimensions. The ASTM standard gives detailed marking requirements for A325 bolts, which serve to distinguish them from other fasteners [1]. For instance, all types are marked A325 on the bolt head in addition to certain radial and other lines that separate Types 1, 2, and 3.

ASTM A449. The material strength of A449 bolts is identical to that of A325 for the diameter range $\frac{1}{4}$ to $1\frac{1}{2}$ in., but drops to 90 ksi ultimate stress for larger sizes. The A449 steel is essentially a carbon–manganese type that has been quenched and tempered to achieve the higher strength.

The quality control requirements for A449 bolts are significantly less stringent than those for A325. This is the primary reason why the LRFD specification does not allow the use of A449 for anything but anchor bolts and threaded rods, and for structural fasteners with diameters larger than $1\frac{1}{2}$ in.

ASTM A490. A490 bolts are the highest strength mechanical fasteners in use for steel structures today, with an ultimate tensile stress of 150 ksi for the bolt material in the range of diameters covered by this standard ($\frac{1}{2}$ to $1\frac{1}{2}$ in.). The high strength is achieved through the choice of alloying elements and a quench and tempering heat treatment, which also produces a very hard, fine-grained crystalline structure. For these reasons A490 is particularly sensitive to high service temperatures and cyclic loads, and should not be hot-dip galvanized or otherwise coated [*23*].

As for A325, A490 bolts are produced as Types 1, 2, and 3, with Type 3 being the corrosion resistant kind that is required for use with structures of A242, A588, or A709 steel. Types 1 and 2 differ in chemical composition in a fashion similar to those of the A325 types, and Type 2 is usually a better choice. The purchaser must specify the kind that is required, or Type 1 will be supplied [*1*]. However, the designer should observe that high-temperature usage of A490 is not advisable and special care should be taken when very low temperatures are present.

The manufacturing quality control criteria for A490 are the same as those of A325. The ASTM marking requirements for A490 follow the system of A325, using the A490 notation in addition to radial and other lines to help distinguish these bolts from others.

There is no grade of structural steel that has properties identical or even close to A490. The most likely comparison is A514.

ASTM A354. A354 bolts are made from steel that is essentially the same as that of the A490 fasteners. However, the quality control criteria are less stringent for the former, and A354 bolts are therefore generally not allowed as substitutions for A490. The A354 comes in three grades, and only A354 BD may be an acceptable alternative to A490. This will be necessary if the strength of A490 is required for bolts with diameters larger than $1\frac{1}{2}$ in., since A354 is produced in sizes up to 4 in. in diameter. On the other hand, the engineer might prefer to redesign the connection in order to be able to use more reasonable bolt sizes. This is particularly important if bolt pretensioning is specified (see Chapter 11), since it will be practically impossible to tighten the large-diameter fasteners to the required level.

Other Fastener Materials.   High-strength bolts are also produced to other material specifications, such as those of the Society of Automotive Engineers (SAE). SAE grades 5 and 8 correspond approximately to ASTM A325 and A490, but threads and other geometric characteristics are different. Many similar examples can be found. It is generally advisable to examine all properties of any such bolt materials and obtain the express approval of all authorities before specifying these bolts.

4.8 STEELS FOR WELDING ELECTRODES

The steel grades that are used for consumable welding electrodes or, as they are officially referred to, as filler metals for the many welding processes that are currently in use, are governed by the specifications of the American Welding Society (AWS). The processes that are most useful in the construction of steel-framed buildings and bridges are discussed in some detail in Chapter 11, and only the material characteristics of the structural filler metal grades are outlined in the following. The electrode classification and designation system is also explained here.

The design of welded connections is governed by the criteria of the LRFD specification and the AWS Structural Welding Code [4]. In Canada the applicable design documents are the Canadian Standards Association (CSA) standard for the design of steel structures and standard for welded steel construction [24]. Common to all of these is the general concept of weld and base metal matching, which essentially implies that the strength of the metal in the joint itself will be equal to or higher than that of the steel in the structural component, the base metal. This design philosophy and the problems and advantages that pertain to it are discussed in Chapter 11. At this point we note only that matching not only means compatibility in chemical and mechanical properties, but also that the behavior of the materials during and after the welding must be compatible. For example, the large contraction strains that accompany the cooling of the weld metal after it has been deposited may give rise to serious connection problems unless they are considered at the design stage. This focuses both on the welding (residual) stresses that will always result from restraining a cooling metal, as well as the cracks that may appear if the degree of restraining is sufficiently high. In other words, the three-dimensional character of the connection is particularly important for welded joints.

As we discuss in the following, the choice of welding electrode also depends heavily on the type of welding process that is involved, in addition to factors such as welding positions, rate of deposition, and cost. The basic filler metal will therefore fit the base metal type, such as when a mild steel (carbon–manganese) electrode is used to weld A36 components. However, once this choice is made, the designer must determine the type of mild steel electrode that is available for the given welding process, where and how the weld will be placed, the environmental conditions during and after placement, the size of the connection welds, and so on. Briefly, therefore, it must be understood that welding is an engineering discipline all by itself, and the proper choices should be made only by an individual that is well versed in all of the different aspects of this method of joining.

The AISC-approved filler metal specifications of AWS are as follows:

A. 1. Mild steel covered electrodes, AWS A5.1
    2. Low-alloy steel covered electrodes, AWS A5.5
B. 3. Bare mild steel electrodes and fluxes, AWS A5.17
    4. Bare low-alloy steel eletrodes and fluxes, AWS A5.23
C. 5. Mild steel electrodes, AWS A5.18
    6. Low-alloy steel filler metals, AWS A5.28
D. 7. Mild steel electrodes, AWS A5.20
    8. Low-alloy steel electrodes, AWS A5.29

The listing sequence does not follow the AWS specification numbers, but rather the type of welding processes for which the individual electrodes are intended. This is indicated by the capital letters to the left, referring to the following processes:

A. Shielded metal arc welding (SMAW)
B. Submerged arc welding (SAW)
C. Gas metal arc welding (GMAW)
D. Flux cored arc welding (FCAW)

A brief description of the various electrode types and their designation system is given below. Note that when the strength of the electrode material is considered, it is the ultimate tensile stress that governs, since these metals do not have a pronounced yield plateau. The weld design criteria are therefore based on this property [4, 24].

4.8.1 Shielded Metal Arc Welding Electrodes (SMAW)

Up until now the most common method of welding, SMAW or "stick electrode welding," makes use of mild steel or high-strength, low-alloy electrodes that are coated with a mineral-based cover. The cover is necessary to prevent impurities from entering the bath of molten metal that is formed when the weld is deposited, and forms a layer of slag on top of the weld bead. The slag must be removed before any additional welding is done.

The chemical composition of the electrode coating may vary from a high cellulose type to one with a high content of iron powder. The exact chemical makeup is normally regarded as confidential by the producer.

However, the coating must perform the task of protecting the weld metal, and the more important and severely loaded the weld will be in service, the more strict the coating requirements will be. Reference [25] gives detailed data on the many types of electrode coverings.

Mild steel electrodes have nominal tensile strengths F_{EXX} of 60 and 70 ksi, making them suitable for use with A36 steel, for example (F_y = 36 ksi, F_u = 58 ksi). HSLA steels require corresponding types of electrodes, and these have similar strength values of 80, 90, 100, and 110 ksi. The choice of electrode for a given grade of steel will also depend on service conditions, deposition rates, and so on.

Figure 4.11 shows a sample of some typical welding electrodes for shielded metal arc welding. The numbers that are seen close to the uncoated end of the electrodes are the AWS classification numbers, which generally take on the form E*AABC* (or E*AAABC*). The E stands for arc welding electrode, as opposed to R, which is used to indicate a welding rod (not the same as electrode), for instance. The letters *AA* (or *AAA*) indicate the nominal tensile strength of the electrode, *B* is the welding position identifier, and *C* is the coating identifier. *AA* (*AAA*) will take on the values of 60, 70, 80, 90, 100, and 110, giving the material strength. The letter *B* will be either 1 or 2, where 1 represents an electrode that may be used for welding in all positions (horizontal or flat (down-hand), vertical and overhead), and 2 is used for electrodes that should be used to weld only in the horizontal/flat position. The letter *C* can be anything from 0 to 8, representing types of electrode coating that go from the high cellulose kind (*C* = 0 or 1), to high titanium (*C* = 2 or 3), iron powder and titanium (*C* = 4), low hydrogen (*C* = 5, 6, 7), and finally to

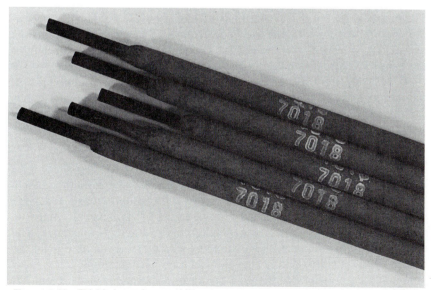

Figure 4.11   Shielded metal arc welding electrodes. Courtesy the Lincoln Electric Co.

iron powder low hydrogen ($C = 8$). For example, the designation E6010 gives an electrode with $F_{EXX} = 60$ ksi that is good for welding in all positions, and whose coating is of the high cellulose-sodium kind [25].

The reader is encouraged to study the electrode classification system in detail, including the characteristics of the coatings. The economy of welded construction rests heavily upon the correct filler metal choices.

4.8.2 Submerged Arc Welding (SAW)

Submerged arc welding is an automated procedure that is dependent on equipment that is unsuitable for field work. It uses bare steel electrodes in the form of wire that is fed continuously from a roll, and therefore needs an additional agent that will protect the molten weld metal from atmospheric impurities. Instead of the coating that is used for SMAW electrodes, SAW utilizes a granular flux that is spread over the weld as soon as it is placed. The AWS specifications that relate to submerged arc welding therefore also specify the type of flux that should be used.

The identification number for SAW electrodes and fluxes follow the system FAB-ECCC (or FAAB-ECCC), where the individual data signify the following:

$$F = \text{flux}$$

$$A \text{ (or } AA) = \text{strength of electrode in 10-ksi increments (e.g., 6 means 60 ksi)}$$

$$B = \text{notch toughness identifier for weld metal}$$

$$E CCC = \text{electrode metal contents classification}$$

The numbers for ECCC are normally specified by the welding engineer at the time the fabrication schedule is determined. The structural engineer is primarily concerned with the numbers for A (AA) and B, which relate to the material's strength and toughness. A (AA) is available as 60 ($=6$), 70 ($=7$), 80 ($=8$), 90 ($=9$), 100 ($=10$), and 110 ($=11$) ksi; B is given as a number from 0 to 4 [25]. A 0 indicates no toughness requirement (which is usually satisfactory for most building structures), and 4 requires a material with a Charpy V-notch toughness of 20 ft-lb at $-60°F$ [25].

4.8.3 Gas Metal Arc Welding (GMAW)

GMAW has become a very prominent structural welding method, for a number of reasons. In addition to working with a continuously fed, bare steel wire as the electrode, the equipment is easily portable, and the welding in itself is less operator-dependent than SMAW, for instance. The role of the coating is taken by a shielding gas, which is emitted from the nozzle of the welding gun during the welding. The shielding gas is most commonly carbon dioxide (CO_2), alone or in combination with argon (Ar) or helium (He) and argon. These have the advantage of being chemically inert, i.e., they will not react with the metal in the weld.

The nominal tensile strength for the GMAW electrodes varies from 60 to 110 ksi, just as the filler metals for SMAW and SAW. The strength grades from 80 ksi and up are low-alloy steels, and should be used accordingly.

The identification numbers for GMAW electrodes follow the layout EAAS-B (or EAAAS-B), where E indicates electrode; AA (or AAA) gives the magnitude of F_{EXX} in ksi; S is a symbol for an uncovered electrode (sometimes S is replaced by U, indicating a specially treated wire); and B relates the chemical composition of the electrode metal. For example, $B = 5$ identifies an electrode with high toughness that contains aluminum and silicon as deoxidizers. A typical GMAW electrode is E70S-1, defining a material with $F_{EXX} = 70$ ksi, to be used with an ArCO$_2$ shielding gas, and whose silicon contents is relatively low. Additional examples and details are given by the respective AWS specifications.

4.8.4 Flux Cored Arc Welding (FCAW)

A method of welding that is now very commonly used in construction projects, FCAW uses electrodes that have a tubular cross section with the flux packed inside the tube. The advantage of this approach is that the electrode can be produced in large lengths, and is therefore available in coils of wire. This allows continuous feeding during welding, with a resulting higher productivity. Flux cored arc welding is therefore replacing SMAW and GMAW in a number of applications, especially when welding has to be done in the field. FCAW is not susceptible to problems associated with the many starts and stops that result from the limited electrode length in SMAW or the minor wind gusts that may blow away the shielding gas of GMAW.

The strength characteristics of the FCAW electrodes are identical to those for the other welding processes. The identification system is similar to that of GMAW, with the exception that a T (for tubular) replaces the S (or U) of the gas metal arc electrodes. Thus, the classification E70T-1 identifies a tubular electrode (i.e., for FCAW) with strength and other properties identical to the E70S-1.

4.8.5 Base Metal and Weld Metal Matching

To facilitate the choice of the proper electrode to go along with a certain base metal and welding procedure, Table 4.8 gives a set of general guidelines on the primary basis of strength. The data in the table are intended for help in choosing the broad type of electrode; other criteria must then be considered to assure good deposition rates, material toughness, and so on. The current issues of the *Welding Handbook* [20, 25] and the *Structural Welding Codes* [4, 24] should be referred to extensively.

TABLE 4.8 ASTM STRUCTURAL STEEL GRADES AND MATCHING WELD ELECTRODES

ASTM Steel Grade	Welding Process			
	SMAW	SAW	GMAW	FCAW
A36 ($t > 8$ in.)	E60XX	F6X-EXXX	E70S-X	E60T-X
A53, grade B	E70XX	F7X-EXXX	E70U-X	E70T-X
A500, grade A				
A36 ($t \leq 8$ in.)				
A441 ($4 \leq t \leq 8$ in.)	E70XX	F7X-EXXX	E70S-X	E60T-X
A500, grade B	E60XX	F6X-EXXX		E70T-X
A501, A529				
A572, grade 42				
A242	E70XX	F7X-EXXX	E70S-X	E70T-X
A441 ($t < 4$ in.)				
A500, grade C				
A572, grade 50				
A588, A618				
A572, grades 60 and 65	E80XX	F8X-EXXX	ER80S-X	E80T-X
A514 ($2\frac{1}{2} \leq t \leq 6$ in.)	E100XX	F10X-EXXX	ER100S-X	E100T-X
A514 ($t < 2\frac{1}{2}$ in.)	E110XX	F11X-EXXX	ER110S-X	E110T-X

Note. Consult AWS welding codes for additional details.

4.9 STEELS FOR STUD SHEAR CONNECTORS

Stud shear connectors are important elements in most types of composite (steel–concrete) construction, and the steel that is used is governed by ASTM standard A108 [*1*], along with the strength requirements of the structural welding code. A108 gives only the chemical composition data for a variety of materials that can be used for cold-drawn bars, and of these only grades 1010, 1015, 1017, and 1020, semi- or fully killed steel, are acceptable for shear connectors.

The A108 steels have no pronounced yield stress due to the cold-forming that has taken place. The welding codes [*4, 24*] therefore specify only that the minimum tensile strength must be 60 ksi.

4.10 BEHAVIOR OF STEEL AT EXTREME TEMPERATURES

The properties of steel that have been described so far are applicable only if the ambient temperature stays within reasonable proximity of 70°F, say, from 30 to 110°F. This range encompasses the service conditions for most structures, but it is still important to understand what will take place if the temperature were to depart significantly from the normal level. It is also necessary to know that some properties improve whereas others deteriorate, depending on the type of extreme environment: The lower strength and higher ductility at elevated temperatures is one typical example.

The characteristics of steel vary a great deal between low-temperature and high-temperature conditions. For that reason these two types of environment are discussed separately.

4.10.1 High-Temperature Behavior

The properties of steel do not change appreciably for temperatures up to approximately 300 to 400°F, although the stress–strain curve shows increasing nonlinearity when the temperature exceeds 250°F. It is therefore fortunate that most structures will never experience heat levels that go past these points, and even in those that do, the high temperatures are normally of very short duration and appear only over a small portion of the structure. A typical example is what may take place in a structure during a fire: The temperatures may reach high levels, but only for a very short time, and normally only in highly localized spots. Exceptions do appear, as has been evidenced by some of the more spectacular fire-related collapses (McCormick Place, in Chicago, Illinois, for example), but these are fortunately few and far between. Realistic conditions are better represented by what took place during the full-scale fire test that was conducted in a parking garage in Scranton, Pennsylvania, in 1972 [26]: Damage was localized, and most of it was easily repaired, for example, by cleaning blackened areas and replacing damaged tiles. Nevertheless, it is important to know how heat affects the material, and, if necessary, to take heat into account in the design process.

The relationships between the temperature and the primary strength and stiffness characteristics of steel are shown in Fig. 4.12. For all

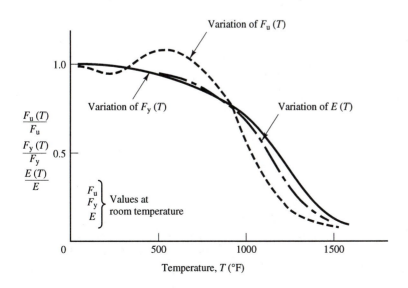

Figure 4.12 Relationship between temperature and primary strength and stiffness characteristics of steel.

practical purposes, F_y, F_u, and E show decreasing values as the temperature increases, although the rate of decrease becomes significant only after the temperature has reached approximately 1000°F. F_u actually exhibits a slight increase between 250° and 600°F, which is due to the phenomenon of strain aging [5]. The yield and ultimate stresses have dropped to about one-half of their room temperature value at 1100–1200°F; at this level E has reached about 60 percent of its original value. The level of E is actually more important for the structure, since deflections are directly proportional to the modulus of elasticity. The phenomenon of creep will also come into play if the loaded structure is subjected to increased temperatures for an extended period [27].

A number of metallurgical changes take place in the steel as the temperature goes from ambient to the upper level of white hot. Figure 4.13 illustrates the primary effects for a typical structural steel and also designates some important manufacturing and fabrication temperature ranges. For example, it is seen that the typical welding preheat ranges from 75 to 200°F, depending on the size of the pieces that are to be welded. Similarly, hot-dip galvanizing takes place at about 875°F; temper-

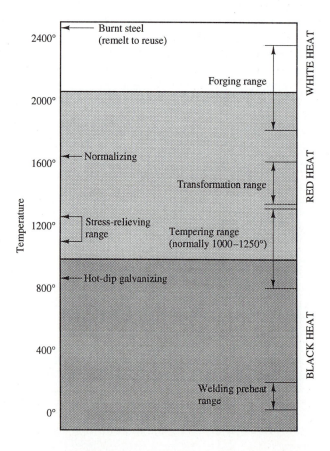

Figure 4.13 Critical temperatures for structural steel.

ing of steel is usually done at temperatures between 1000 and 1250°F; normalizing is done at approximately 1650°F; and stress-relieving (of which annealing is one method) is achieved by heating the material to between 1100 and 1250°F.

The importance and proper use of welding preheat are discussed in the part of Chapter 11 that deals with welded connections. Briefly, however, it is necessary to use this treatment to reduce the impact of placing molten metal on steel of a much lower temperature, as well as to reduce the rate of cooling after welding. Both of these influences will help the steel and weld metal to avoid problems such as hot cracking.

Steel that is allowed to remain at elevated temperatures for an extended time, or steel that cools very slowly may experience grain growth. Although this usually results in a material with higher ductility, the strength and toughness may be lowered. Normalizing is done to refine the grain size as well as to improve the toughness, and is achieved by heating the steel to above the upper transformation temperature, as indicated in Fig. 4.13. At this level the molecular structure of steel goes from alpha-iron to gamma-iron, which allows the subsequent grain size changes to take place relatively easily.

Tempering of steel is normally done in conjunction with quenching, such as when the high-strength quenched and tempered steels are produced. The rapid cooling of the steel due to quenching will produce the hard, fine-grained structure called martensite [5]. Although very strong, martensite is also very brittle, and the tempering is done to reshape the crystalline structure only to the extent that ductility and toughness are increased, but strength is maintained.

Any amount of heat input and subsequent cooling will produce a certain level of built-in or residual stresses, due to the restraints of the material and the structure to the contractions that must take place. This occurs very prominently in welded joints; it will also occur throughout any structure or part of it that has been heated. If the heat has been applied unevenly, and the contractions are not restrained, a certain amount of distortion is bound to appear. This may make structural fitup difficult, but appropriate application of heat and controlled cooling may relieve such problems. Reference [28] gives extensive guidelines for dealing with these types of difficulties.

The designer that is concerned about welding and other residual stresses in joints with high degrees of restraint may choose to redesign the connection geometry to avoid problems. Reference [29] details the solution to such problems as they pertain to lamellar tearing, but the recommendations are excellent advice for the design and fabrication of welded connections in general. It is also possible to use stress relieving in some form, as was done for a number of the beam-column-diagonal connections in the exterior frames of the John Hancock Center in Chicago, Illinois. However, stress relieving is usually an expensive procedure that is limited by the size of the heating ovens and the structural pieces that are to be treated. In most cases it is probably preferable to use other means of reducing the alleged problems of built-in stresses.

Hot-dip galvanizing of steel is sometimes used to enhance the long-term corrosion resistance of the structure. It is still specified for certain applications (some transmission line towers, for example), but the weathering steels have eliminated much of the galvanizing that used to be performed. However, it must be remembered that heat is applied during this process, and certain steels may be adversely affected. For example, the high-strength quenched and tempered material in A514 and in the bolt grades A490 and A354 may not respond properly after galvanizing.

Oxygen or flame cutting is a common fabrication operation that introduces intense heat over a localized area of the material. The rapid cooling after the cutting leaves the material at the edge very brittle and hard. Localized heat treatment or edge machining may be necessary in some cases, especially if the oxygen-cut plate is intended for service under high-cycle tensile loads.

Local quenching effects may appear as a fire is extinguished in a structure, and water from the fire hoses hits heated steel. However, it is rare for local quenching effects to occur over anything but a minor area. The structural effect is therefore minimal, but heat treatment can be applied to remove any problem spots, if the owner of the structure is leery of doing nothing. Naturally, if the members have been deformed badly they may require replacement anyway. However, the material in itself does not usually suffer irreparable damage due to a fire.

4.10.2 Low-Temperature Behavior

In general terms, the strength of steel increases as the temperature is reduced below the ambient level. However, ductility diminishes very rapidly, which means that the material may fail simply because it has lost its ability to deform before fracturing. Loss of ductility is normally demonstrated through the temperature transition or TT curve, which shows the energy that is absorbed in common impact tests of the steel for a range of temperatures. The most common of these tests is the Charpy V-notch test; Fig. 4.14 gives an illustration of a TT curve for a typical structural steel based on a set of such tests. The details of the testing method are prescribed by ASTM A370 [1].

Failure of steel that takes place with little or no permanent (i.e., plastic) deformation is called brittle fracture, and is influenced by a combination of the effects of low temperature, tensile stress, loading rate, and joint restraint in the region around the failure initiation spot. For example, points of stress concentration, sometimes referred to as "stress risers," may indicate potential starting areas for brittle fracture.

Brittle fracture and the related problems of structural steel fatigue are discussed more fully in the next section. In the low-temperature context, the service conditions for some structures necessitate analysis of the potential for brittle fracture. This applies particularly to structures

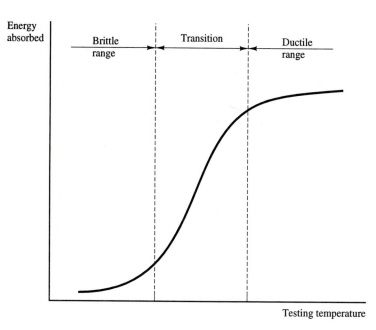

Figure 4.14 Temperate transition curve for structural steel.

such as bridges and offshore oil-drilling platforms in some geographical regions, where the temperatures may fall well below the "normal." However, loading and joint restraint play an important role as well, and it is therefore possible to get a brittle fracture even if the ambient temperature is at a reasonable level.

4.11 FATIGUE AND FRACTURE

Reference [22] is currently one of the most extensive and authoritative sources on the problems and remedies of fatigue and fracture in steel structures. The two phenomena are closely related, yet sufficiently different to warrant individual design procedures. The following gives but a brief introduction to fatigue and fracture. The interested reader is advised to study [22] and the many references that are given in that book for a complete documentation of causes, effects, and solution procedures.

Brittle fracture occurs with little or no prior warning, because of the absence of plastic deformations. In addition, it also proceeds at very high speed once it has started, often leading to a disastrous failure for the structure. Thus, the failures of the Kings Bridge in Melbourne, Australia, in 1962 and the Point Pleasant Bridge in West Virginia in 1967 are but two examples of the consequences of brittle fracture. Naturally, the failures of the "Liberty" ships during World War II are well-known examples of the same phenomenon.

Some fractures that have occurred over the last several years have emphasized the need for the structural engineer to gain a better understanding of the brittle fracture phenomenon. On several occasions fractures have been found in heavy wide-flange shapes that have been used as tension members. Normally associated with fully welded splices, the failures have been found to develop as a result of the poor fracture toughness of the steel in the area of the intersection of the flange and the web, as well as in the web. The problems were compounded by difficult welding conditions, high restraint to weld contraction deformations, high hardness of rough, flame-cut surfaces, and the existence of cracks in the base metal before any welding was done. A rather complex problem that now has been resolved through the addition of material selection criteria and detailed fabrication requirements, it has focused on the needs of the practicing engineer to play an active role in the early stages of the design and fabrication process. Mostly, this problem has pointed out that brittle fracture is not limited to low-temperature, dynamic loading service conditions.

References [30–32] detail the investigations that led to the resolution of the "jumbo" wide-flange shape fracture problem. The reader is encouraged to study these sources, so as to make appropriate use of the information as it is needed.

Basically, fracture is related to a single or just a few applications of the load, whereas the fatigue failures are uniquely tied to loads that cycle a great many times during the life of the structure. The brittle fracture occurs due to a combination of low ductility, high load application rate (=impact), and the degree of restraint in the material. A fatigue failure occurs at cyclic load levels that may be substantially below the full capacity of the structure, yet come about because the presence of sharp notches promotes crack initiation and propagation in the material. A brittle fracture, therefore, is a spontaneous event; fatigue failure takes place over a certain length of time.

The material mechanism of fatigue is not completely understood, although the failure is obviously related to the ductility and toughness of the steel. Many fatigue cracks may be initiated while they are microscopic, such that by the time they are discovered, they have already gone through a number of load cycles. This is illustrated in Fig. 4.15, which shows the relationship between the size of the crack a and the number of cycles N. Because of the cumulative nature of the damage, by the time the flaw has been discovered a significant fraction of the life of the structure may have been expended. Some theories have been developed to predict the fatigue life of a structure as the result of the cumulative damage. The best known and most used approach is that of the Miner-Palmgren hypothesis, which is given mathematically as

$$\sum_{i=1}^{n} \frac{n_i}{N_i} = 1 \tag{4.4}$$

In this equation, n_i is the number of load cycles at level i, and N_i is the fatigue life (i.e., number of load cycles) of the structure if the load were maintained at level i until failure. Although Miner's formula, as Eq. (4.4)

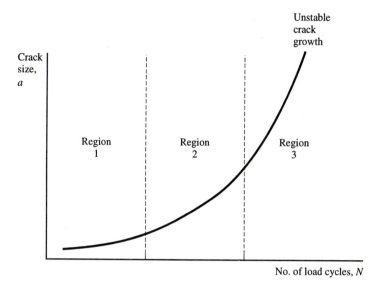

Figure 4.15 Growth of a
fatigue crack.

is known, is burdened with inacurracies in attempting to represent the
fatigue life of a structure, it is still probably the simplest and most
accurate tool for a number of applications.

The magnitude of the largest applied stress, or rather the range of
stress between the largest and smallest values in a structural component
or detail, exerts a pronounced influence on the fatigue life of a structure.
To that end it is frequently useful to display the relationship between the
stress range S_r and the number of cycles to failure N in what is commonly
referred to as an S-N diagram. Figure 4.16 illustrates this concept, and the
curve immediately points out two features of most structural steels and
their details:

1. The fatigue life increases with a lower stress range.

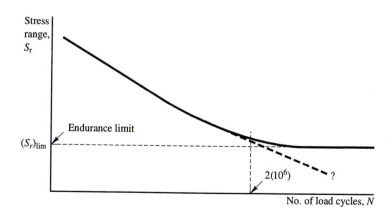

Figure 4.16 An S-N diagram.

2. Many metals reach a fatigue or endurance limit. In Fig. 4.16 this
   is the value $(S_r)_{lim}$. A stress range below this level is not regarded
   as causing fatigue.*

Previous fatigue design criteria were based on the magnitude of the maximum stress, and the ratio between the maximum and minimum stresses or
stress ratio [6, 22]. This has now been abandoned in favor of the stress
range, which has been found to be a more representative factor [22, 30,
31].

The strength of the structure is rarely impaired if the number of load
cycles is less than approximately 100,000. For that reason most building
structures do not have to be analyzed for fatigue conditions. In bridges
and other structures where load cycles of 2,000,000 or more may be
common during their lifetime, the design for fatigue is a very real and
important part of the overall evaluation. The LRFD specification gives
detailed requirements for these analyses.

The fatigue strength of a structure depends heavily on the many
structural connection details that are used. Many of these are related to
welded joints, since the probability of having some form of flaw in a weld
can be appreciable. In any case, some details influence the fatigue life
more severely than others, and the LRFD specification and other design
criteria therefore provide extensive information on the fatigue life classifications of most structural connections and fastening systems. Appendix
G in the LRFD specification gives all the fatigue design data.

In the context of LRFD, fatigue is really an anomaly, since it is an
ultimate limit state that is reached at service load levels. It is currently
treated as a ULS, since the material or member actually fails, but bear in
mind that this happens at service loads.

The conditions that govern the occurrence of a brittle fracture have
already been outlined. Much research work has been done to develop
fracture mechanics [22] into a realistic design tool, and some success has
been achieved through fracture control plans, fitness-for-purpose criteria,
and so on. These developments are continuing, but it is difficult to estimate when these concepts may be universally accepted and used for
design in civil engineering structures.

4.12 CORROSION RESISTANCE
AND STRUCTURAL MAINTENANCE

The corrosion of steel is one of the major difficulties associated with the
use of steel-framed structures, not so much because it may actually lead
to some loss of strength, but mostly because of economics and aesthetics.

* There is still some dispute among fatigue experts as to whether endurance
limits exist at all for any materials. The interested reader is advised to follow the
developments in this branch of engineering closely.

Although an advanced state of corrosion can be a serious structural problem, as has been found in some of the older bridge structures in the United States and elsewhere, such neglect of the need for maintenance is the exception rather than the rule. The long-term economy of a project demands that provisions be made to maintain the structure such that it can continue to serve its intended purpose. For large, exposed bridges and the like these costs can be substantial. The life-long contract of a French painting company to maintain the Eiffel Tower in Paris is a case in point: Work crews are constantly painting one part or other of the structure.

Corrosion control is normally achieved in one of two ways: (1) through painting of the structure in the shop and the field, and (2) through the use of corrosion-resistant (weathering) steels. Weathering steels have already been discussed to some extent in the evaluation of the properties of the weathering steels, A242 and A588. Their corrosion resistance is two or four times that of the mild structural steel, depending on the chemical composition. For structural applications where the material may be left exposed (open parking garages, bridges, etc.), the use of weathering steel can be an excellent choice. However, it is important to bear in mind that the weathering process may take several years, and the final coloring may not be uniform. Exposure of the details of the structure will vary, as do the chemical elements in the atmosphere that influence the oxidation of the steel. Thus, if some color variations are acceptable, and adequate provisions are made to protect the structure itself and those around it from the stains due to rain and other runoff, weathering steel will probably prove to be the best choice.

Although weathering steel is intended for use without paint, some designers and agencies require that the material be painted anyway. While this increases the corrosion resistance, it also increases the cost, and the final benefits may be dubious. Even if the structure will not continue to be painted, the uneven weathering that takes place as the coats of paint wear off will leave a less than attractive impression.

Painting of structural steel is normally done in several stages, from the initial coats of primer that are applied in the fabricating shop, to the final application that is done in the field. The cost of these operations can be substantial,* and it is therefore important to specify the proper types of paint and surface preparation methods. The Steel Structures Painting Council (SSPC), headquartered in Pittsburgh, Pennsylvania, has developed extensive manuals for surface preparation and painting. Reference [32] gives an extract of these requirements as they pertain to shop painting.

It is reasonable to require corrosion protection for structural members that will be exposed to environmental conditions that have such

---

* Five to 15 percent of the material cost is not an unusual figure for painting.

effects. On the other hand, for elements that are to be fully covered and kept within the protected atmosphere of an office building, for example, corrosion protection in the form of paint or similar treatment should be waived. The degree of corrosion that will take place under such conditions is minimal, and certainly does not warrant full-scale painting.

4.13 COST FACTORS

The cost of the material for a typical steel structure normally amounts to between 25 and 75 percent of the structural budget, which in turn may be 15 to 40 percent of the cost of the entire project. These numbers depend on many factors, and there are no fixed rules about how to choose materials and structural systems for optimum economy. However, certain ground rules can be followed:

1. Higher grades of steel are more expensive than mild structural steel.
2. Higher grades of steel are normally less readily available from suppliers or steel mills.
3. A strength-controlled design may benefit from the choice of a higher grade of steel, despite the higher cost.
4. A deflection- or stability-controlled design will probably not benefit from the use of a higher grade of steel, since deflections depend on E, which is independent of the steel strength. Short columns may benefit; longer ones may not, since the buckling strength of a long compression member is most heavily influenced by E, not F_y.
5. Most structures can benefit by using more than a single grade of steel for the primary members. This may not be the case for smaller structures.
6. The types of connections that are chosen will influence the project economy significantly. It is generally preferable to use shop-welded, field-bolted connections.
7. The base selling price of the steel is set by the producer at the time the order is placed. This will take into account the size of the order, delivery time, and so on. The user should consult with the structural steel fabricator before an order is placed.
8. A number of extras are added to the base selling price of the steel, covering the cost of items such as
   (a) Chemistry requirements
   (b) Testing requirements
   (c) Shape/size requirements
   (d) Quantity requirements
   (e) Quality requirements
   (f) Additional requirements

For example, A36 may be specified as semi- or fully killed, necessitating chemical additives (a) and certain tests (b). Under (f), the cost of special shipping or handling procedures warrant handling extras. Many similar examples can be given.

The pricing picture is complex, and the potential user is well advised to consult with fabricators and steel mill representatives alike to obtain as complete data as possible. Mutual trust and the exchange of all necessary information are prerequisites for a successful project.

REFERENCES

[1] American Society for Testing and Materials (ASTM), *1989 Annual Book of ASTM Standards, Parts 3, 4, and 5.* ASTM, Philadelphia, Pennsylvania, 1989. (*Note:* ASTM standards are reissued every year, sometimes with revisions included. The user should obtain the most recent standards available.)

[2] Canadian Standards Association (CSA), *General Requirements for Rolled and Welded Structural Quality Steel,* CSA Standard No. CSA G40.20-M, CSA, Rexdale, Ontario, Canada, 1986.

[3] Canadian Standards Association (CSA), *Structural Quality Steels,* CSA Standard No. CSA G40.21-M, CSA, Rexdale, Ontario, Canada, 1986.

[4] American Welding Society (AWS), *Structural Welding Code,* AWS Specification No. D1.1-88, AWS, Miami, FL, 1988.

[5] United States Steel Corporation, *The Making, Shaping and Treating of Steel,* 9th ed., U.S. Steel Corporation (now USS Division, USX Corporation), Pittsburgh, PA, 1974.

[6] Tall, L., Editor-in-Chief, *Structural Steel Design,* 2d ed., Ronald Press, New York, 1974.

[7] Princeton University Art Museum, *The Eads Bridge,* a publication prepared in connection with a commemorative exhibition, Princeton University, Princeton, NJ, October 1974.

[8] Canadian Institute of Steel Construction (CISC), *General Information on Structural Steel,* rev. 2d ed., CISC, Willowdale (Toronto), Ontario, Canada, 1975.

[9] American Society for Testing and Materials (ASTM), *Estimating the Average Grain Size of Metals,* ASTM Standard No. E112-77, Part 11, 1982 Annual Book of ASTM Standards, ASTM, Philadelphia, PA, 1982.

[10] American Iron and Steel Institute (AISI), *Specification for the Design of Cold-Formed Steel Structural Members,* AISI, Washington, DC, 19 August 1986, with Addendum of 11 December 1989.

[11] American Iron and Steel Institute (AISI), *Load and Resistance Factor Design Specification for Cold-Formed Steel Structural Members,* AISI, Washington, DC, September 1991.

[12] Canadian Standards Association (CSA), *Cold-Formed Steel Structural Members,* CSA Standard No. S136-1987, CSA, Rexdale, Ontario, Canada, 1987.

[13] Steel Deck Institute (SDI), *Steel Deck Institute Diaphragm Design Manual,* 2d ed., SDI, Canton, OH, 1987.

[14] American Institute of Steel Construction (AISC), *Manual of Steel Construction—Load and Resistance Factor Design,* First Edition, AISC, Chicago, IL, 1986.

[15] Sherman, D. R., *Tentative Criteria for Structural Applications of Steel Tubing and Pipe,* American Iron and Steel Institute, Washington, DC, August 1976.

[16] Canadian Institute of Steel Construction (CISC), *Handbook of Steel Construction,* 3d ed., CISC, Willowdale (Toronto), Ontario, Canada, 1980.

[17] Bjorhovde, R., and Tall, L., *Survey of Utilization and Manufacture of Heavy Columns,* Fritz Engineering Laboratory Report No. 337.7, Lehigh University, Bethlehem, PA, October 1970.

[18] Blodgett, O. W., Arc welded steel columns: A progress report, *AWS Welding Journal*, Vol. 58, No. 6, June 1979 (pp. 15–29).

[19] American Institute of Steel Construction (AISC), *Iron and Steel Beams 1873–1952*, AISC, Chicago, IL, 1952.

[20] American Welding Society (AWS), *Metals and Their Weldability*, AWS Welding Handbook, Vol. 4, 7th ed., AWS, Miami, FL, 1982.

[21] Mendelson, A., *Plasticity: Theory and Application*, MacMillan, New York, 1968.

[22] Rolfe, S. T., and Barsom, J. M., *Fracture and Fatigue Control in Structures*, 2d ed., Prentice-Hall, Englewood Cliffs, NJ, 1987.

[23] Kulak, G. L., Fisher, J. W., and Struik, J. H. A., *Guide to Design Criteria for Bolted and Riveted Joints*, 2d ed., Wiley, New York, 1987.

[24] Canadian Standards Association (CSA), *Welded Steel Construction (Metal Arc Welding)*, CSA Standard No. W59-1987, CSA, Rexdale, Ontario, Canada, June 1987.

[25] American Welding Society (AWS), *Welding Processes*, AWS Welding Handbook, Vol. 2, 8th ed., AWS, Miami, FL, 1991.

[26] American Iron and Steel Institute (AISI), *Automobile Burn-Out Test in Open-Air Parking Structure*, Final Report, Gage-Babcock & Associates, Inc., Westchester, IL, January 1973.

[27] United States Steel Corporation (USS), *Steel for Elevated Temperature Service*, USS Publication No. ADUSS 43-1089-05, Pittsburgh, PA, December 1974.

[28] British Welding Institute, *Control of Distortion in Welded Fabrications*, 2d ed., The Welding Institute, Abington, Cambridge, UK, 1975.

[29] American Institute of Steel Construction (AISC), Commentary on highly restrained welded connections, *AISC Engineering Journal*, Vol. 10, No. 3, 1973 (pp. 61–73).

[30] Fisher, J. W., and Pense, A. W., Experience with the use of heavy W-shapes in tension, *AISC Engineering Journal*, Vol. 24, No. 2, 1987 (pp. 63–78).

[31] Barsom, J. M., *Non-Column Applications of Wide-Flange Shapes*, Proceedings, 1988 National Steel Construction Conference, American Institute of Steel Construction, Miami Beach, FL, June 1988.

[32] Bjorhovde, R., *Solutions for the Use of Jumbo Shapes*, Proceedings, 1988 National Steel Construction Conference, American Institute of Steel Construction, Miami Beach, FL, June 1988 (pp. 2-1 to 2-20).

[33] Fisher, J. W., Frank, K. H., Hirt, M. A., and McNamee, B. M., *Effect of Weldments on the Fatigue Strength of Steel Beams*, NCHRP Report No. 102, Highway Research Board, Washington, DC, 1970.

[34] Fisher, J. W., Albrecht, P. A., Yen, B. T., Klingerman, D. J., and McNamee, B. M., *Fatigue Strength of Steel Beams with Welded Stiffeners and Attachments*, NCHRP Report No. 147, Highway Research Board, Washington, DC, 1974.

[35] American Institute of Steel Construction (AISC), *A Guide to Shop Painting of Structural Steel*, Joint Publication, AISC and the Steel Structures Painting Council (SSPC), AISC-SSPC, Chicago, IL, 1972.

CHAPTER FIVE

Tension Members

5.1 INTRODUCTION

It is relatively rare to find a pure axially loaded member in a structure, although there are many applications that approach this condition. The structural elements considered in this chapter are subjected to a concentric tensile load as the primary force. The influence of secondary effects, such as load misalignment, are explored in the overall context of strength and behavior. Members subjected to a combination of tension and moment, for example, may have to be analyzed as beams or combined force (beam-column) members. This depends on how the moment is applied. Conditions such as these are treated in Chapters 7, 8, and 9.

Attention is focused on the behavior of the member, and how this is reflected in its strength as well as in the design criteria for strength and serviceability. The nature of the tension member is such that the service conditions rarely control the size or shape of the member. However, it is important that the designer check the appropriate serviceability conditions.

The type of end connections used for a tension member will have a significant influence on its behavior and strength. The analysis and design of the connections is dealt with in Chapter 11; their effect on the tension element is examined here.

5.2 TENSION MEMBERS IN STRUCTURES

A variety of tension member types may be found in building structures. Among the more important ones are bracing members, hangers, tie rods, sag rods, truss chords and webs, and guy wires.

Tension members used to brace structures are normally long and slender. Because they are very flexible, it is important that they be erected with a preload of some magnitude. Otherwise, the structure that they are intended to brace may have to move an unacceptable amount before the brace becomes effective. The preload may be provided by fabricating the member with a "draw," that is, with a detailed length dimension that is less than the theoretical one, or for rods by use of turnbuckles. The member, therefore, is taut when erected.

Other examples of tension members are the hangers that connect the cables to the floor system of a suspension bridge, and the sag rods that support the purlins in the roof structure or the girts in the walls of a steel-framed building. Hangers are clearly critical members, since they are essential to the integrity, strength, and behavior of the bridge. The influence of sag rods may not be quite as critical. Their failure may produce unsightly displacements in the walls, and could result in stability problems for the purlins or the girts, especially during the construction phase. Figure 5.1 shows an example of sag rods in a building application.

Tension members are also found as chords, diagonals, and verticals of trusses. Figure 5.2 illustrates typical simply supported trusses, with the tension members indicated.

Figure 5.1 Sag rods in a roof system.

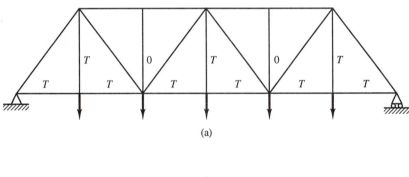

(a)

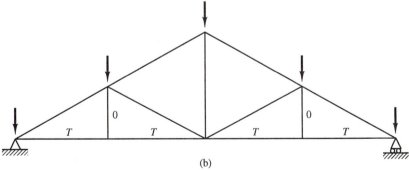

(b)

Figure 5.2 Tension members in a typical truss: (a) Warren Truss; (b) Howe Truss.

5.3 CROSS-SECTIONAL SHAPES
FOR TENSION MEMBERS

The treatment of tension members in this book is limited to those that utilize structural steel shapes, plates, and combinations thereof. High-strength steel cables and wire rope are not considered, although these are important for a number of special types of structures [1, 2].

Figure 5.3   (a) Eye bar; (b) pin-connected
link

(a)

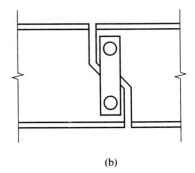

(b)

Two other types of tension members used in some structures are eyebars and pin-connected links, as illustrated in Fig. 5.3. The eyebar is not in practical use in structures that are being designed today, but can be found in older bridges, trusses, and similar structures where it is used in lieu of cables or wire rope. It was normally fabricated by forging from a single piece of steel, using a constant thickness for the full length of the member. The pin-connected link that is shown in Fig. 5.3b is in use in a number of industrial structures today; such members usually require some form of reinforcement of the material around the hole [3].

The more common shapes used for tension members are shown in Fig. 5.4, and some typical built-up cross sections are given in Fig. 5.5. The solid circular bar is frequently used, either as a threaded rod or welded to other members. The threaded end provides a simple connection to the structure, but the design must take into account the reduction of the cross-sectional area that is caused by the threads. Upset rods are occa-

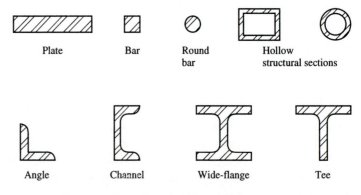

Figure 5.4   Shapes used as
tension members.

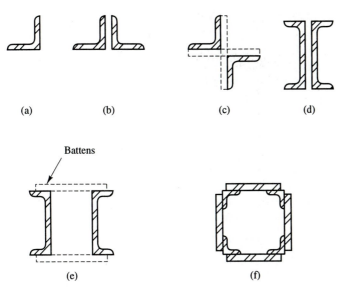

(a)   (b)   (c)   (d)

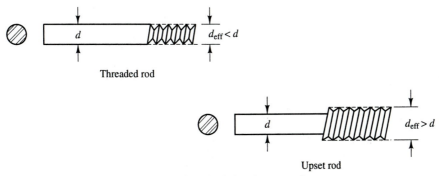

Battens

Figure 5.5 Typical built-up
tension members.

(e)   (f)

sionally used instead of the normal rods; the enlarged end permits thread-
ing without reducing the cross-sectional area below that of the main
portion of the rod. The differences between these two types of rods can be
seen in Fig. 5.6.

Square, rectangular, and circular tubes have become more commonly used for tension members over the past few years, largely due to their
attractive appearance and ease of maintenance. However, the end connections may become complicated and expensive, depending on the particular application. The tubes are especially useful for longer tension
components, where slenderness and related serviceability considerations
may be important.

Single angles are used extensively in transmission towers. Double
angles and double channels, as shown in Fig. 5.5b and d, are probably the
most popular tension members for planar trusses, due to the fact that
gusset plates can be conveniently placed in the space between the individ-

d $d_{eff} < d$

Threaded rod

d $d_{eff} > d$

Upset rod

Figure 5.6 Threaded and upset rods.

ual shapes. The end connections for these members are therefore easy to design and fabricate, and allow for symmetry in the vertical plane.

Large tensile forces usually require cross sections that may dictate that the member be made from a wide-flange shape, a tee, double channels, or built-up shapes, such as those given in Fig. 5.5e and f. Built-up cross sections were more commonly used in the past, when the cost of labor was lower; today, large-force tension members would probably be made from rolled shapes. Current structural applications of such elements are found in truss bridges, bracing members in power plants, and similar industrial structures.

5.4 BEHAVIOR AND STRENGTH OF TENSION MEMBERS

To perform a truly complete design of a structure, it is imperative for a structural engineer to have a thorough understanding of the strength as well as the behavior of all components, connections, and the overall assembly. Unfortunately, the behavior aspect is often neglected or down played. However, it is crucial to appreciate how the structure will respond to service as well as ultimate loads, if it is to be proportioned adequately. This requires an understanding of material behavior in addition to that of the structure itself, and further emphasizes the importance of the choice of the grade and type of steel for a particular application. In other words, it is at least as important to bear in mind the ductility of the member and the material, since their strength and service load computations alone tend to obscure these considerations.

5.4.1 Load–Deformation Relationship

It is often useful to compare the behavior of a full-size tension member under load to that of a tension test specimen. In many ways the member is an extension of the small-scale specimen, although their differences underline the fact that the tension test displays the properties of the material per se, while the full-size member incorporates the influence of size and shape and many other parameters. This is further illustrated by the load–deformation curves shown in Fig. 5.7, where Fig. 5.7a gives the stress–strain curve for a typical mild steel tension test, and Fig. 5.7b shows the load–deformation data for a full-size tension member. Simplistically speaking, the curve in Fig. 5.7a represents the behavior of an individual "fiber" of the steel, whereas the curve in Fig. 5.7b reflects the load–deformation relationship of a member that is made up of a large number of such fibers. The differences between the curves in Fig. 5.7a and b may be attributed to a number of factors:

1. The load P is not applied concentrically to the member.
2. The full-size shape has certain built-in or residual stresses.

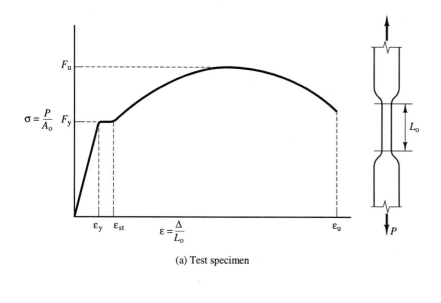

(a) Test specimen

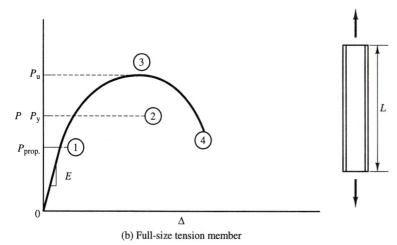

(b) Full-size tension member

Figure 5.7 Load–deformation curves.

3. The tension member is not perfectly straight.
4. The cross-sectional shape and size vary to some extent along the member length.

Items 1 and 3 both imply that there will be a certain amount of bending in the member, in addition to the effect of the axial load. This is demonstrated schematically in Fig. 5.8, from which it is clear that the moments that are produced create a nonuniform stress distribution in the cross section. Therefore, as the load is increased, the fiber that is stressed the

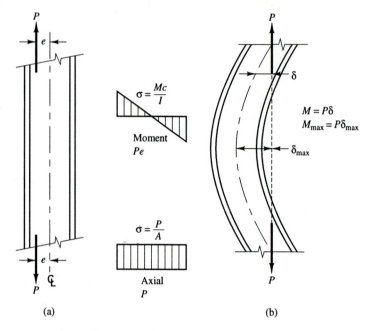

Figure 5.8 Bending of the tension member: (a) Eccentric load; (b) member not straight.

most in tension due to bending plus axial load will yield before any other fiber. That fiber reaches the yield plateau of the stress–strain curve (see Fig. 5.7a) before all of the others, which are still strained in the elastic range.

The phenomenon of residual stress (item 2) is discussed in detail in Chapter 6, which deals with the behavior and strength of columns. Suffice it to observe here that the uneven cooling that takes place in the shape after any kind of heat input (e.g., heating due to rolling, welding, or flame-cutting) results in varying degrees of restraint to contraction of the fibers in the shape. As a result, some parts of the cross section will end up in compression and others will be in tension, and these stresses exist in the member before any load has been applied. Residual stresses therefore effectively act as a preload. When the tensile residual stresses are combined with those due to a tensile force (and possibly also a bending moment, as discussed previously), yielding will start earlier in these fibers than in those that are stressed in compression.

The effect of residual stresses on a tension member is essentially limited to producing early yielding in some parts of the shape. This means that the member will exhibit larger longitudinal deformations than would appear if no residual stresses existed. Although it is possible that the fibers that are the first to reach yield strain will have been strained into the strain-hardening region and beyond by the time the last fibers reach F_y, for all practical purposes the limit state of yielding may be defined by the yield load:

$$P_y = F_y A \qquad (5.1)$$

where A is the nominal cross-sectional area of the shape. The presence of residual stresses in compression members results in a significant lowering of the actual strength (see Chapter 6). For tension members, on the other hand, the only effect is a shift in the load–deformation curve that reflects the larger deformations, but the strength stays the same.

Figure 5.9 illustrates the summation of the effects due to residual stresses, uniform axial stresses, and bending. The resulting distribution emphasizes the aspect of gradual yielding in the cross section. The data that have been used to develop the individual distributions shown in Fig. 5.9 are realistic, but not necessarily representative of all shapes or steel grades. They are intended only as a demonstration of concepts.

The last of the effects that produce a load-deformation curve that differs from that of the tension specimen is a member cross-sectional area that is not a constant along the length (item 4). Although this variation in area might result from a change in the actual shape dimensions, a more important factor is the presence of holes in the cross section, such as will be found when bolted end connections are used for the tension member. The treatment of this aspect of tension member behavior is dealt with in Section 5.5. Recall that a hole creates a stress concentration, such as that illustrated in Fig. 5.10. As a result, the application of an increasing tensile force will cause yielding at the edges of the hole first. As the load continues to increase, all fibers will eventually reach the yield stress, F_y, and the reduction in the strength will therefore be only a reflection of the absence of material. On the other hand, while stress concentrations may not reduce the static yield strength of the member, they do have a significant effect on the dynamic or fatigue-related structural performance.

The various ranges of the load-deflection curve for the full-size member delineate certain aspects of the behavior that are important (refer again to Fig. 5.7b):

1. *Range 0–1:* All fibers in the cross section are stressed to less than the yield stress. The member is therefore fully elastic, and all deformations are recoverable upon unloading.

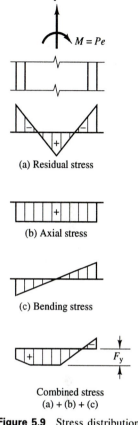

(a) Residual stress

(b) Axial stress

(c) Bending stress

Combined stress
(a) + (b) + (c)

Figure 5.9   Stress distribution due to different load effects.

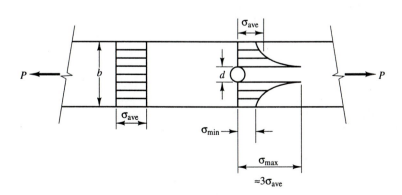

Figure 5.10   Stress concentration due to holes in a member.

2. *Range 1–2:* First yield occurs for $P = P_{prop}$, which is the proportional limit. All subsequent load increases produce yielding in
larger and larger parts of the cross section. The member is
therefore partially elastic and partially plastic, and unloading to
zero load will result in a certain permanent deformation. At point
2, all fibers have reached or exceeded the yield strain, and the
applied load equals or exceeds the yield load F_yA.

3. *Range 2–3:* Certain fibers that yielded early are bound to reach
strain-hardening, in some cases even before $P = F_yA$. As the
load increases past point 2, a significant portion of the cross
section may strain-harden, which accounts for the continuing
increase in the load to point 3. At this stage P equals F_uA, the
ultimate load.

4. *Range 3–4:* As the tension member is deformed past the displacement that corresponds to point 3, a number of fibers will already
have reached the descending portion of the stress–strain curve,
as seen in Fig. 5.7a. The load on the member therefore must be
reduced to maintain equilibrium. At some stage between points 3
and 4, some fibers will begin to fracture (i.e., to reach the rupture
strain ε_u), which rapidly produces the tensile fracture of the
member.

This describes the preferred behavior of a tension member. Large deformations take place before fracture, and there is ample evidence of the
impending collapse before it occurs. Certain combinations of cross section, holes, and types of steel may not exhibit this kind of behavior. For
these it is possible that failure may take place with little or no warning; the
material may fracture at one or more sections before extensive yielding
has occurred. This has important design implications, as we demonstrate
in the following.

Particular attention must be paid to the case of combined axial load
and bending, as alluded to earlier in this chapter. The minor bending
effects that are created by initial out-of-straightness, for example, can be
accommodated by the member itself without any special design modifications. This is due to the ductility of the material and the member, and its
consequent ability to redistribute load from highly stressed to less highly
stressed regions. However, it is important to recognize that bending in a
tension member may result from two very different kinds of loading
conditions.

If the bending is due to slight eccentricities from the axial loading, as
illustrated in Fig. 5.11a, it will probably be of no concern. This is because
the most highly stressed remote fibers will yield and the less stressed
fibers will then pick up more load, as described above. In this case, the
yield load and the fracture capacity of the member are dependent only on
the cross-sectional area and the grade of steel. However, if the bending
results from a lateral load, as shown in Fig. 5.11b, the member may fail as

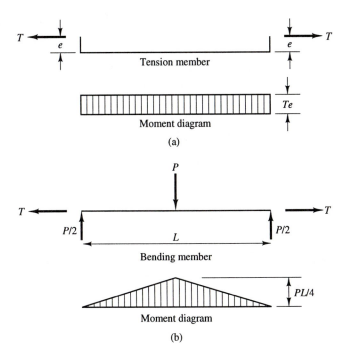

Figure 5.11 Bending resulting from (a) an axial load eccentricity and (b) a lateral load.

a beam. Since the axial tension will reduce the bending capacity, it is important that it be designed using an interaction approach. This is discussed in detail in Chapter 9.

5.4.2 Limit States of Strength

In the design of tension members, the following cross-sectional area definitions are particularly important:

1. *Gross cross-sectional area, A_g*
2. *Net area, A_n*
3. *Effective net area, A_e*

For a steel plate with width b, thickness t, and a single hole of diameter d, the gross and net areas of the plate will be

$$A_g = bt$$

and

$$A_n = (b - d)t \qquad (5.2)$$

The effective net area is a function of the cross section and the method of end connection. It is related to the phenomenon of shear lag, and is discussed in detail in Section 5.5.4.

The behavior of the tension member under increasing load has already been discussed, to the extent that the gradual spread of yielding due to residual stresses, cross-sectional properties, and loading pattern is understood. Considering the gross and net cross-sectional areas of the member, this can now be phrased in terms of the two modes of failure:

1. Yielding on the gross area
2. Fracture on the effective net area

Both of these represent limit states of strength that must be taken into account in the design of the tension member. The resistance factors are $\phi = 0.9$ for yielding and $\phi = 0.75$ for fracture.

Yielding on the Gross Area. If the size of the gross area is such that the condition

$$\sigma = \frac{P_n}{A_g} = F_y \tag{5.3}$$

is reached before the ultimate stress F_u has been developed anywhere in the member, it is said to have failed by yielding on the gross area. This is a limit state of strength that is accompanied by large overall deformations in the tension member, and therefore gives ample warning of an impending failure. Naturally, since $A_n < A_g$, the yield stress will have been reached in all fibers of the net cross section before Eq. (5.3) is satisfied. However, since the net area exists only at one or a few specific and limited locations along the tension member, the deformations that are associated with the condition $F_y = (P_n/A_n)$ will be small and essentially confined to these regions. Because the total deformation of the member is proportional to its length, these small regions will not be critical. Consequently, the only yielding limit state of significance is yielding on the gross area.

Uncontrolled yielding of a tension member can be about as disastrous for the structure as an actual member fracture. For instance, if the bottom chord of a truss yields without limit, the truss will deflect to a point where it will, in fact, collapse, even though the chord itself does not fracture. This is obviously a large deflection/deformation phenomenon.

Fracture on the Effective Net Area. If the size of the effective net area is such that the condition

$$\sigma = \frac{P_n}{A_e} = F_u \tag{5.4}$$

is reached before yielding has developed on the gross area, the member is said to have failed by developing the ultimate tensile stress on the effective net area. From our discussion of the deformations in a tension member, we know that failure according to Eq. (5.4) is accompanied by small overall deformations. Since this also incorporates the actual fracture of the steel on the effective net section, this ultimate limit state is a disastrous failure that involves the loss of the integrity of a structural member. The small deformations give little or no warning of the impending failure, which occurs suddenly and offers limited opportunities to take corrective action before the fracture.

On the basis of these two definitions, the tension member strength is a function of A_e, A_g, F_y, and F_u. It has already been noted that the cross-sectional properties and end connections are important, because of their influence on the effective net area. The ratio of F_y to F_u is also a significant parameter. It is conceivable that the average stress on the effective net section may reach F_u before the average stress on the gross area reaches F_y. When these stresses are defined as

$$\sigma_{\text{gross}} = \frac{P_n}{A_g} = F_y \quad \text{and} \quad \sigma_{\text{eff}} = \frac{P_n}{A_e} = F_u \tag{5.5}$$

then Eqs. (5.5) may be rewritten as

$$P_n = \sigma_{\text{gross}} A_g = F_y A_g \tag{5.6a}$$

and

$$P_n = \sigma_{\text{eff}} A_e = F_u A_e \tag{5.6b}$$

The two limit states occur simultaneously if

$$F_u A_e = F_y A_g \tag{5.7}$$

or

$$\frac{A_e}{A_g} = \frac{F_y}{F_u} \tag{5.8}$$

The limit state of yielding on the gross section governs if the right side of Eq. (5.7) is less than the left side, or, using Eq. (5.8),

$$\textit{Yielding on } A_g: \quad \frac{A_e}{A_g} > \frac{F_y}{F_u} \tag{5.9}$$

Conversely, fracture on the effective net section governs when

$$\textit{Fracture on } A_e: \quad \frac{A_e}{A_g} < \frac{F_y}{F_u} \tag{5.10}$$

The LRFD design criteria consider both of the above failure modes. Note that a steel with a small (F_y/F_u) value, such as ASTM A36 and similar mild structural steels, will allow more of the cross section to be removed in the form of bolt holes before the fracture limit state will govern. In other words, the redistribution of the stresses from the higher stressed to the lower stressed regions of the cross section is facilitated in the more ductile steels with large reserve capacity.

These comparisons are applicable only for normal bolted connections and the corresponding areas. Equations (5.5)–(5.10) are not intended to cover the case of large cutouts in members. These require special design considerations, and are beyond the scope of this book, since they are uncommon features of tension members in most building structures.

Welded connections remove no material from the cross section, which means that $A_n = A_g$ for these cases. However, depending on the placement of the welds and the type of cross section that is used, shear lag may play a role, making the effective net area less than the net area. Details of the treatment of tension members with welded connections are given in Section 5.5.4.

5.5 COMPUTATION OF AREAS

The criteria that govern the computation of the cross-sectional areas of tension members are given in Sections B1, B2, and B3 of the LRFD specification. We discuss them in detail, including data on some of the studies that led to their development.

5.5.1 Gross Area

The computation of the gross area of a member is straightforward. A section is made perpendicular to the longitudinal axis of the element, along which the tensile force is acting, and the gross area is found as the area of that cross section. No holes or other area reductions can be present where the section is made.

In the case of plates, bars, and solid circular shapes, the value of A_g is found directly as the value of bt for plates and bars and $\pi d^2/4$ for circular shapes, where d is the diameter of the shape. For the structural steel shapes that are used in construction, the *LRFD Manual* gives the data for their gross areas. However, in lieu of using the tabulated values, the LRFD specification, Section B1, allows A_g to be computed as

$$A_g = \Sigma w_i t_i \tag{5.11}$$

where w_i and t_i are the width and thickness, respectively, of the rectangular cross-sectional element i of the shape. Equation (5.11) obviously applies only to shapes that are made up of flat plate components, such as the wide-flange and the channel. The calculation for hollow circular shapes or tubes is similarly simple.

A special procedure may be used for areas of angles. This is particularly useful in the determination of the net area of such shapes, as we demonstrate later in this chapter. The angle may be treated as an equivalent flat plate, where the dimensions are:

$$Width:\qquad w_e = l_1 + l_2 - t \qquad\qquad (5.12a)$$

$$Thickness:\quad t_e = t \qquad\qquad (5.12b)$$

and the gross area is found from

$$Gross\ area:\ A_g = w_e t_e = (l_1 + l_2 - t)t \qquad (5.13)$$

In these expressions, l_1 and l_2 are the nominal leg lengths of the angle, and t is the leg thickness. The equivalent flat plate concept is further illustrated by Fig. 5.12.

5.5.2 Net Area

As noted in Section 5.4.2, only the reduction in cross-sectional area that results from typical connections is considered here. Thus, holes resulting from mechanical fasteners (bolts, rivets) and certain types of welds (plug welds, slot welds) are included. The fastening elements are normally used only to connect the tension member to the adjacent parts of the structure, and the reduced areas therefore appear mostly at the ends.

Plug and slot welds are relatively uncommon in most structures where the framing members are made from structural shapes and plates. Cold-formed elements and metal building components use this fastening method to a much greater extent.

The types of welded joints that are used in regular structural steel frames are essentially limited to fillet welds and full and partial penetration groove welds. The characteristics of these kinds of joints are discussed in detail in Chapter 11. Such welds do not reduce or increase the cross-sectional area of the member, and for that reason the net area is

$$Welded\ joints:\quad A_n = A_g \qquad\qquad (5.14)$$

Figure 5.12 Equivalent flat plate.

which applies to tension members with welded end connections. Although the welds leave A_n equal to A_g, it is still possible to have the effective net area A_e less than A_g, as a result of the shear lag effects.

In the computation of the net area for a tension member with bolted end connections, the size of the individual hole and how that is determined are important. The criteria for standard bolt holes and the geometries of oversize holes and slotted holes are covered in the LRFD specification, Section J3.7.

Size of Standard Holes.   In fabricated steel construction it is not the size of the holes that is specified, but rather the nominal diameter of the fastener. The hole is then sized according to that dimension, as well as the way the hole is made. Exceptions to this procedure will naturally occur, but the design almost always will be based on the bolt size that is chosen.

Recognizing the needs for fabrication and erection tolerances, standard bolt holes are made $\frac{1}{16}$ in. larger in diameter than that of the bolt. Thus, a $\frac{3}{4}$-in. bolt will have a hole that is $(\frac{3}{4} + \frac{1}{16}) = \frac{13}{16}$ in. in diameter. Since hole making may damage some of the material immediately adjacent to the hole, that material may not be considered fully effective in transmitting load. This is schematically illustrated in Fig. 5.13 for the case of punched holes. As the punch is applied to the material, the edges around the hole are deformed, as shown in Fig. 5.13b. This region is discounted and the effective hole size that is used for the net area calculations in tension members is therefore increased by another $\frac{1}{16}$ in.

Provided the fabricator has punching equipment of sufficient capacity, the LRFD specification, Section M5, allows punching to be used for material thicknesses up to $\frac{1}{8}$ in. larger than the nominal bolt diameter. Otherwise the holes must be drilled, or subpunched and reamed. Special rules apply to quenched and tempered steels, such as ASTM A514. In design of tension members, therefore, it is standard practice to deduct for a hole whose diameter is given by the expression

$$\textit{Effective hole diameter:} \quad d_e = d + \tfrac{1}{16} + \tfrac{1}{16} \qquad (5.15)$$

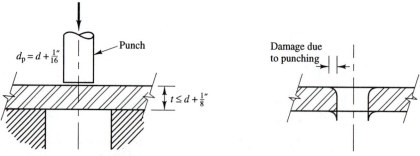

Figure 5.13   Damage caused by hole punching.

   The following examples demonstrate typical net area calculations. Figure 5.14 gives the size of plates with a single line (a) and a double line (b) of holes.

EXAMPLE 5.1

Given: A single line of standard holes for $\frac{3}{4}$-in. bolts are placed in a $6 \times \frac{1}{2}$ plate, as shown in Fig. 5.14a. Determine the gross and net areas.

Solution

Gross area (section 2-2): $A_g = 6(\frac{1}{2}) = 3.0$ in.^{2}

Net area (section 1-1): $A_n = (b - d_e)t$

For a $\frac{3}{4}$-in.-diameter bolt, the effective hole size is

$$d_e = \tfrac{3}{4} + \tfrac{1}{16} + \tfrac{1}{16} = \tfrac{7}{8} \text{ in.}$$

Consequently,

$$A_n = (6.0 - \tfrac{7}{8})(\tfrac{1}{2}) = 2.56 \text{ in.}^2$$

EXAMPLE 5.2

Given: A double line of standard holes for $\frac{7}{8}$-in. bolts are placed in a $10 \times \frac{3}{4}$ plate, as shown in Fig. 5.14b. Determine the gross and net areas.

Solution

Gross area (section 2-2): $A_g = 10(\frac{3}{4}) = 7.5$ in.^{2}
Effective hole size for a $\frac{7}{8}$-in.-diameter bolt:

$$d_e = \tfrac{7}{8} + \tfrac{1}{16} + \tfrac{1}{16} = 1.0 \text{ in.}$$

Net area (section 1-1): $A_n = (10.0 - 2(1.0))(3.4)$
$$= 6.00 \text{ in.}^2$$

Figure 5.14 also indicates two measurements that are important in the design of tension members as well as connections. The center-to-center distance between adjacent holes in the direction parallel to the primary applied force (here: tension) is defined as the pitch s. Similarly, the center-to-center distance between adjacent holes in the direction perpendicular to the primary applied force is defined as the gage g. The latter

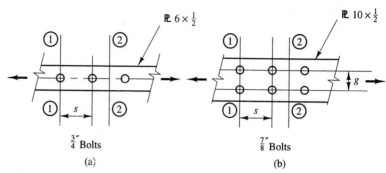

Figure 5.14 Plates with (a) a single line of holes and (b) a double line of holes (Example 5.1).

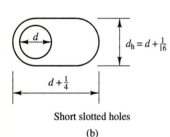

Oversize holes

(a)

$d_h = d + \frac{3}{16}$

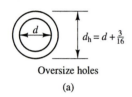

$d_h = d + \frac{1}{16}$

$d + \frac{1}{4}$

Short slotted holes

(b)

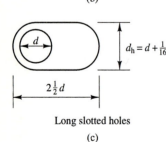

$d_h = d + \frac{1}{16}$

$2\frac{1}{2}d$

Long slotted holes

(c)

Figure 5.15 Criteria for normal bolt diameters in oversize and slotted holes.

obviously applies only if the member has more than one line of holes, such as that of Fig. 5.14b.

Oversize and Slotted Holes. The LRFD specification, Section J3.7, also gives the required measurements for larger than standard or oversize holes, as well as for short slotted and long slotted holes. Figure 5.15 illustrates the criteria that apply for nominal bolt diameters less than or equal to $\frac{7}{8}$ in.; the reader is referred to the specification for data for larger bolts. Basically, these types of holes are used in many instances to facilitate the erection of the structure, and in other cases to allow certain rotations or deformations to take place in the frame for specific loading levels. The specification also gives detailed rules for the use of oversize and slotted holes. These are covered in Chapter 11.

When the tension member is a short connecting element, such as a stiffener or a gusset plate, the LRFD specification, Section J5.2, specifies that the net area cannot be assumed to be larger than 85 percent of the gross area. The reason is that the calculation for the fracture limit state assumes that the fracture stress is uniformly distributed across the net area. When the member is short, and the net area and the gross area are close to equal, there may not be sufficient length for the entire cross section to yield uniformly. In this case the area that is the first to yield may reach fracture at an early stage, and the fracture limit state is therefore reached prematurely. This is an undesirable mode of failure, primarily because it is not ductile, and because it occurs suddenly, with little or no warning.

5.5.3 Influence of Hole Placement

The sample computations of net area that are given in the preceding part of this chapter represent simple, clear-cut cases, where the net area is taken through the section that produces the largest reduction. It is obvi-

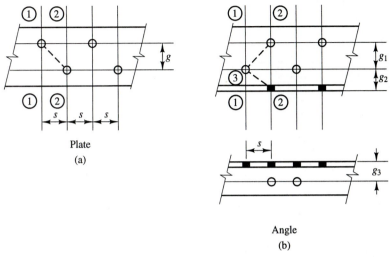

Figure 5.16   A pattern of staggered holes.

ous for both of the examples where the governing section or chain must be located. However, hole placement does not always follow simple patterns such as those of Fig. 5.14. For several reasons it is sometimes practical to utilize a pattern of staggered holes, as indicated in Fig. 5.16. Thus, Fig. 5.16a shows an application of staggered holes for a plate, and Fig. 5.16b demonstrates an example for an angle. Pitch and gage dimensions are also indicated.

The location of the governing net section is not obvious for either of the two cases shown in Fig. 5.16. For the plate, perpendicular sections 1-1 and 2-2 give identical A_n values, where only a deduction for one hole has been taken. For the angle, it is clear that section 2-2, which incorporates one hole deduction for each of the legs, for a total of two, is more critical than section 1-1. However, the question for both the plate and the angle is whether the presence of a hole on one section may influence the strength and behavior of a different, but adjacent, one. For example, the force flow in the plate on section 2-2 will be different from that on section 1-1, as illustrated by the stress distribution and stress resultants for the two sections. This is shown in Fig. 5.17, which indicates the approximate elastic distributions. It is intuitively clear that if Sections 1-1 and 2-2 are close enough together, the behavior of either will not be the same as when there is no hole in the vicinity. On this basis, it is possible that a section or chain such as Section 1-2 of Fig. 5.16a or Section 2-3-2 of Fig. 5.16b may be the governing one for the tension member. In other words, holes on adjacent sections will interact, although not as severely as when they are located on the same perpendicular section.

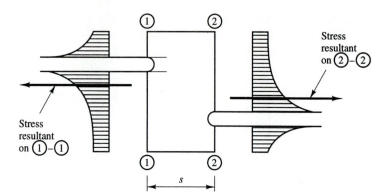

Figure 5.17 Stress distribution and stress resultants for sections 1–1 and 2–2 of Fig. 5.16 (a).

Early studies of these phenomena were made by Cochrane and Smith [5–7], and were later expanded by the findings of Bijlaard [8, 9]. McGuire [4] gives an excellent and thorough discussion of the approaches and the results. The following, therefore, presents only the essential criteria and the procedures that are specified by AISC.

In the original work it was assumed that the object of the design was to arrive at a balanced solution, and this was thought to have been achieved if the maximum principal stresses on perpendicular and inclined sections through the member were the same [5, 7]. In other words, the Tresca failure criterion was used, which implies that the yield stress in shear is equal to one-half of the tensile yield stress.

As a final simplification, it was found that the influence of the staggered hole could be accounted for by using the following expression for the net width of the section:

$$w_n = w_g - d - xd \tag{5.16}$$

where w_n is the net width, w_g is the gross width, d is the diameter of the hole, and xd represents the strength-reducing influence of the staggered hole. Cochrane and Smith showed that xd could be expressed as

$$xd = d - \frac{s^2}{4g + 2d} \tag{5.17a}$$

This ratio reduces to $s^2/4g$, if it is assumed that $2d$ is significantly less than $4g$. This is a reasonably realistic assumption which leads to

$$xd = d - \frac{s^2}{4g} \tag{5.17b}$$

When Eq. (5.17b) is substituted into Eq. (5.16), the net width of a two-hole section, with one-hole offset with a pitch of s and a gage of g, becomes

$$w_n = w_g - d - \left(d - \frac{s^2}{4g} \right)$$

or

$$w_n = w_g - 2d + \frac{s^2}{4g} \tag{5.18}$$

Effectively, Eq. (5.18) states that the net width is found by deducting fully for all holes in the chain, but since this overestimates the effect of the staggered holes, a certain width contribution is added.

Equation (5.18) applies directly in the case of a pair of holes that are staggered by a pitch of s and a gage of g. In general, one $(s^2/4g)$ term is introduced for every stagger in the chain. Therefore, there will be $(m - 1)$ staggers for a chain with m holes, and the generalized form of the net width equation becomes

$$w_n = w_g - md_e + \Sigma \left(\frac{s_i^2}{4g_i} \right) \tag{5.19}$$

where d_e has been used to emphasize that the effective hole diameter should be accounted for. Naturally, the net area is found as $A_n = w_n t$, or as $A_n = w_{n1} t_1 + w_{n2} t_2 + \cdots$, where the latter applies to the case of a cross section that is made up of plates of different thicknesses and widths.

Bijlaard [8, 9] expanded the earlier developments by basing his analysis of the plate model on the maximum distortion energy or von Mises failure criterion. This gives a shear yield stress of $\tau_y = F_y/\sqrt{3}$, and is generally regarded as the most appropriate failure model for ductile materials like steel. The findings demonstrated that Cochrane's result was correct for a pitch of $s = 0$; it obviously would not hold for the case of a gage of $g = 0$ (single line of holes). On the whole, however, the practical ranges of s and g values are such that the Cochrane and the Bijlaard results differ only by about 10 to 15 percent. This is the reason why Eq. (5.19) continues to be used in design specifications.

Examples 5.3 and 5.4 (Fig. 5.18) demonstrate practical applications of the staggered hole criterion, and Example 5.4 also shows the use of the equivalent flat plate approach for angles.

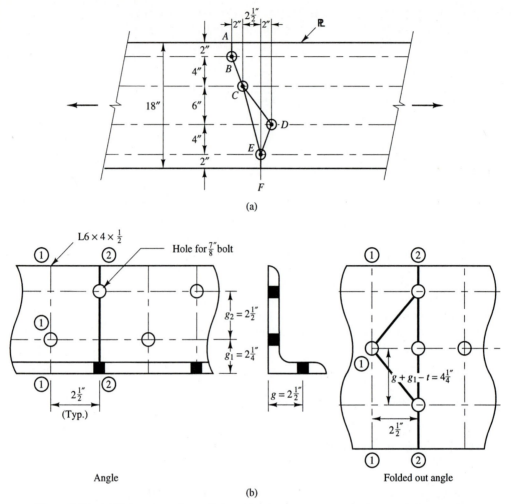

Figure 5.18 (a) Hole pattern for an 18-in.-wide plate that is loaded in tension (Example 5.3); (b) placement of holes in an angle tension member (Example 5.4).

EXAMPLE 5.3

Given: The hole pattern for an 18-in.-wide plate that is loaded in tension is shown in Fig. 5.18a. Determine the net width that governs the design.

Solution

Step 1. Chain *A B C E F*

Deduct for 3 holes @ $(\frac{3}{4} + \frac{1}{8})$ $= -\ 2.63$ in.

For BC, add $s^2/4g = 2.0^2/(4)(4)$ $= +\ 0.25$ in.

For CE, add $s^2/4g = 2.5^2/(4)(10)$ $= +\ 0.16$ in.

Total deduction $= -\ 2.22$ in.

Step 2.   Chain $A\ B\ C\ D\ E\ F$

Deduct for 4 holes @ $(\frac{3}{4} + \frac{1}{8})$ $= -\ 3.50$ in.

For BC, add as in step 1 $= +\ 0.25$ in.

For CD, add $s^2/4g = 4.5^2/(4)(6)$ $= +0.85$ in.

For DE, add $s^2/4g = 2.0^2/(4)(4)$ $= +\ 0.25$ in.

Total deduction $= -\ 2.15$ in.

Step 3.   Net width $= 18.0 - 2.22 = 15.78$ in.

Step 4.   Check maximum allowable net area:
Maximum allowable net width $= 18.0(0.85) = 15.3$ in.
Since 15.3 is less than 15.78, the maximum allowable value of step 4 governs:
15.3 in.

EXAMPLE 5.4

Given: A $6 \times 4 \times \frac{1}{2}$ angle is used as a tension member. It has holes for $\frac{7}{8}$-in.
bolts that are placed as shown in Fig. 5.18b.*

Solution

Step 1.   Determine the width of the folded-out angle "plate":

$$w_e = l_1 + l_2 - t = 9.50 \text{ in.} \qquad \text{(from Eq. (5.13))}$$

Step 2.   Gross area of plate:

$$A_g = w_e t = 9.50(0.5) = 4.75 \text{ in.}^2$$

(*Note:* This happens to be exactly the same as the area that is given for the angle in
the *LRFD Manual.* That will not always be the case for such computations, since
the "plate" is an artificial one.)

* The placement of holes in angles and other structural shapes generally
follows accepted practice. The *LRFD Manual* gives detailed data (e.g., see
p. 5-166 for angles).

Step 3. The gages for the holes are shown in Fig. 5.18b. The gage between the holes closest to the heel of the angle in the two legs is found as $(g + g_1 - t) = 2.50 + 2.25 - 0.50 = 4.25$ in.

Step 4. Computation of net area: The governing net section will be section 2-2 or section 2-1-2.
Section 2-2: $A_{n2} = [9.50 - 2 (\frac{7}{8} + \frac{1}{8})] 0.50 = 3.75$ in.^{2}
Section 2-1-2: There are two staggers of the bolt holes in this chain, and both have the same pitch ($s = 2.50$ in.). The gages are different, with one of 4.25 in. and the other of 2.50 in. The net area for this chain becomes

$$A_{n3} = \left[9.50 - 3(1) + \frac{2.50^2}{4(2.50)} + (2.50^2/4(4.25)) \right] 0.50 = \underline{3.50 \text{ in.}^2}$$

Step 5. Governing net area: Since $A_{n3} = 3.50$ in.^{2} is less than A_{n2} as well as $A_{n.max} = 0.85 A_g = 4.04$ in.^{2}, it governs the net area of the given angle and hole configuration.

5.6 SHEAR LAG

When only a portion of a tension member is attached to the connecting element, the net area for calculating the fracture case must be reduced for the effect of *shear lag*. This phenomenon occurs because the stresses must flow from a uniform distribution across the area at some distance from the connection to the more restricted area where the connection is located. The portion of the area that is participating effectively in the transfer of the force is smaller than the full net area, which therefore must be reduced. The reduced area is the *effective net area A_e*.

Through an extensive research program at the University of Illinois, Munse and Chesson [*10, 11*] showed that the reduction can be approximated by multiplying the net area by the factor U:

$$U = 1 - \frac{x}{L} \qquad (5.20)$$

where

x = distance between the centroidal axis of the tension member and the face of the connection element

L = length of the connection between the tension member and the connection element

Figure 5.19 illustrates the conceptual basis for Eq. (5.20). Essentially, the effective length of the connection is reduced by the distance x and is assumed to be $(L - x)$. The reduced length is found by assuming a 45° angle of transfer, as shown in Fig. 5.19. The net area of the tension member therefore is reduced by the factor $(L - x)/L$, or $1 - x/L$.

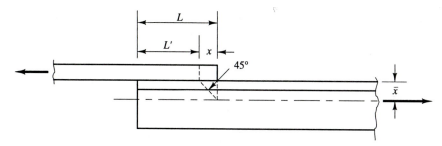

Figure 5.19   Conceptual basis for Eq. (5.20).

A typical example of the calculation of the shear lag factor is shown in Fig. 5.20 for both a bolted and a welded case. For the bolted case, N is the total number of bolts, and s is the pitch. The LRFD specification, Section B3, lists the values of the shear lag factor U for a number of cases. In general, they are based on Eq. (5.20). However, in some cases the specified values are more liberal, because subsequent tests indicated that the equation gives unnecessarily conservative results.

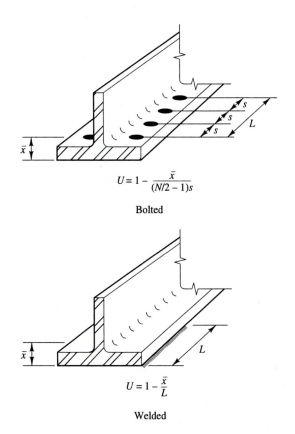

$$U = 1 - \frac{\bar{x}}{(N/2 - 1)s}$$

Bolted

$$U = 1 - \frac{\bar{x}}{L}$$

Welded

Figure 5.20   Calculation of the shear lag factor for a bolted and a welded connection.

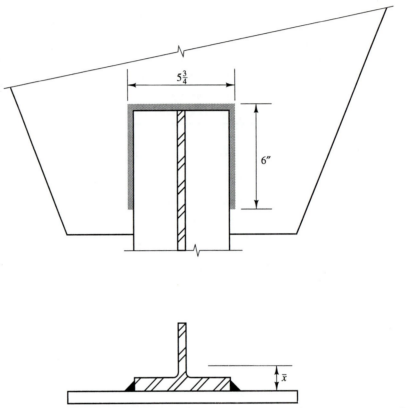

Figure 5.21   A structural tee that has been welded to a gusset plate, transverse as well
as parallel to the tensile force.

There is a second major type of tension member connection where
only a portion of the cross section is attached. Figure 5.21 shows a
structural tee (WT) that has been welded to a gusset plate, transverse as
well as parallel to the tensile force. The expression $1 - x/L$ is a good
estimate of the reduction factor due to the connectors (bolts or welds) that
are oriented parallel to the force. However, the LRFD specification,
Section B3, states that in the case of a transverse weld, only the portion of
the tension member that is actually connected is effective in carrying the
load. Two separate reduction factors therefore apply to the tee in
Fig. 5.21. This is not specifically stated in the specification, but it is
recommended that the weighted average U value should be used for such
cases. A numerical solution is given in Example 5.7.

5.7 BLOCK SHEAR

Block shear failure in a connection occurs when a portion of the member
tears out in a combination of tension and shear. Figure 5.22 shows a
typical case. The resistance to the tear-out is provided by a combination

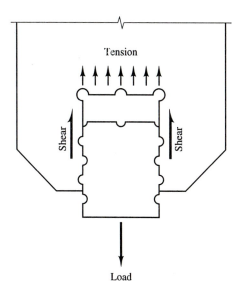

Figure 5.22 Block shear failure.

of shear on one plane and tension on a plane normal to it. Previous design rules utilized a combination of shear and tensile fracture to reflect this phenomenon. However, it is now clear that this solution is not realistic, to the effect that as soon as one plane fractures, the other plane, which is still at the yield stage, is then assumed to pick up all of the load and reach fracture.

The LRFD specification design criteria are based on research work of Birkemoe and Gilmor [*12*], Ricles and Yura [*13*], and Hardash and Bjorhovde [*14*]. The approach is detailed in the following.

For block shear failure actually to occur, two possibilities exist: Failure may take place through a combination of shear yield and tensile fracture, or through a combination of tensile yield and shear fracture. The LRFD specification addresses this in Section J5.2. The controlling case is the one that gives the larger load. Since the computations are easily done, and it is not always clear which of the modes of behavior will govern, both limit states are usually checked. The expressions to use are the following:

$$\text{Shear yield} \; + \; \text{Tensile fracture}$$
$$\phi(0.6F_y A_{vg} \; + \; F_u A_{nt}) \tag{5.21}$$

$$\text{Tensile yield} \; + \; \text{Shear fracture}$$
$$\phi(F_y A_{tg} \; + \; 0.6F_u A_{ns}) \tag{5.22}$$

where

$$\phi = \text{resistance factor for fracture} = 0.75$$
$$A_{vg} = \text{gross area in shear}$$
$$A_{tg} = \text{gross area in tension}$$

A_{ns} = net area in shear

A_{nt} = net area in tension

The reason the controlling case is the larger of the two failure loads that represents the actual ultimate limit state can be determined from an examination of Fig. 5.23. In Fig. 5.23a, the resistance to block shear is primarily afforded by shear. Since the shear fracture load is larger than

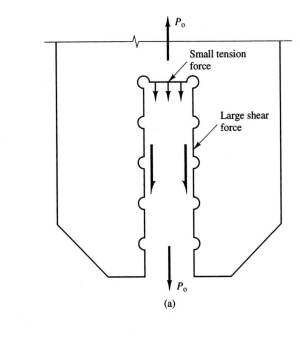

(a)

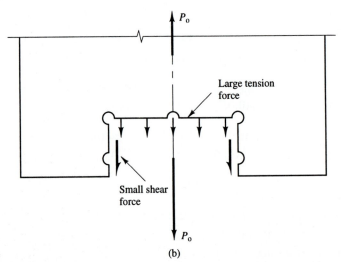

(b)

Figure 5.23 Ultimate limit states governed by (a) shear fracture and tensile yield and (b) tensile fracture and shear yield.

the shear yield load, failure will be governed by shear fracture in combination with tensile yield. On the other hand, Fig. 5.23b illustrates a case where tension is the primary stress resultant; in this case the limit state is that of combined tensile fracture and shear yield.

EXAMPLE 5.5

Given: For the gusset plate in the heavy bracing connection that is shown in Fig. 5.24, check whether the plate thickness of $\frac{1}{2}$ in. is adequate to resist block shear. The steel grade is A36 ($F_y = 36$ ksi, $F_u = 58$ ksi), and the holes are drilled, for $\frac{7}{8}$-in.-diameter A325 high-strength bolts.

Solution

It is assumed that the total factored load of 225 kips is distributed evenly between all of the bolts, such that the load is transferred from the diagonal into the gusset

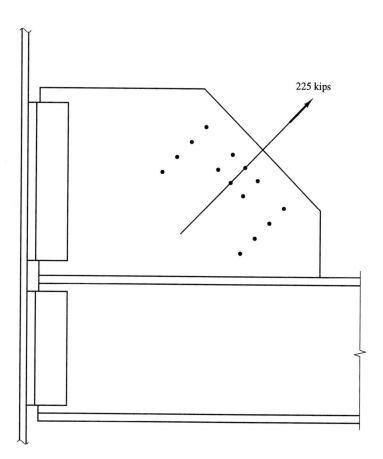

225 kips

Figure 5.24 Gusset plate in a heavy bracing connection (Example 5.5).

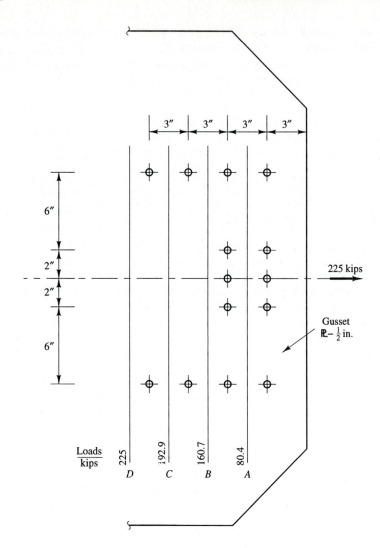

Figure 5.25 Transfer of load from the diagonal into the gusset plate (Example 5.5).

plate, as shown in Fig. 5.25. The loads listed are those that are in the gusset plate at each plane. For example, the load on plane B is 160.7 kips.

The following calculations check planes A and C for shear fracture and tensile yield. Assume that planes B and D have already been evaluated and found to be acceptable.

Plane A

A_{ns} has two shear planes

$$A_{ns} = 2[3.00 - (\tfrac{7}{8} + \tfrac{1}{16})\,0.50]\,\tfrac{1}{2} = 2.53 \text{ in.}^2$$

A_{tg} has one tension plane

$$A_{tg} = [2(6.00) + 2(2.00)]\,\tfrac{1}{2} = 8.00 \text{ in.}^2$$

Block shear capacity:

$$P_{bs} = 0.75(0.6F_uA_{ns} + F_yA_{tg})$$
$$= 0.75[0.6(58)(2.53) + 36(8.00)] = 261 \text{ kips} > 80.4 \text{ kips} \quad \underline{\text{OK}}$$

Plane C

$$A_{ns} = 2[9.00 - 2.50(0.938)]\tfrac{1}{2} = 6.66 \text{ in.}^2$$
$$A_{tg} = 8.00 \text{ in.}^2$$

Block shear capacity:

$$P_{bs} = 0.75[0.6(58)(6.66) + 36(8.00)] = 390 \text{ kips} > 192.9 \text{ kips} \quad \underline{\text{OK}}$$

Since all planes are adequate, it is not necessary to analyze the other limit state, shear yield plus tension fracture.

EXAMPLE 5.6

Given: A W14×43 wide-flange shape is connected by flange plates, as shown in Fig. 5.26. The bolts are $\tfrac{7}{8}$-in.-diameter A490-X, high-strength bolts. For a single flange the design strength of the bolts in shear is 211 kips; and the design strength of the bolts in bearing is 548 kips. (See Chapter 11 for the details on bolt shear and bearing strength computations.) Determine the design strength of the member.

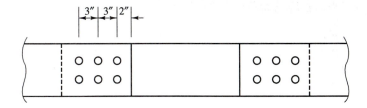

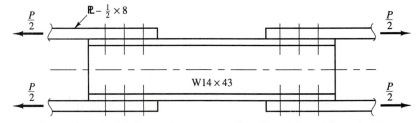

Figure 5.26   A wide-flange shape connected by flange plates (Example 5.6).

Solution

Yield Limit State of Shape:
  Gross area of W14×43 = A_g = 12.60 in.²
  Design yield strength of the member = $\phi F_y A_g$ = 0.90(36)(12.60) = <u>408 kips</u>

Fracture Limit State of Member:
  Step 1.   Hole area to be deducted in the flange at each connection:

$$2(\tfrac{7}{8} + \tfrac{1}{8})0.53 = 1.06 \text{ in.}^2$$

  Step 2.   Determine x and L to be used in the shear lag calculation: see Fig. 5.27. The W14×43 is considered as two tee sections, each a WT7×21.5. The x of each WT is found in the *LRFD Manual* as 1.31 in. The value of L is

$$L = (N/2 - 1)s = (6/2 - 1)(3.00) = 6.00 \text{ in.}$$

  Step 3.   Calculate the shear lag factor U:

$$U = 1 - x/L = 1 - 1.31/6.00 = 0.78$$

  The LRFD specification, Section B3, notes that for this case, a value of $U = 0.85$ may be used.

  Step 4.   The design strength of the member is twice the capacity of each flange connection. The gross area of each tee is 6.31 in.².

Design fracture strength = $2\phi A_n F_u U$ = 2(0.75)(6.31 − 1.06)(58)(0.85)

$$= \underline{388 \text{ kips}}$$

Block Shear of Section Flanges:
The block shear limit state must be checked for tear-out of the flange, as shown in Fig. 5.28. The design strength is provided by the larger of the two basic block shear combinations: (1) Shear yield + Tension fracture; (2) Tension yield + Shear fracture.

  Step 1.   Fracture on shear plane:

$$P_{bs} = 2(0.6)F_u A_{ns}$$

$$= 2(0.6)(58)[8.00 - 2.50(\tfrac{15}{16})](0.53) = 209 \text{ kips}$$

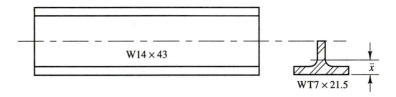

Figure 5.27 Shear lag calculation for Example 5.6.

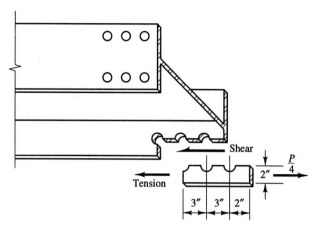

Figure 5.28 Block shear (Example 5.6).

Step 2.   Yield on tension plane:

$$P_{bs} = 2F_y A_g$$
$$= 2(36)(2)(0.53) = 76.3 \text{ kips}$$

Step 3.   Design strength for condition 1:

$$(P_{bs})_{total} = \phi(\text{Shear} + \text{Tension})$$
$$= 0.75(209 + 76.3) = 214 \text{ kips}$$

Step 4.   Fracture on tension plane:

$$P_{bs} = 2F_u A_n$$
$$= 2(58)[2.00 - (\tfrac{1}{2})(\tfrac{15}{16})](0.53)$$
$$= 94 \text{ kips}$$

Step 5.   Yield on shear plane:

$$P_{bs} = 2(0.6)F_y A_g$$
$$= 2(0.6)(36)(8.00)(0.53) = 183 \text{ kips}$$

Step 6.   Design strength for condition 2:

$$(P_{bs})_{total} = \phi(\text{Tension} + \text{Shear})$$
$$= 0.75(94 + 183) = 208 \text{ kips}$$

Step 7.   Governing block shear strength:
The largest capacity from steps 3 and 6 is condition 1; thus,

$$\text{Block shear design strength} = 214 \text{ kips.}$$

SUMMARY FOR CONNECTION:

Bolt design shear strength (given)	211 kips
Connection bolt bearing capacity (given)	548 kips
Gross section yielding of member	408 kips
Effective net section fracture of member	388 kips
Block shear capacity of flanges	214 kips

Conclusion: The bolt design shear strength of the connection of 211 kips governs.

EXAMPLE 5.7

Given: A WT5 × 11 in A36 steel is connected to a gusset plate with the ends completely welded and each side welded along 6 in., as was shown in Fig. 5.21. Determine the effective net area of the member. From the *LRFD Manual*, A_g = 3.24 in.², $\bar{x}$ = 1.07 in. and t_f = 0.36 in.

Solution

For the 6-in. welds, the value of $U = 1 - 1.07/6.00 = 0.822$. For the end weld, the value of $U = 5.75(0.36)/3.24 = 0.639$. The effective reduction coefficient is now found as the weighted average of the above:

$$U_{\text{eff}} = \frac{0.822(12) + 0.639(5.75)}{2(6.00) + 5.75} = 0.76$$

The effective net area is then

$$A_e = U_{\text{eff}}A_n = 0.76(3.24) = 2.46 \text{ in.}^2$$

EXAMPLE 5.8

Given: The truss diagonal member in Fig. 5.29 consists of a pair of angles L4 × 3 × ⅜ that are loaded in tension. Determine the design strength T of one angle. The bolts that will be used are ¾-in. A325-N, and the steel is A36. The bolt design strengths for the connection in one angle are 46.5 kips in shear and 88.1 kips in bearing.

Solution

Step 1.   Determine the angle design strength for the limit state of yielding on the gross cross section.

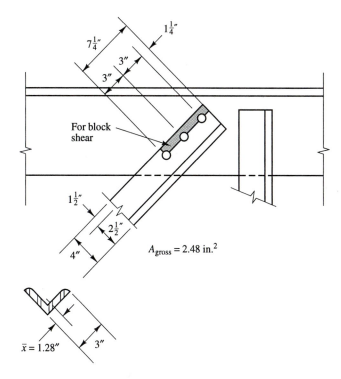

Figure 5.29 Truss diagonal member (Example 5.8).

$$A_\text{g} = 2.48 \text{ in.}^2$$

$$T = \phi F_y A_\text{g} = 0.9(36)(2.48) = 80.4 \text{ kips}$$

Step 2.   Determine the angle design strength in tension fracture.

$$A_\text{n} = 2.48 - \tfrac{3}{8}(\tfrac{3}{4} + \tfrac{1}{8}) = 1.23 \text{ in.}^2$$

Reduction coefficient:

$$U = 1 - x/L = 1 - 1.28/6.00 = 0.787$$

$$T = U\phi F_u A_\text{n} = 0.787(0.75)(58)(1.23)$$

$$= 42.1 \text{ kips}$$

Step 3.   Check angle design strength in block shear, for the tension yield and shear fracture combination.
  Tension yield load:

$$P_\text{bs} = \phi F_y A_\text{tg} = 0.75(36)(\tfrac{3}{8})(1.5) = 15.2 \text{ kips}$$

Shear fracture load:

$$P_{bs} = \phi(0.6)F_u A_{ns}$$

$$= 0.75(0.6)(58)(\tfrac{3}{8})[7.25 - 2.5(\tfrac{7}{8} + \tfrac{1}{16})] = 48.0 \text{ kips}$$

Total block shear capacity:

$$(P_{bs})_{total} = 15.2 + 48.0 = 63.2 \text{ kips}$$

Since the lowest tensile design strength is that of the angle in tension fracture, the load 42.1 kips controls. Further, since the larger of the two block shear modes controls, additional computations are not needed.

Conclusion: The design tension strength of the angle is 42.1 kips.

5.8 EYE-BARS AND PIN-CONNECTED MEMBERS

Figure 5.30 illustrates an eyebar and a pin-connected member. Historically, eye-bars were forged and used primarily as chords and web members for trusses. They have not been used in new construction for many years. Pin-connected members are used mainly for special applications, such as hangers in suspension structures; they are usually flame-cut.

    The LRFD specification, Section D3, details the design rules for these members. It is important to realize that the rules for the various dimensions are maxima, to be used in the calculations. There is no reason why an actual member cannot be fabricated larger than the sizes that are shown. In real structures, this is normally the case.

    Eye-bar tension members are designed only for the limit state of

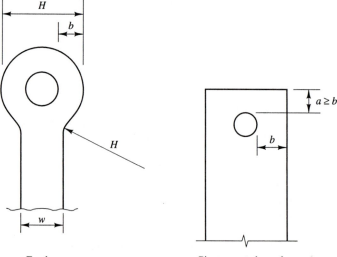

Figure 5.30   An eye-bar and a pin-connected member.

Eye-bar   Pin-connected member

yielding on the gross section, while pin-connected members are designed for this limit state as well as fracture on the net section. The reason for this is that the dimensional restrictions for the eye-bar preclude the possibility of fracture at a factored load that is less than the yield limit state.

EXAMPLE 5.9

Given: Design an eye-bar as a tension member. The dead load is 30 kips and the live load is 70 kips. The steel has a yield stress of 50 ksi.

Solution

Step 1.   Factored load $= 1.2(30) + 1.6(70) = 148$ kips

Step 2.   The LRFD specification, Section D3, stipulates that eye-bars are to be designed for the yield limit state, using Eq. (D1-1). The resistance factor is 0.90.

$$\phi P_n = \phi F_y A_y = 148 \text{ kips}$$
$$A_y = 148/(0.9)(50) = 3.29 \text{ in.}^2$$

Try using an eye-bar body width of $W = 5.0$ in., and a thickness of $t = \frac{3}{4}$ in. This gives an area of 3.75 in.²

Step 3.   Determine the dimensions of the eye-bar head:

$$W = 5.0 \text{ in.}; t = \tfrac{3}{4} \text{ in.}$$

It is required that: $W < 8t$

Here: $5.0 < 8(\tfrac{3}{4}) = 6.0 \text{ in.}^2$ <u>OK</u>

It is also required that $\frac{3}{4}W > b > \frac{2}{3}W$.

Here: $3.75 > b > 3.33$.

Conclusion: Use $b = 3.50$ in.
The pin diameter d_p must satisfy $d_p > \frac{7}{8}W$, *or:* $d_p > \frac{7}{8}(5.0) = 4.375$ in.

Conclusion: Use a pin with diameter $4\frac{3}{8}$ in.
Pin hole $= 4\frac{3}{8} + \frac{1}{32} = 4\frac{13}{32}$ in.
Diameter of head, H:
Pin hole $+ 2b = 4\frac{13}{32} + 2(3\frac{1}{2})$
$\qquad\qquad = 11\frac{13}{32}$ in.

Conclusion: Use a head with a diameter of $H = 11\frac{1}{2}$ in.

EXAMPLE 5.10

Given: Design a pin-connected member for the same conditions as in the above example.

Solution

Step 1. Same as previous example: the factored load is 148 kips.

Step 2. Check bearing against the base metal. Assume a pin diameter of 4.0 in., and a material thickness of $t = \frac{3}{4}$ in. Then:

$$A_{pb} = dt = 4.0(\tfrac{3}{4}) = 3.0 \text{ in.}$$

$$\phi P_n = \phi A_{pb} F_y = 1.0(3.0)(50.0) = 150 \text{ kips} > 148 \text{ kips} \qquad \underline{\text{OK}}$$

Step 3. Check for shear fracture of the member. Assume the value of $a = 2$ in.

$$A_{sf} = 2t\left(a + \frac{d}{2}\right)$$

$$= 2(\tfrac{3}{4})\left(2.0 + \frac{4.0}{2}\right) = 6.0 \text{ in.}^2$$

$$\phi P_n = \phi A_{sf} F_y = 0.75(6.0)(50)$$

$$= 225 \text{ kips} > 148 \text{ kips} \qquad \underline{\text{OK}}$$

Step 4. Check for tension fracture of the member.

$$b_{eff} = 2t + 0.63 = 2(\tfrac{3}{4}) + 0.63$$

$$= 2.13 \text{ in.}$$

$$\phi P_n = \phi(2)tb_{eff} F_u$$

$$= 0.75(2)(\tfrac{3}{4})(2.13)(60)$$

$$= 311 \text{ kips} > 148 \text{ kips} \qquad \underline{\text{OK}}$$

PROBLEMS

5.1. Determine the gross and net areas, A_g and A_n, for an $8 \times \frac{3}{4}$-in. plate with a single line of standard punched holes for $\frac{7}{8}$-in. bolts.

5.2. Determine the gross and net areas for a $10 \times \frac{1}{2}$-in. plate with a single line of standard punched holes for $\frac{3}{4}$-in. bolts.

5.3. Determine the gross and net areas for a $4 \times 4 \times \frac{1}{2}$ L with two lines of standard punched holes for $\frac{3}{4}$-in. bolts.

5.4. Determine the gross and net areas for a $5 \times 3 \times \frac{3}{8}$ L with two lines of standard punched holes for $\frac{7}{8}$-in. bolts.

5.5. Determine the gross and net areas for a WT8×20 with three lines, one in each projecting element, of standard punched holes for $\frac{7}{8}$-in. bolts.

5.6. Determine the gross and net areas for a C15×40 with five lines, three in the web and one in each flange, of standard punched holes for $\frac{7}{8}$-in. bolts.

5.7. Determine the net width for a 10 × ½-in. plate with ¾-in. bolts placed in three lines as shown.

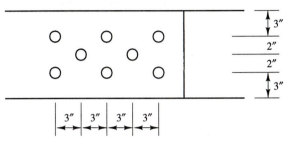

5.8. Determine the net width for a 10 × ½-in. plate with ¾-in. bolts placed in three lines as shown.

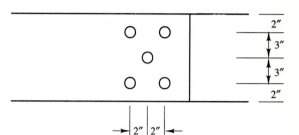

5.9. Determine the net width for the 6 × 4 × ⅝ L shown with ⅞-in. bolts.

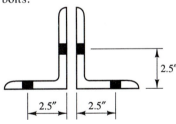

5.10. Determine the net area for the double 4 × 4 × ½ L's shown with ¾-in. bolts.

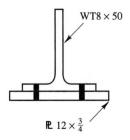

5.11. For the WT8 × 50 attached through the flange to a 12 × ¾-in. plate with two rows of ⅞-in. bolts as shown, determine the shear lag factor and the effective net area if the pitch is 4 in.

WT8 × 50

℞ 12 × ¾

5.12 A single 6 × 6 × 1 L is used as a tension brace in a multistory building. One leg of the angle is attached to a gusset plate with a single line of ⅞-in. bolts. Determine the shear lag factor and the effective net area for three bolts with a spacing of 3 in.

5.13. The WT8 × 50 and 12 × ¾-in. plate of Problem 5.11 are welded along the tips of the flange over a length of 12 in. on each flange. Determine the shear lag factor for this connection.

5.14. Determine the design strength of the 12 × ½-in.

A36 plate connected to two 12-in. plates as shown with two lines of $\frac{3}{4}$-in. bolts.

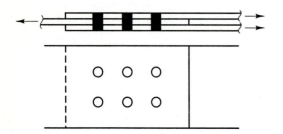

5.15. Determine the design strength of an 8 × $\frac{1}{2}$-in. A529 plate connected with three lines of $\frac{7}{8}$-in. bolts.

5.16. Determine the design strength of a 6 × 4 × $\frac{3}{4}$ L attached through the long leg to a gusset plate with 10 $\frac{7}{8}$-in. bolts in two lines. Use A36 steel.

5.17. Determine the design strength of a WT7×24 attached to a gusset plate with 8 $\frac{3}{4}$-in. bolts in two lines in the flange. Use A36 steel.

5.18. Design a 10-ft-long, single-angle tension member to support a live load of 49.5 kips and a dead load of 16.5 kips ($L/D = 3$). The member is to be connected through one leg and only one bolt hole will occur at any cross section. Use A36 steel and limit slenderness to length/300 as indicated in the LRFD specification, Section B7.

5.19. Design a 10-ft-long, single-angle tension member as in Problem 5.18 with the same total service load, 66 kips. Use a live load of 7.3 kips and a dead load of 58.7 kips, $L/D = 0.125$.

5.20. Design a 10-ft-long, single-angle tension member as in Problem 5.18 with the same service load using a live load of 55.0 kips and a dead load of 11.0 kips, $L/D = 5$.

5.21. Design the 27.0-ft tension wind brace for a multi-story building to resist a wind force of 380.0 kips. Assume that $\frac{7}{8}$-in. bolts will be used so that no more than 4 holes will occur at any cross section. Use an A36 WT section.

REFERENCES

[1] Bethlehem Steel Corporation, *Cable Roof Structures*, Bethlehem Steel Corporation Booklet No. 2318-A, Bethlehem, PA, November 1968.

[2] Jensen, J. J., *Eine Statische und Dynamische Untersuchung der Seil- und Membrantragwerke* (A Study of the Static and Dynamic Characteristics of Cables and Cable Nets), Publication No. 70-1, Division of Structural Mechanics, Technical University of Norway, Trondheim, Norway, September 1970.

[3] Johnston, B. G., Pin-connected plate links, *Transactions, ASCE*, Vol. 104, 1939 (pp. 314–336).

[4] McGuire, W., *Steel Structures,* Prentice-Hall, Englewood Cliffs, NJ, 1968.

[5] Cochrane, V. H., Calculating net section of riveted tension members and fixing rivet stagger, *Engineering News*, April 1908.

[6] Cochrane, V. H., Rules for riveted hole deduction in tension members, *Engineering News-Record*, 16 November 1922.

[7] Smith, T. A., Diagram of net section of riveted tension members, *Engineering News*, May 1915.

[8] Bijlaard, P. P., Theory of local plastic deformation, *Publications, IABSE*, Vol. 6, 1940/41.

[9] Bijlaard, P. P., Discussion of "Investigation and limit analysis of net area in tension" by W. G. Brady and D. C. Drucker, *Transactions, ASCE*, Vol. 120, 1955 (pp. 1156–1163).

[10] Chesson, E., and Munse, W. H., Riveted and bolted joints: Truss-type tensile connections, *Journal of the Structural Division, ASCE*, Vol. 89, No. ST1, February 1963 (pp. 67–106).

[11] Munse, W. H., and Chesson, E., Riveted and bolted joints: Net section design, *Journal of the Structural Division, ASCE*, Vol. 89, No. ST1, February 1963 (pp. 107–126).

[12] Birkemoe, P. C., and Gilmor, M. I., Behavior of bearing critical double-angle beam connections, *AISC Engineering Journal*, Vol. 15, 4th Quarter, 1978 (pp. 109–115).

[13] Ricles, J. M., and Yura, J. A., Strength of double row bolted web connections, *Journal of the Structural Division, ASCE*, Vol. 109, No. ST1, January 1983 (pp. 126–142).

[14] Hardash, S. G., and Bjorhovde, R., New design criteria for gusset plates in tension, *AISC Engineering Journal*, Vol. 22, 1985 (pp. 77–94).

CHAPTER SIX

Columns

6.1 THE STRUCTURAL COLUMN

The column is one of the basic elements found in building structures. It performs the key function of transmitting compressive loads from one point of the structure to another. Thus, a more appropriate name is compression member. In this book the terms column and compression member are used interchangeably, without a preference for either.

Columns are normally used as vertical elements, typified by the beautiful examples seen in Greek and Roman temples. These ancient

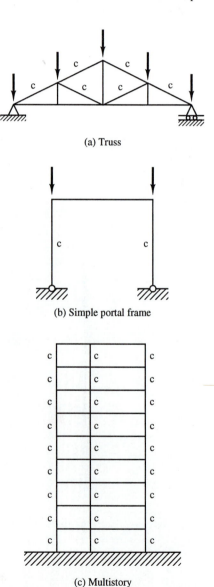

(a) Truss

(b) Simple portal frame

(c) Multistory
frame

Figure 6.1 Typical compression members in structures.

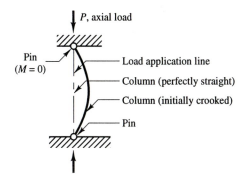

Figure 6.2   The basic column.

columns generally had large cross-sectional dimensions, and the loads they carried were probably far below their potential capacity. They were works of art, based on little or no engineering design in the way it is understood today.

As structures and materials changed, more slender elements evolved. Most significant was the introduction of steel as a construction material,which led to the development of a host of shapes and structural applications. It also demanded increased analytical ability on the part of the designer and a better understanding of the strength and behavior of the member and its material. Column stability became a primary concern, and much work was done to resolve the questions related to this phenomenon.

Figure 6.1 illustrates some typical examples of compression members in structures. Although much of the discussion of column strength and behavior in this chapter relates to the individual member, the structural column is a part of the assembly that makes up the complete structure. In the design process, therefore, it is vital to take into account the elements that surround and frame into the column in question.

Figure 6.2 gives a schematic illustration of the basic column. It is a single member, supported at the ends by perfect hinges or pins. Thus, it is also referred to as a pinned column. The load is axial, applied concentrically at the ends. This basic column is a special case of the more generalized structural member, the beam-column, which is subjected to axial loads, bending moments, and other stress resultants. Although the basic column rarely exists in actual structures, it is essential in the later study of beam-columns to have a clear understanding of the response of the pure, pinned-end compression member.

6.2 FACTORS INFLUENCING COLUMN STRENGTH

6.2.1 The Complexity of Column Behavior

Although the structural column is a seemingly simple structural member, the interaction between the responses and characteristics of the material, the cross section, the method of fabrication, the imperfections and other geometric factors, and the end conditions, make the column one of the most complex individual structural members. The following parameters have been shown to have an effect on column strength:

1. *Material properties*
   (a) Stress–strain relationship
   (b) Yield stress
2. *Manufacturing method*
   (a) Hot-rolled shape
   (b) Welded built-up shape
      (i) Using flame-cut plates
      (ii) Using universal mill (UM) plates
   (c) Cold-straightened shape
      (i) Rotorizing (continuous straightening)
      (ii) Gag (point) straightening
3. *Shape of cross section*
   (a) Area of steel
   (b) Cross-section geometry (W, C, WT, etc.)
   (c) Bending axis (x vs. y)
4. *Length*
5. *Initial out-of-straightness*
   (a) Maximum value
   (b) Distribution along column length
6. *End support conditions*
   (a) Without sway, pinned or otherwise
   (b) With sway, pinned or otherwise
   (c) Restrained ends, with or without sway

A brief discussion of these parameters is given in the following. The most important factors are discussed in more detail later in this chapter.

Material Properties. Yield stress and modulus of elasticity are clearly the most important material properties. For short columns, the strength is directly proportional to F_y. For longer columns, yield stress plays less of a role, since the capacity will be shown to be influenced more by stiffness, which is a function of the magnitude of the modulus of elasticity. The ultimate limit state of *very* short columns may reflect a strength increase due to strain-hardening, but for hot-rolled structural shapes this is almost always overridden by other factors.

Manufacturing Method. The method of manufacturing is one of the primary factors since it influences the magnitude and distribution of residual stress in the cross section. The various forms of heat input, from rolling, welding, flame cutting, and heat straightening, along with heat dissipation by controlled or uncontrolled cooling in the steel mill, cooling after welding or cutting, normalizing, and stress relief annealing, all affect the residual stress.

Cold-straightening has the effect of redistributing the residual stresses, due to local plastification, but the influence is significant only when the straightening is applied to the full length of the member. This process is usually referred to as rotorizing. Concentrated load or gag straightening alters the residual stress distribution only in the immediate vicinity of the load application point.

Cross Section.   The size of a shape is obviously important, since the larger the area, the larger the load-carrying capacity for the same stress level. In addition, size and geometry of shape may influence the residual stresses which tend to be higher for larger thickness material [1]. The geometry also influences the residual stress distribution, due to its influence on the dissipation of the heat stored in the steel. The geometry of the cross section influences the principal moments of inertia which will impact on the capacity of all but the shortest columns. The complex topic of residual stress has been the subject of research efforts for many years. The characteristics of strong and weak axis buckling and the influence of moment of inertia are covered in depth later.

Length.   Length has long been understood to be an important factor in determining column strength. Although other factors may be important, their influence may be hidden in the design approach, whereas the influence of column length is readily seen in design calculations. Regardless of design approach, columns are generally referred to as either short columns, intermediate columns, or long columns. Column shape also plays a role in these definitions; however, length is primary.

Initial Out-of-Straightness.   The initial out-of-straightness is also a major column strength parameter. The shape of the distribution, i.e., its variation along the length of the column, is not significant, although it has been found to differ from the commonly assumed half sine wave [2]. The amount of crookedness in any member is limited by manufacturing tolerance limits set by ASTM, which reflects standard mill practice. Thus, the maximum value of eccentricity e_i for rolled wide-flange shapes is approximately $L/1000$, with some minor modifications for longer lengths and certain geometries.

End Support Conditions.   As was the case for the influence of length, the effect of well-defined support conditions has been recognized for a long time. Recently, studies of the strength of columns as influenced by actual beams and beam-to-column connections have been carried out. End restraint is a significant factor, and is regarded as one of the primary column strength parameters. It is addressed in detail in Section 6.8.

6.2.2 Residual Stress

Common structural steel generally has a carbon content of 0.3 percent or less, which results in steel with a "melting point" of approximately 2700°F (1500°C). Hot-rolling of ingots or other products takes place at temperatures of 1650°F (900°C) or higher. Heat input associated with any of the arc welding processes or flame cutting an edge results in liquid metal in the weld bead or cut edge. The production of steel and the fabrication of components involves substantial heat, which leads to the development of residual stresses in the cross section.

Development of Residual Stress.   During hot-rolling of steel, the cross section of the shape or plate is heated to a uniform temperature. Once the rolling is completed, the heat must dissipate. The cooling process is normally not uniform; thus, all of the individual "fibers" in the shape cool at a different rate so that some of the fibers reach room temperature well before others. This is schematically illustrated in Fig. 6.3. With different parts of the shape or plate cooling at different rates, residual stresses are developed. In Fig. 6.3a and b, points A and B represent fibers with significantly different cooling characteristics. Point A is located close to the edge or flange tip; the heat can escape fairly readily. Point B is centrally located in the mass and the dissipation of heat requires more time, since the paths for heat escape are more complex. The fiber at A, therefore, will reach ambient or room temperature before the fiber at B.

The cooling process results in contraction of the fibers. Once the fibers at A, the first to cool, have reached ambient temperature, they have also reached what otherwise would be their final length. However, since the fibers in the immediate vicinity of A still have to release some heat and thus contract, they will pull on the fibers at A to shorten them. In other words, the fibers that are the first to cool will end up having to contract more than necessary for their own cooling in order to accommodate the cooling of the surrounding material. With the fibers thus maintained in a

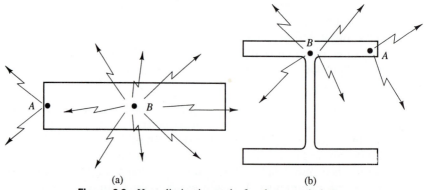

(a)   (b)

Figure 6.3 Heat dissipation paths for shapes and plates.

state of forced shortening, the effect is that they are being subjected to a state of compressive stress.

The amount of shortening due to cooling is independent of the steel yield stress. Since the modulus of elasticity is essentially a constant for all grades of steel, regardless of the yield stress level, the magnitude and distribution of the residual stress is independent of the steel grade. Thus, whether the steel has a yield stress of 36 or 100 ksi is immaterial; residual stress magnitudes and distribution will be the same. This explains why residual stresses are generally more detrimental to lower strength steels, since these steels are stressed to a much higher percentage of their yield stress.

When the fibers at B, which are the last to cool, are ready to reach ambient temperature, all of the surrounding material has already reached this level. Consequently, these fibers are restrained from completing the amount of contraction that would occur if they were not surrounded by other steel fibers. Therefore, fibers at B are held in a state of forced elongation, resulting in a tensile residual stress.

The entire process is significantly more complex than described; however, the result of the differential cooling rates within the cross section may be understood from this simplified presentation. A range of residual stresses will exist, from tension to compression, in the absence of any applied, external load. Since the shape is a continuum of material, the residual stresses will vary continuously, satisfying overall equilibrium and strain compatibility. Figure 6.4a and b provides two simple examples of distributions of residual stress, using a common rolled (universal mill, UM) plate and a medium-size hot-rolled shape, W12×65 [1, 3]. The maximum compressive and tensile values are σ_{rc} and σ_{rt}, respectively.

Extensive data on residual stress distributions in a variety of hot-rolled shapes and plates and welded built-up shapes are available in the literature [1-7]. For columns and similar members, for which compressive residual stresses are the most important, σ_{rc} usually assumes a value between 10 and 30 ksi for hot-rolled shapes and plates, with the higher values for large thickness material (e.g., flange thickness of a W shape of 4 to 6 in.). For welded built-up shapes with universal mill plates as flanges, the magnitude of σ_{rc} is higher than that found for the plate by itself, for smaller thickness elements, due to the effect of the heat of welding [3]. For welded shapes with flame-cut plates the picture is much more complicated, but generally more favorable than for UM plates [2-4, 6, 7].

In summary, then, the residual stress principle can be stated as follows:

The fibers that cool first will have residual compression; those that cool last will have residual tension.

Localized Heat Effects.   The effects of localized heat input, such as from welding and flame cutting, depend on the level and duration of the

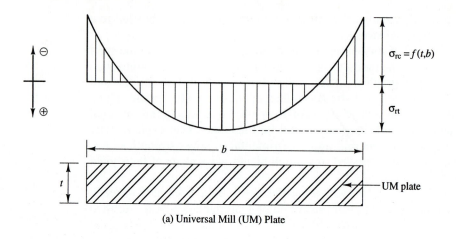

(a) Universal Mill (UM) Plate

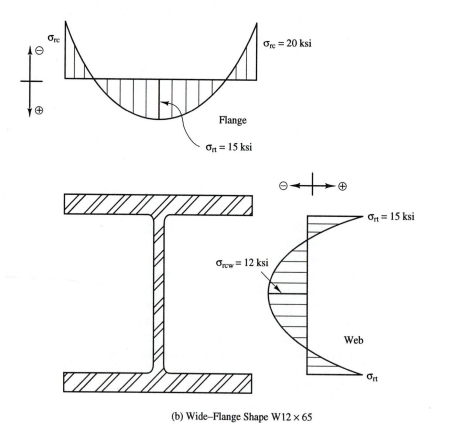

(b) Wide–Flange Shape W12 × 65

Figure 6.4   Residual stress distributions in rolled plates and shapes.

heat involved. Basically, the portion of the steel in the cross section that is affected will only be substantial for a shape or plate of small area; for others, the heat influences the residual stress distribution only in the immediate vicinity of the weld or flame-cut edge. Further, since the heat is intense and local, the cooling tends to be very rapid, unless it is slowed by the use of pre- or post-heat. The normally high rate of cooling associated with these processes causes changes in the metallurgical structure of the base metal close to the weld or the flame-cut edge (the heat-affected zone), primarily by promoting development of the fine-grained crystalline structure of martensite. This "new" material is stronger and much harder than the original steel, but less ductile.

The basic principle of "cools first—compression; cools last—tension" also applies to the residual stresses that develop as a result of localized heat input. Thus, high tensile residual stresses are found at weld bead and flame-cut edges, generally around 60 to 70 ksi, which is close to the yield level of the material that has developed at the weld or flame-cut edge [*1–4, 6*].

Figure 6.5 shows typical residual stress distributions in a flame-cut plate and a welded built-up shape using such plates [*1, 3*]. Considering the tensile stress at the flange tips, it is clear why shapes with flame-cut plates are preferable for column usage. Instead of having early yielding in the parts of the cross section that contribute the most to the moment of inertia, as is the case when flange tip compression exists, the tensile stress delays the local plastification. This phenomenon emphasizes the importance of heat as a means of influencing structural capacity.

Residual Stress Distribution Principles. The final residual stress distribution in a shape or plate cannot be determined by superposition of the stresses due to each of the individual effects. However, it can be assessed realistically by utilizing the following key points, in addition to the "residual stress principle":

1. The residual stress distribution must satisfy axial force and moment equilibrium in the absence of any applied external stress resultants.
2. Residual stress magnitudes and distribution are independent of the base metal yield stress.
3. Maximum compressive residual stress in a plate or the flange of a hot-rolled shape is inversely proportional to the width factor of the plate or flange [*3*], given as

$$\text{WF} = \frac{2(b + t)}{b^2 t} \qquad (6.1)$$

where b is the plate or flange width and t is the thickness, both expressed in inches. As a reference value, an average σ_{rc} of 14 ksi for a width factor of 0.10 may be used [*3*].

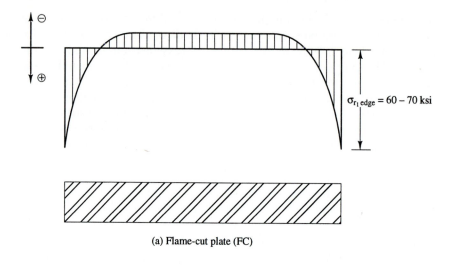

(a) Flame-cut plate (FC)

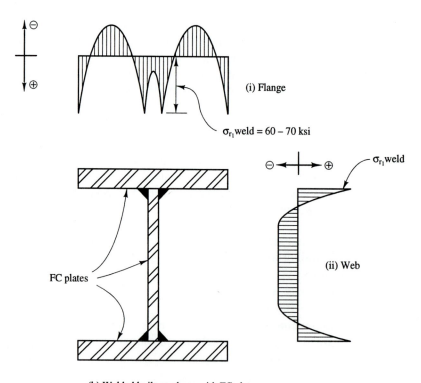

(b) Welded built-up shape with FC plates

Figure 6.5   Residual stress distributions in flame-cut (FC) plate and welded built-up shapes.

4. In general, the center one-half of the width of a hot-rolled plate or flange will exhibit tensile residual stress. The distribution is typically parabolic, but other patterns are also common.

5. The tensile residual stress at a weld or flame-cut edge can conservatively be set equal to approximately 60 ksi.

6. The residual stress distribution in a welded built-up box cross section can be represented by four flame-cut plates, assembled as a box. The corner stresses are therefore approximately 60 ksi in tension.

Residual Stresses Due to Cold-Forming.   Cold-forming of steel shapes or plates has many applications. For example, when shapes come from the steel mill initially curved in some fashion, it is customary to cold-straighten the members to bring them into agreement with the applicable materials delivery standard. In the production of rectangular tubular shapes, a high degree of local cold-forming is used when the 90° corners are formed. Finally, cold-forming is used extensively in the steel deck and pre-engineered building (metal building) industry, where shapes of many types are formed from sheet and strip [8, 9].

Cold-straightening is applied in either continuous (rotorizing) or localized (gag straightening) fashion, as discussed briefly earlier. Gag straightening affects the residual stress distribution only in the area surrounding the load application points. Rotorizing, however, is applied continuously along the full length of the member and, therefore, permanently changes the residual stress distribution for the column as a whole. The magnitude of the straightening moment is somewhere between the yield and plastic moments, creating local plastification in the cross section. Upon elastic unloading, the residual stresses are redistributed such that the compressive and tensile peak stresses are reduced significantly [4, 10]. An example is shown in Fig. 6.6. The column strength of a rotorized member is therefore higher than that of an unstraightened one.

In the forming of tubes, the localized deformation that occurs in the corners results in a material with properties that are different from those of the base metal. For example, the yield stress is higher and the ductility is lower. This is taken into account in the strength criteria for cold-formed structures [8, 9].

6.2.3 Initial Crookedness

Perfectly straight members exist only in theoretical solutions. Real structural shapes are delivered from the steel mill in a more or less crooked form. Although not perfectly straight, the product always is expected to satisfy the straightness requirements of the applicable materials delivery standard. Column design criteria of the past covered the effects of any initial crookedness through the factor of safety, and used the strength of the straight member as the actual criterion. With limit states philosophy,

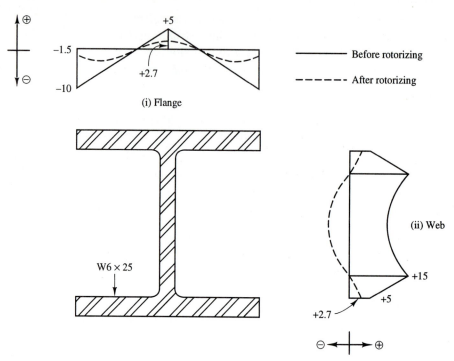

Figure 6.6   Residual stress distributions before and after straightening.

however, all of the major parameters have to be accounted for. The strength equations therefore include the influence of initial crookedness.

Initial crookedness develops as a result of the cooling conditions for the shape, once it has been rolled to final dimensions. To use the wide-flange as an example, the shape emerges from the last set of rolls in the steel mill, uniformly heated to about 1650°F (900°C). The length of the member depends on the specific mill, but 40 ft or more is normal. The shape is then left on the cooling bed to cool in air over a period of time. Other members are also placed on the bed, and the heat dissipation for the member, therefore, is uniform neither throughout the cross section, nor along the length or around all sides of the member. As a result, heat is usually retained longer in the midlength portion of the column and in the parts of the cross section exposed most directly to the heat of the adjacent shapes on the cooling bed.

The resulting nonuniform cooling leads not only to residual stresses, but also to a length of steel that naturally ends up in a curved configuration along its length. The amount of curvature or initial out-of-straightness is difficult to predict because of the number of uncontrollable factors. Depending on how the shape has been placed on the cooling bed, the curvature may appear in both orthogonal directions of the cross section, meaning that the member is initially curved about both principal axes. Crookedness about the major axis is called camber and curvature about the minor axis is designated sweep. Figure 6.7a illustrates camber and Fig. 6.7b illustrates sweep.

Figure 6.7 shows a length of steel that is bent in single, symmetric curvature. Although this is a common form of initial out-of-straightness, actual shapes often display variations from this simple form. For column strength, however, the single curvature shape (often modeled as a half sine wave [2]) has the greatest effect and is used in the basic column model.

The maximum out-of-straightness is limited by materials delivery standards (ASTM A6 in the United States, CSA G40.12 in Canada, Euronorm in western Europe, etc.). It is typically based on the length of the member, and for most industrialized nations is set as $e_{i,max} \approx L/1000$ for wide-flange shapes. The value is different for other kinds of cross sections. A column that does not meet this straightness requirement is rotorized or gag-straightened to bring it into compliance with the code.

As an example, ASTM A6 gives the following basic requirement for the permissible camber and sweep for W-shapes, $e_{i,max}$ [7]:

$$e_{i,max} = \frac{\frac{1}{8} \text{ in. (Number of feet of total length)}}{10} \tag{6.2}$$

which works out to $L/960$. For structural purposes this has been conveniently rounded to $L/1000$.

It is important to note that the combined effects of residual stress and initial crookedness cannot be obtained merely by combining the two terms. In some cases and for certain slenderness ratio ranges, the strength

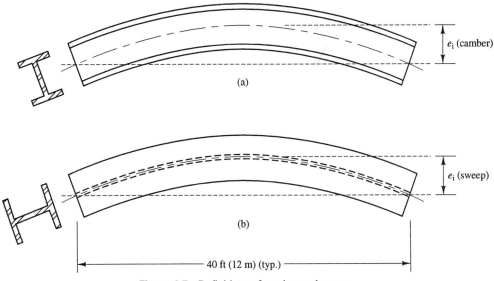

Figure 6.7 Definitions of camber and sweep.

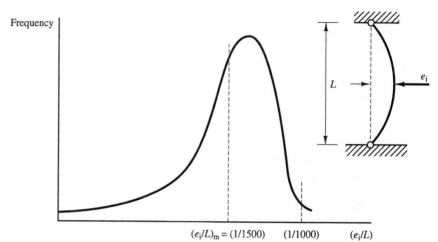

Figure 6.8   Random variation of initial crookedness.

of a column with residual stress and initial curvature is less than what would be found if the effects of both were added; in other cases it is not as critical as the sum would seem to indicate [*2, 4, 11–13*].

Wide-flange shapes have initial crookedness values that range from zero to the common maximum of $L/1000$ or somewhat larger. Analyses of test specimens as well as columns in actual structures show a random variation of e_i as indicated in Fig. 6.8, with an average of approximately $L/1500$ [*2*]. Other hot-rolled shapes (C, T, etc.) show the same basic pattern. The values of e_i for cross sections such as round and rectangular tubes tend to be smaller, with an average around $L/6000$ [*14*].

The out-of-straightness used for the basic column in limit states design codes varies, depending on the fundamental approach to structural reliability. It is equal to the maximum of $L/1000$ for the Canadian standard [*15*] and Eurocode 3 [*16*]. The LRFD specification bases its column curve on the average of $L/1500$ [*2, 11*], on the premise that all of the other column strength parameters of the LRFD development reflect mean values. Both concepts have their advantages.

6.3 CROSS-SECTIONAL SHAPES FOR COLUMNS

6.3.1 Basic Selection Criteria

The basic column is subjected to a compressive axial force; thus, the area of the cross section is of prime importance. With buckling as the critical failure mode, the geometry of the section as reflected by the moment of inertia is also a major factor. The various structural shapes described in Chapter 4 offer a wide range of readily available alternatives. Since residual stresses play an important role in column strength, the method of fabrication also becomes critical with rolled as well as built-up shapes.

Finally, the material strength will play a role. Therefore, the choice of member can have a major impact on the success of any design.

The position of the column in the structure ties the column response to the structure. Therefore, the designer must consider the buckling load in both directions to determine whether the in-plane or out-of-plane behavior yields the lowest value.

The designer should insure that there is a measure of compatibility between the sizes of a column and a beam that are joined as well as the details of the connection. Although not necessarily critical for the strength of the axially loaded member, it is important when considering constructability and structural performance during service.

6.3.2 Doubly Symmetric Cross Sections

Wide-flange shapes are by far the most common sections used for columns in buildings and other structures. This is in part historical, but also prompted by the fact that the W-column tends to make connections to adjacent members easier and therefore less costly than other doubly symmetric sections. For compressive axial loads, sections with properties about each axis that are similar in value will yield strengths about those axes that are in close agreement. Sections with lower I_x/I_y ratios are normally preferred. Based on this, preferred shapes are those in the W14 series with unit weights of 43 lb/ft and heavier, the W12 series from 40 through 336 lb/ft, the W10 series from 33 through 112 lb/ft, the W8 series from 24 through 67 lb/ft, and most of the W4, 5, and 6 series. Naturally, any W-shape can be used as a column, although a large difference between the principal moments of inertia may mean that there is little, if any, "balance" between the axes of buckling.

Square, rectangular, and circular tubes (hollow structural sections and pipes) have obvious advantages as far as column usage is concerned. First, the square cross sections have identical I_x and I_y values and the circular tubes have an infinite number of principal axes. All, including the majority of rectangular tubes, therefore have very favorable moment of inertia ratios. Second, the closed form of the tubes makes them architecturally attractive, as well as especially appropriate for use in structures that require "clean" environments, such as buildings for the electronics and food industries. Finally, the residual stresses in these shapes, in particular the ones that have undergone a measure of post-production heating, are low, and the column strength is therefore relatively high. The disadvantage for these shapes lies in the difficulty of connecting to beams that might be framing into the column face.

Welded built-up columns are suitable for a number of structures, especially when large axial loads have to be accommodated, such as in high-rise buildings, or when there are particular architectural demands that have to be met. Figure 6.9 gives some typical built-up column examples. Fully welded wide-flange and box shapes are shown in Fig. 6.9a and b. Figure 6.9c and d shows columns that utilize rolled W-shapes

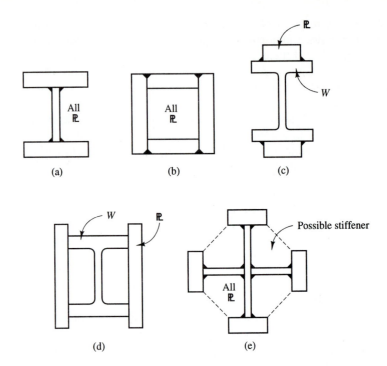

Figure 6.9 Examples of welded built-up column shapes.

along with welded plate attachments, and Fig. 6.9e shows a cruciform cross section that may use horizontal stiffeners, as indicated. Among other considerations, some built-up shapes will have significantly higher relative column strength, depending on the fabrication methods [2, 4].

6.3.3 Singly Symmetric and Asymmetric Cross Sections

For lower axial loads as well as for a number of common structural types, singly symmetric or asymmetric cross sections may be the most suitable. Typical shapes include equal- and unequal-legged angles, the tee, and the channel. Some examples are shown in Fig. 6.10. For larger axial forces in truss members, double channels placed back to back are also common.

For compression members with a single or no axis of symmetry, special considerations must be given to the limit states of torsional buckling or flexural-torsional buckling [17]. Sometimes referred to as twist buckling, this failure mode could also occur in doubly symmetric shapes

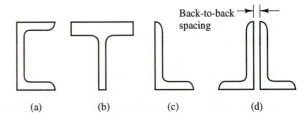

Figure 6.10 Singly symmetric and asymmetric column cross sections.

where the width-to-thickness ratio of the plate elements in the cross-section are high, and the plates, therefore, are not fully effective in carrying the axial load [*8*].

6.3.4 Cross-Sectional Areas

Since the axial force is compressive, the net cross-sectional area that is critical for tension members does not apply to columns. It is assumed that bolts will eventually fill any holes that are made for connections and that whatever local overstress may occur in their vicinity will be adequately accommodated by the section. Thus, the full or gross area of the column shape is used to determine the axial load capacity. For example, the yield load P_y and the nominal axial load P_n for a column are given by

$$P_y = F_y A_g \tag{6.3}$$

$$P_n = F_{cr} A_g \tag{6.4}$$

where F_{cr} is the critical buckling stress.

   For members with large cutouts, such as may be found when openings are made in the web to allow pass-through of a duct or similar element, consideration of the hole may become important. This is a relatively rare occurrence for compression members in building structures and the specification does not address the design of such members. If encountered, it is recommended that a local strength analysis be carried out, using a nonlinear finite element method, or some other appropriate method.

   In the past, built-up compression members in the form of laced or battened columns were often used in industrial buildings and bridge superstructures. Schematic examples are given in Fig. 6.11. Labor costs make such members prohibitive for today's steel structures. The specification applies for the case of renovation or retrofitting; additional information can be found in Ref. [*4*].

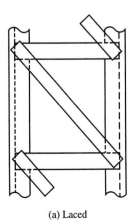
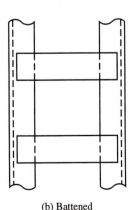

(a) Laced   (b) Battened

Figure 6.11 Laced and battened columns.

6.4 COLUMN BEHAVIOR

6.4.1 Basic Column

The fundamental requirement of any column strength theory is that it must be based on the basic principles of statics and strength of materials, while also taking into account the stress–strain relationship of the materials and reflecting geometric imperfections of all relevant kinds.

The simplest case to be considered is the compression member whose length is sufficiently short that it will not impact on the strength of the member. The strength of these short columns is directly related to their yield strength. Although they rarely occur in actual structures, the capacity of a short column provides an upper bound for column strength calculations, which usually is referred to as the crushing strength.

6.4.2 Elastic Behavior

The centrally loaded, pinned-end column of some length has been the subject of extensive research for more than 250 years. Although the problem of the stability of an elastic, axially loaded member had been examined to some extent by earlier scholars, it is generally accepted that the 1744 and 1759 publications of Leonard Euler form the historical starting point [4, 18].

The elastic buckling solution for a pinned-end column is well known. The initially straight, concentrically loaded column in which all fibers remain elastic is shown in Fig. 6.12a. If the column is of sufficient length, failure will occur by the member taking on a displaced configuration, as shown in Fig. 6.12b. This is the buckling shape for the pure column.

Examination of the free body diagram shown in Fig. 6.12c, and assuming that the deflection remains small, yields the following equations:

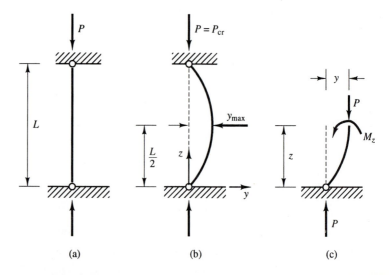

Figure 6.12 Stability conditions for elastic columns.

(a)   (b)   (c)

$$\textit{External moment:} \quad M_z = Py \tag{6.5a}$$

$$\textit{Internal moment:} \quad \frac{d^2y}{dz^2} = y'' = -\frac{M_z}{EI_x} \tag{6.5b}$$

The x and y axes are the centroidal axes and the z axis defines the position along the length of the member. When we combine Eqs. (6.5a) and (6.5b) and define $k^2 = (P/EI_x)$, the differential equation for the deflected column becomes

$$y'' + k^2y = 0 \tag{6.6}$$

which has the solution

$$y = A \sin kz + B \cos kz \tag{6.7}$$

where A and B are constants of integration. Applying the boundary conditions at each end of the column ($y = 0$ for both $z = 0$ and $z = L$) to determine A and B, we find that $B = 0$ and

$$A \sin kL = 0 \tag{6.8a}$$

Equation (6.8a) has a nontrivial solution only for

$$\sin kL = 0 \tag{6.8b}$$

which leads to

$$kL = n\pi \quad (n = 1, 2, \ldots) \tag{6.9}$$

The minimum value of Eq. (6.9) is found when $n = 1$. Substituting into Eq. (6.9) for $k = \sqrt{(P/EI_x)}$ and solving for P leads to a solution for the buckling load P_{cr}, also known as the Euler load P_E, such that

$$P_{cr} = \frac{\pi^2 EI_x}{L^2} = P_E \tag{6.10}$$

Thus, P_{cr} is the maximum load that will cause the column to take on the shape defined by Eq. (6.5). That shape is a half sine wave when $n = 1$; however, the maximum amplitude y_{max}, as shown in Fig. 6.12b, cannot be determined. That is, any value of A will satisfy Eq. (6.8a) as long as the original assumptions are satisfied.

The buckled column shown in Fig. 6.12b represents the first mode. For higher values of n, the buckling configurations will appear as in Fig. 6.13. Thus, $n = 2$ represents the second mode; it has an associated critical load of $n^2 = 4$ times that of the first mode. Similarly, the third mode buckles at a load $n^2 = 9$ times P_E. It is often convenient to consider

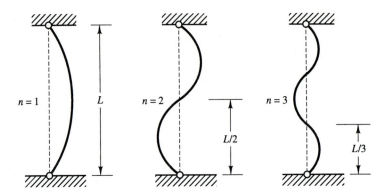

Figure 6.13 Elastic column buckling modes.

a physical interpretation of these various modes. Each deflected portion of the buckled column may be treated as an individual, pinned-end column whose length is the distance between inflection points. On this basis, the associated load of mode 2 is found using $L/2$ for the column length in the buckling load equation, and the load of mode 3 is found using $L/3$.

This approach leads to the concept of the effective length. The effective length is equal to the distance between the inflection points of the buckled member. Just as the pins of the column ends can support no moment, neither do the inflection points. Therefore, the column behaves as if it were a pure column whose length was equal to the distance between the inflection points.

To incorporate effective length into the column equation, a new term, the effective length factor K, is defined. For mode 2, $K = 0.5$, while for mode 3, $K = \frac{1}{3}$. The physical arrangement of connections and bracing will impact the effective length and thus the strength of the column. The arrangement of connections and bracing is addressed in detail in Section 6.8.

The Euler equation is one of the best known equations in mechanics and structural engineering. Euler was the first to recognize that columns could fail through bending rather than crushing. He showed that the column would remain straight until the critical load was reached, at which time it could either continue to remain straight or assume a half sine wave deflected shape. Figure 6.14 illustrates the concepts through a load–deflection diagram.

The column may sustain a load larger than the buckling load, as indicated by the vertical branch of the curve above P_{cr}. However, this is an unstable state of equilibrium, and an infinitesimal perturbation will cause the column to seek a stable, or lower energy level configuration. This is represented by the horizontal branch of the load-deflection curve, either left or right of the vertical. The magnitude of the deflection δ is indefinite, although small, since any value represents an equilibrium position. The behavior exhibited at the critical load, where multiple paths may be followed for equilibrium, is referred to as bifurcation.

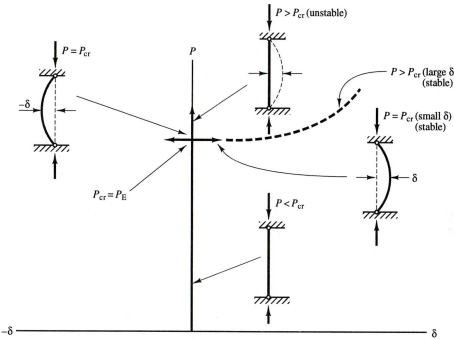

Figure 6.14   Load–deflection curves for an elastic column.

This type of failure of axially loaded members is termed flexural buckling, indicating that the column assumes a "bent" position when the critical load is reached. No bending occurs prior to failure, since the member experiences only axial deformations. Buckling will often happen such that bending is introduced about the principal axis with the smaller moment of inertia. This is referred to as buckling about the weak (or minor) axis. Buckling about the strong (or major) axis could occur if other factors dictate. This is addressed in detail later in this chapter.

Euler's solution showed that column strength is directly proportional to E and I, and inversely proportional to the square of the length. Figure 6.15 illustrates the relationship between the buckling load and the column length in the form of a column curve. This particular column curve is commonly known as the Euler curve; it is asymptotic to $P_{cr} = 0$ as L approaches infinity, and asymptotic to $P_{cr} = \infty$ as L approaches zero.

The Euler equation can also be given in terms of critical stress. This is obtained by dividing Eq. (6.10) by the area A, yielding

$$\sigma_{cr} = \frac{P_{cr}}{A} = \frac{\pi^2 E}{L^2} \frac{I}{A} = \frac{\pi^2 E}{L^2} r^2 \qquad (6.11)$$

where r is the radius of gyration, given by $r = \sqrt{I/A}$. Rewriting Eq. (6.11), we find the critical or Euler stress formula:

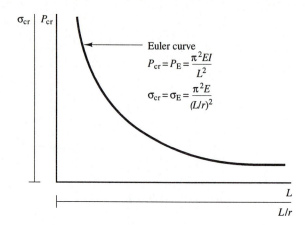

Figure 6.15 Column curve for elastic buckling.

$$\sigma_{cr} = \frac{\pi^2 E}{(L/r)^2} = \sigma_E \tag{6.12}$$

where L/r is known as the slenderness ratio. Equation (6.12) is also reflected in Fig. 6.15, using the stress and slenderness ratio axes shown.

6.4.3 Columns with Initial Curvature or Eccentric Axial Load

Over the 100 to 150 years following Euler's work, studies of column stability focused on developing design curves, partly based on test results and partly based on elastic strength analyses of members with or without various forms of geometric imperfections. The Rankine and Gordon equations (now usually referred to as the Rankine-Gordon column curve), the Johnson parabola, the Perry-Robertson formula, the secant formula, and the Tetmajer straight line equation [4, 17–21] are all such examples.

Figure 6.16a illustrates a pinned-end, initially curved column, and Fig. 6.16b shows a member with the axial load applied eccentrically at the ends. In both cases it is clear that since the axial load introduces a bending moment, the member is no longer a pure column. That is, part of the resistance is used to carry the axial load, and part is used to carry the bending moment. The response is therefore no longer that of a classical stability phenomenon, where no displacements occur until the buckling load is reached. Imperfect elements such as these reflect load–deflection problems, and Fig. 6.16c shows some typical P-δ relationships.

The early studies of such columns focused on determining the maximum stress in the cross section, since the concept of safety was based on having an adequate margin between the service stress and the strength of the material. Limiting the maximum stress to the yield value, divided by a factor of safety, was a key concept in these developments. Indeed, this approach was the forerunner of allowable stress design, as it is known today.

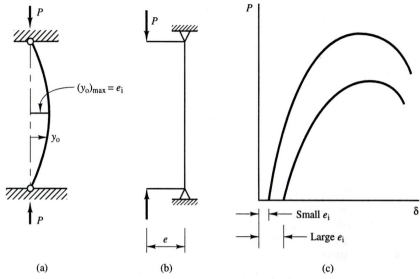

Figure 6.16   Columns with imperfections.

For the initially crooked column in Fig. 6.16a, and ignoring signs, the maximum stress is found from

$$\sigma_{max} = \frac{P}{A} + \frac{M_{max}c}{I} \tag{6.13}$$

where M_{max} is the maximum moment, and c is the distance from the neutral axis to the extreme fiber in the cross section. The maximum moment is the amplified value of the initial crookedness moment Pe_i, computed as [20]

$$M_{max} = \frac{Pe_i}{1 - (P/P_E)} \tag{6.14}$$

where e_i is the maximum value of the initial out-of-straightness. Substituting for M_{max} from Eq. (6.14) into Eq. (6.13), and using $I = Ar^2$, yields the maximum stress in the column:

$$\sigma_{max} = \frac{P}{A}\left[1 + \left(\frac{e_i\,c}{r^2}\right)\left(\frac{1}{1 - (P/P_E)}\right)\right] \tag{6.15}$$

Equation (6.15) is the Perry-Robertson formula, which was used in several design specifications for many years.

For the column with the eccentric axial load in Fig. 6.16b a similar evaluation of the maximum stress leads to the expression

$$\sigma_{max} = \frac{P}{A} \left[1 + \left(\frac{e_i c}{r^2} \right) \sec[(\pi/2)\sqrt{(P/P_E)}] \right] \qquad (6.16)$$

which is the well-known secant formula. It was also used in design specifications for many years, and was particularly useful when column test results formed the basis for code criteria. Specifically, the eccentricity factor ec/r^2 was treated as an imperfection factor, and the value of ec/r^2 was calibrated to produce good correlation between the test results and the column strength as determined by the equation.

The limitations of the Perry-Robertson, secant, and similar formulas arise from the fact that they cannot capitalize on the ductility and the stress redistribution characteristic of steel, two of its greatest assets. Although adequate for very long and slender columns, which truly behave elastically, their semiempirical nature makes them unsuitable for limit states design philosophies such as LRFD.

6.4.4 Inelastic Buckling Concepts

Late in the 19th century, independent research efforts by Engesser and Considère [4, 17, 18] recognized the limitations of the Euler theory, particularly that it was based on elastic material response. It was known, even then, that many materials exhibited distinctly nonlinear characteristics, either through continuously curving stress–strain diagrams, or curves with plateaus in the manner of yielding, as is the case with today's mild structural steel.

The contributions of Engesser's original work (1889) were twofold. First, a theory was developed that assumed that the modulus of elasticity at the instant of buckling was equal to the slope of the material stress–strain curve at the level of the buckling stress. This is indicated by the value E_- in Fig. 6.17, where the buckling stress and strain are σ_1 and ε_1, respectively. The corresponding stress distribution is shown in Fig. 6.18a, which implies that all of the fibers in the cross section respond according to the tangent modulus. That is, as the column goes from being perfectly straight to infinitesimally curved, the Engesser solution assumed that loading and unloading fibers alike would behave according to E_T. As a result, the corresponding buckling load of the column was defined as the tangent modulus load P_T, and it was shown that it could be computed from an Euler-like expression with E_T replacing E:

$$P_T = \frac{\pi^2 E_T I}{L^2} \qquad (6.17)$$

Second, the theory postulated that unloading would take place in the cross section once buckling had occurred and that during this unloading all fibers would respond according to the tangent modulus. It was subsequently thought that unloading would be elastic, using the full value of E

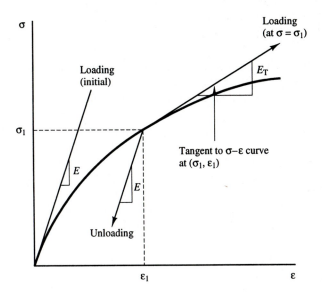

Figure 6.17 Definition of tangent modulus for nonlinear material.

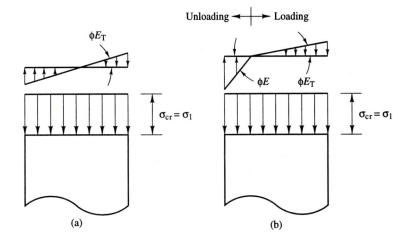

Figure 6.18 Stress distributions for inelastic buckling: (a) tangent modulus theory; (b) reduced modulus theory.

as indicated in Fig. 6.17, with the corresponding stress distribution shown in Fig. 6.18b.

Following this line of reasoning, Engesser and Considère independently developed buckling solutions based on elastic unloading [*4, 17, 18*]. Since this involved dealing with a material whose fibers would respond according to E or E_T, depending on their location within the cross section, the associated critical load was designated the double modulus or reduced modulus load P_R. Again, it could be calculated from an Euler-like expression, with E_R replacing E:

$$P_R = \frac{\pi^2 E_R I}{L^2} \tag{6.18}$$

where E_R is the reduced modulus for the given material and cross section. Since E_R must be greater than E_T, due to the fact that part of the cross section is elastic, the reduced modulus load is also larger than the tangent modulus load.

    Carefully conducted tests by von Kàrmàn and others over many years [4, 17, 18] have shown that actual buckling loads tended to be much closer to those predicted by the tangent modulus equation than those from the reduced modulus equation. It was not until 1947, that Shanley [22] was able to demonstrate that there would be no stress reversal in the cross section as the column reached the tangent modulus load, since the initial deflection associated with the onset of buckling was infinitesimal and the radius of curvature of the buckled member was very large. Thus, Shanley proved that the tangent modulus load was the largest one for which the column would remain straight. He also showed that once the deflection reached a finite quantity, there would be a stress reversal, and elastic unloading would take place. Thus, an increase in the load-carrying capacity would result as predicted by the reduced modulus equation, Eq. (6.18). This phenomenon is illustrated in Fig. 6.19, which shows the load-deflection curve for an initially perfectly straight member, buckling at P_T, and reaching a subsequent maximum value of $P_{max,T}$, generally 5 to 10 percent above P_T [23].

    The resolution of the tangent modulus vs. reduced modulus question had a significant impact on the understanding of steel column behavior. Studies in the 1940s and 1950s, particularly at Lehigh University [5], focused on the presence of residual stresses in hot-rolled and other steel members as the reason for inelastic response that was predicted by the tangent modulus equation.

    It was shown in Section 6.2 that some parts of a cross section will be subjected to residual tension, while other parts are subjected to residual compression. With the application of a compressive axial force, as in a column, Osgood, Yang et al., Johnston, Beedle, and others [4, 17, 18, 21]

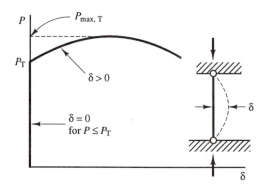

Figure 6.19 Load–deflection behavior for tangent modulus buckling.

demonstrated that local yielding would develop in the shape, as illustrated in Fig. 6.20, when the sum of the applied stress and the residual stress in any area was equal to or larger than the yield stress. Since a portion of the section is stressed beyond the yield stress, this was referred to as a partially elastic, partially plastic state of stress, now simply called inelastic.

Using a very short column, referred to as the stub column, to develop the stress–strain characteristic for a column cross section, it was shown that for a given level of applied stress, the slope of the tangent to the stub column curve was the tangent modulus for the shape. A stress–strain curve for a stub column is shown in Fig. 6.21.

The tangent modulus E_T can be thought of as the average modulus of elasticity for the complete cross section, knowing that some fibers are responding elastically in accordance with the full E, while others respond with a modulus of zero. For steel, with a linearly elastic, perfectly plastic σ-ϵ curve, the yielded (plastic) regions deform with no increase in stress; hence, $E_T = 0$. For the interior of the shape, where the residual stress plus the applied stress is less than F_y, the material is elastic, and $E_T = E$. Thus, it can be seen that the maximum compressive residual stress is of great importance, since it establishes at what applied stress level the cross section will cease being fully elastic.

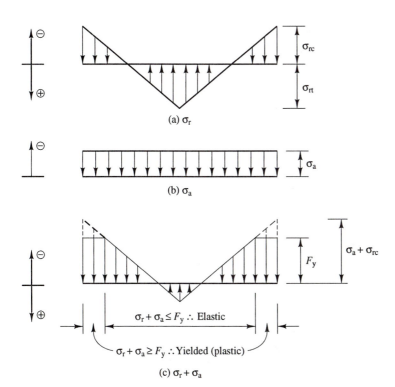

(a) σ_r

(b) σ_a

$\sigma_r + \sigma_a \leq F_y$ ∴ Elastic

$\sigma_r + \sigma_a \geq F_y$ ∴ Yielded (plastic)

(c) $\sigma_r + \sigma_a$

Figure 6.20 Superposition of residual stress and uniform axial stress.

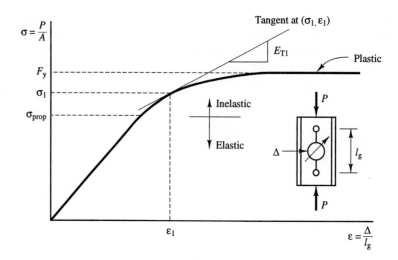

Figure 6.21   Stub column curve.

The tangent modulus load became understood as the buckling capacity of a perfectly straight member of nonlinearly elastic material, with residual stresses. The results had a major impact on design specifications around the world, especially since the strength of a column whose cross section is only partly elastic will be less than that of a fully elastic one. Theoretical evaluations and scores of column tests proved the accuracy of this finding [4, 5].

A tangent modulus column curve is plotted in Fig. 6.22. On the same diagram, the Euler curve and the crushing load plateau are shown. Strength reductions of 20 to 40 percent are common in the intermediate region of the curve. Note that for buckling stresses below the proportional limit σ_{prop}, the tangent modulus and Euler curves coincide, since both reflect the s'rength of perfectly straight elastic columns.

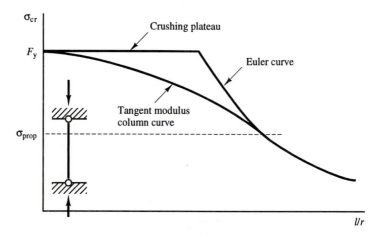

Figure 6.22   Tangent modulus column curve.

6.4.5 Maximum Strength Theory

The studies that led to the development of the Perry-Robertson and secant formulas were based on initially crooked and eccentrically loaded column models, along with an assumed elastic response. These analyses were done without knowledge of the presence of residual stresses. Following the development of the tangent modulus model, however, it was recognized that a complete column evaluation needed to take into account both residual stress and initial out-of-straightness. For many years, the column formulas found in the most common specifications were strictly tangent modulus expressions, where an increased factor of safety covered the effects of initial crookedness and unintentional load eccentricity [24].

The solution to the inelastic, initially curved column problem is primarily a question of numerical modeling. The solution method developed in some of the key studies of the time [2, 4, 11, 25] was based on an incremental, iterative numerical routine which established the load–deflection curve for the column. For each deflection increment, the corresponding equilibrium value of the axial load would be found, taking into account residual stress, magnitude of initial out-of-straightness, yield stress level and variations throughout the shape, and local yielding as well as elastic unloading in elemental areas of the cross section. In this fashion the P-δ curve was determined, including the all-important peak or maximum strength of the member. The correlation with full-scale tests was good, and correlation with computation results was generally within 5 percent of the test values [2].

Figure 6.23 shows three typical P-δ curves, including the special case of a perfectly straight member, which is governed by the tangent modulus load P_T. The nonlinear response is evident, as is the fact that the maximum strength P_{max} may be substantially lower than P_T. Research has demonstrated that for initial crookedness values around the common maximum of $e_i = L/1000$, where L is the column length, and the mean of $e_i = L/1500$, the drop from P_T to P_{max} could be as much as 25 to 30 percent, with the magnitude of the differences dependent on the slenderness ratio [2, 4, 11].

Figure 6.24 provides a comparison between the maximum strength column curve P_{max} and the tangent modulus P_T and Euler P_E curves. As may be seen from the figure, it is in the middle range of slenderness ratios where the difference between P_{max} and P_T is the greatest. This is, of course, the slenderness range for the majority of columns in actual structures.

Maximum strength studies have established that the two major column strength factors are residual stress and initial out-of-straightness. Although a number of other influences play a role, these two outweigh all others. Thus, the initially curved or geometrically imperfect member has generally been accepted as the basic column model [4]. This has led the way toward the use of limit states design principles as presented in the LRFD specification.

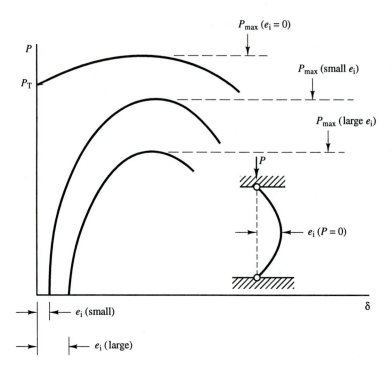

Figure 6.23 Load–deflection curves for columns.

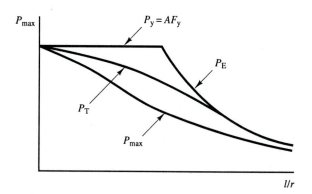

Figure 6.24 Maximum strength column curves.

6.4.6 Limit States Design Concepts

Having determined that the maximum strength model provided the most accurate representation of column strength, studies then proceeded to consider the impact of additional variables. The random nature of strength parameters such as yield stress and residual stress had been known for a number of years. Some studies also examined construction effects, in the form of out-of-straightness and out-of-plumb (i.e., deviation from the vertical) of columns in actual structures [2, 26].

All in all, the variability of the column strength parameters was reflected in significant column strength variations. As an example, Fig. 6.25 shows the upper and lower envelope curves for a collection of maximum-strength column curves, representing the full range of steel grades and shapes used in construction in North America [2]. The range of variation is substantial; the distribution is typically quite skewed toward the lower envelope curve, as indicated by the frequency curve shown for an arbitrary slenderness ratio.

The data in Fig. 6.25 are indicative of a concern that can be addressed only through LRFD. The strength variation is so large that using an "average" curve to represent the strength of all types of columns will be conservative for some and unconservative for others. The variation has always been there; it was only through the maximum strength and LRFD studies that it was actually quantified.

Several studies have addressed various ways of improving the accuracy of column design through modifications of the design approach or curve that is used. One such approach uses multiple column curves. Sets of such curves have been developed in the United States [2, 4] and Europe [25], the latter now part of Eurocode 3 [16]. The Canadian steel design standard uses two curves, based on the American research results [2, 15]. Other studies have proposed different ways of arriving at improved reliability [27, 28], and work along these lines is likely to continue for some time. Although the multiple column curve concept has definite advantages, it also has drawbacks, in the form of increased code complexity. It is expected that the U.S. structural steel design specifications will continue to use a single curve for the foreseeable future.

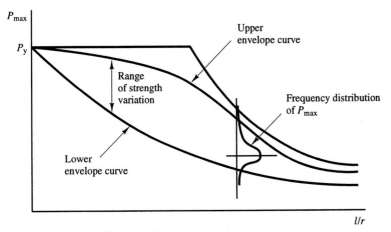

Figure 6.25   Column strength variation.

6.5 COLUMN LENGTH

6.5.1 The Influence of Column Length

The primary column strength parameters have been detailed in preceding sections. Here, the capacity of the compression member is examined in the light of these parameters and how they interact with the length of the column. Behavior and strength vary significantly as a function of the length, and it is therefore common and convenient to subdivide compression members into representative categories of behavior. The ultimate limit states are different for each category. The following categories and ultimate limit states are used to define column behavior:

LENGTH CATEGORY	LIMIT STATE
Short columns	Crushing (yield load)
Intermediate columns	Inelastic flexural buckling
Long columns	Elastic flexural buckling

6.5.2 Short Columns

Known most commonly in the research literature as stub columns, short columns fail in pure axial compression or crushing of the member. Since it is so short, flexural stability is not a viable limit state: Yielding of the material in the full cross section is reached before overall buckling can take place. The representative load–deformation curve is therefore an axial load–axial shortening relationship, a stub column curve, as shown in Fig. 6.21.

The stub column curve is obtained by testing a full-size column cross section, with a length that is usually no more than about three to four times the largest cross-sectional dimension (depth of a W-shape, for example). The load is applied concentrically, and the longitudinal deformation Δ is measured over a certain gage length l_g, as indicated in Fig. 6.21. Technical Memorandum No. 3 in Ref. [4] gives details of the testing and data acquisition procedures for stub column tests, and also discusses the interpretation of the findings.

The behavior is fully elastic for an applied stress less than the proportional limit σ_{prop}, and the full cross-section modulus of elasticity is E. Once σ_{prop} is reached, all subsequent response is inelastic, since the applied stress plus the residual stress will equal or exceed the yield stress in certain areas of the cross section. The tangent modulus of elasticity applies to the gross cross section, given as

$$0 \le E_{\text{T}} \le E \qquad \text{for } \sigma_{\text{prop}} \le \sigma = \frac{P}{A_g} \le F_y \qquad (6.19)$$

Once the yield stress is reached in all fibers, the axial capacity of the stub column is exhausted. The corresponding load is the yield load P_y.

The yield load is a benchmark capacity for compression members in that it represents the upper bound of useful strength for a member under pure axial load. For some shapes and certain types of steel, however, it is possible that strain hardening may be reached after the stub column has gone through a substantial plastic deformation. This is indicated by the dashed branch marked ST in Fig. 6.21.

Similarly, for shapes with medium to high width–thickness ratios of the elements in the cross section (diameter to wall thickness ratio for circular tubes), local buckling may occur after a certain amount of plastic deformation. This is accompanied by a drop in the axial load on the stub column, as indicated by the dashed branch LB in Fig. 6.21.

6.5.3 Intermediate-Length Columns

Columns of intermediate-length reflect the majority of compression members in structures. In terms of slenderness ratio, these columns have KL/r values in the approximate range of 40 to 120. The controlling limit state is inelastic flexural buckling, which reflects the influence of residual stress and initial crookedness along with length.

Typical load–deflection curves for intermediate-length columns were shown in Fig. 6-23. The perfectly straight member reaches the tangent modulus load; columns with initial crookedness e_i reach the maximum strength at the peak of the curve. For small magnitudes of e_i, say around $L/10{,}000$, the maximum strength will be close to P_T, but this is generally not the case for realistic structural members, where the initial out-of-straightness lies within the range $L/2000$ to $L/1000$.

Depending on the yield stress, the residual stress distribution, the crookedness, and the slenderness of the column, the peak of the load–deflection curve, the maximum strength, is reached at a stage where substantial local yielding has occurred. Thus, inelastic buckling is the primary response of intermediate-length compression members.

It is often useful to present column strength curves in a nondimensional form, thereby removing the material yield strength from the picture. If a new term, λ^2, is defined as the ratio of the yield stress to the Euler stress F_y/F_E, the nondimensional slenderness may be represented by

$$\lambda = \frac{1}{\pi}\sqrt{\frac{F_y}{E}}\frac{KL}{r} \tag{6.20}$$

Since the yield stress is the useful upper bound of the column strength, it is helpful to note that the Euler load will equal the yield load at a value $\lambda = 1.0$. This can be seen as the intersection of the yield plateau with the Euler elastic buckling curve in Fig. 6.26.

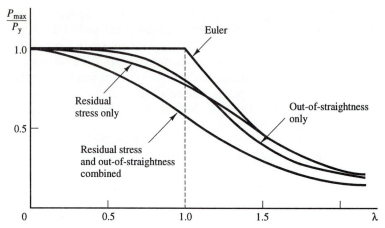

Figure 6.26 Column curves and imperfection effects.

Figure 6.26 also shows maximum strength column curves for a typical member, depicting the influence of residual stress alone (i.e., a tangent modulus curve), out-of-straightness alone, and the combined effect of the two parameters. The reduction of the capacity in the intermediate slenderness range is substantial when both terms are incorporated into the solution. The synergism of residual stress and initial crookedness is well established [2, 4, 13], although it is only relatively recently that the individual term influences have been separated and documented for certain types of columns [12].

A study presented by Chernenko and Kennedy [12] has shown that for welded built-up H-shapes, the synergism between residual stress and initial crookedness in some cases works to the advantage of the column. That is, some columns with both imperfections are actually somewhat stronger than if only one or the other is present. It is likely that this can be attributed to the unique residual stress distributions that are found in built-up shapes with flame-cut plates, since these have very high tensile residual stress at the most critical locations within the shapes. For the usual hot-rolled shapes, however, the documentation of the negative (i.e., strength-reducing) synergism of combining residual stress and initial out-of-straightness is clear.

The limiting slenderness for intermediate columns is generally defined as the boundary between elastic and inelastic buckling. Figure 6.27 shows a typical tangent modulus column curve, using the nondimensional representation of (P_T/P_y) vs. λ.

Due to the effects of residual stress, the tangent modulus column curve falls below the yield plateau and the Euler curve in the inelastic response region. The extent of that region is defined by the level of the maximum compressive residual stress σ_{rc} or the proportional limit stress, given as $\sigma_{prop} = F_y - \sigma_{rc}$. In Fig. 6.27 this is indicated by the horizontal

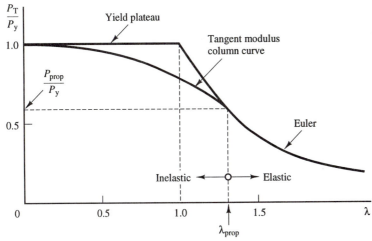

Figure 6.27   Tangent modulus column curve.

dashed line, at the load ratio of (P_{prop}/P_y). The slenderness that defines the end of the inelastic region and the beginning of the elastic region is found by setting the Euler stress equal to the proportional limit stress and solving for the corresponding slenderness ratio:

$$\left(\frac{KL}{r}\right)_{prop} = \pi\sqrt{\frac{E}{F_y - \sigma_{rc}}} \qquad (6.21)$$

and therefore

$$\lambda_{prop} = \frac{1}{\pi}\sqrt{\frac{F_y}{E}}\left(\frac{KL}{r}\right)_{prop}$$

which gives

$$\lambda_{prop} = \sqrt{\frac{F_y}{F_y - \sigma_{rc}}} = \frac{1}{\sqrt{1 - \sigma_{rc}/F_y}} \qquad (6.22)$$

The proportional limit slenderness in structural shapes and the most common structural steel grades will typically vary between 1.12 (50-ksi yield stress steel, 10-ksi maximum compressive residual stress) and 2.45 (36-ksi yield stress steel, 30-ksi maximum compressive residual stress). The latter case reflects a very heavy rolled shape or a small welded built-up shape using rolled plates (universal mill) for flanges.

6.5.4 Long Columns

Even though residual stresses are present, long columns still fail in elastic flexural buckling. Length is the overriding factor since the critical stress is so small that the sum of the maximum compressive residual and applied stress is less than the proportional limit. Thus, the entire cross section behaves elastically. The maximum strength is still less than the Euler value, however, due to the initial out-of-straightness, although it will be close to P_E for very long columns. This can be seen in Fig. 6.26 by the closeness of the Euler curve and the column curves that indicate the presence of initial crookedness alone as well as initial crookedness plus residual stress, as the slenderness becomes large.

6.6 COLUMN STRENGTH APPROXIMATIONS AND DESIGN CRITERIA

6.6.1 Development of Design Criteria

As is the case with all transformations from theory to practice, the development of accurate yet usable design criteria for columns is a complex undertaking. As has been shown, strength of a compression member is influenced by many factors, and it is essential to settle on which factors are the most important and thus need to be reflected in the end result.

Clearly, residual stress, initial out-of-straightness, and column length must enter into the design equation. Further, to maintain the pinned-end member as the basic element, end restraint is better covered through the overall frame analysis. In comparison to these factors, other parameters are of secondary importance, and are preferably taken into account through the value of the resistance factor.

6.6.2 Limit States

The column design curve must be able to reflect all of the relevant limit states, i.e., crushing and inelastic and elastic flexural buckling. For very low slenderness ratios, the ultimate limit state is the yield load P_y, although the steel and shape that are used may actually permit the maximum strength to exceed P_y due to strain hardening. For columns with high slenderness ratios, the ultimate limit state is elastic buckling. For the majority of columns in actual structures, the ultimate limit state will be inelastic buckling.

In addition to the limit states associated with column buckling that have already been presented, three additional limit states must be considered: torsional buckling, flexural–torsional buckling, and local buckling. The first two limit states are associated with the overall cross-section shape, while local buckling is a limit state associated with the individual elements that make up the column shape. If these elements are themselves slender, they will not be able to fully participate in carrying the column load. These limit states must also be considered in establishing the strength of a column and are treated in Sections 6.9 and 6.10.

6.6.3 Variation of Column Strength

The variation of the strength of columns has been discussed at some length throughout this chapter. It is generally agreed that a single column curve cannot adequately represent the strength of all types of columns without some form of recognition of this great variability. Some design standards utilize multiple column curves, others base the requirements on a single equation and use a somewhat lower resistance factor. The aim, obviously, is to have as uniform and satisfactory levels of reliability as possible.

  The final decision rests with the code writers; the implementation rests with the design community. Ultimately, it is not code complexity that will control the outcome; rather, the forces of economy will undoubtedly come to bear. Both approaches have their advantages and disadvantages.

6.7 LRFD COLUMN DESIGN

Although the behavior of columns as discussed up to this point is seen to be quite complex, the actual representation of that behavior as found in the LRFD specification is straightforward. For doubly symmetric shapes whose elements meet the minimum width thickness ratios required by Section B5.1 of the specification, the critical stress is defined as [29]

for $\lambda_c \leq 1.5$

$$F_{cr} = (0.658^{\lambda_c^2})F_y \qquad (6.23)$$

for $\lambda_c > 1.5$

$$F_{cr} = \left(\frac{0.877}{\lambda_c^2}\right)F_y \qquad (6.24)$$

where the nondimensional slenderness as defined in Section 6.5.3 is

$$\lambda_c = \frac{KL}{\pi r}\sqrt{\frac{F_y}{E}} \qquad (6.25)$$

The capacity of the column, then, is given as

$$\phi P_n = \phi A_g F_{cr} \qquad (6.26)$$

where $\phi = 0.85$.

  Members whose elements do not meet the width thickness requirements, singly symmetric members, and nonsymmetric members are treated in Sections 6.9 and 6.10.

The ratio of critical stress to yield stress F_{cr}/F_y, as found from Eqs. (6.23) and (6.24) is plotted in Fig. 6.28 against the nondimensional slenderness λ_c. Columns whose slenderness is less than $\lambda_c = 1.5$ are generally considered intermediate-length columns, while columns with slenderness greater than 1.5 are called long columns. It is seen that the short column, discussed in Section 6.5.2, does not actually exist as a separate entity in the code equations.

Another way to look at column strength is presented in Fig. 6.29 where critical stress is plotted versus slenderness ratio, based again on Eqs. (6.23) and (6.24). Here it can be seen that for large slenderness ratios, the actual value depending on F_y, the critical stress is independent of the value of F_y. This results from the elastic buckling as predicted by Eq. (6.24).

The determination of the capacity of a given column section with a given effective length is straightforward. The first step is to determine the critical slenderness ratio. From this ratio the critical stress is determined and from that stress, the critical load. Although tables are available in the *LRFD Manual,* the following examples will not take advantage of these aids.

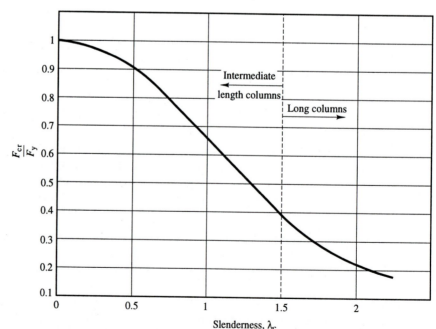

Figure 6.28   Ratio of critical stress to yield stress vs. nondimensional slenderness.

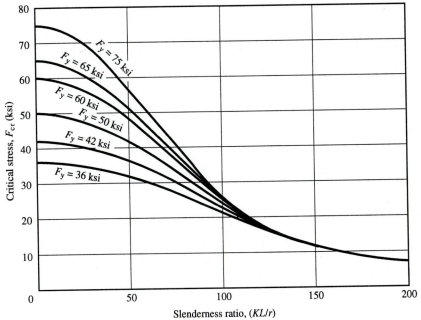

Figure 6.29   Critical stress vs. slenderness ratio.

EXAMPLE 6.1

Given: Determine the capacity of a W8×58 with an effective length about the y axis of 12 ft. F_y = 50 ksi.

Solution

For the W8×58, A = 17.1 in.², r_y = 2.10 in.

$$\frac{KL}{r} = \frac{12.0(12.0)}{2.10} = 68.57$$

$$\lambda_c = \frac{68.57}{\pi} \sqrt{\frac{50.0}{29,000}} = 0.906$$

since λ_c < 1.5,

$$F_{cr} = (0.658^{0.906^2})50.0 = 35.46 \text{ ksi}$$

and

$$\phi P_n = 0.85(17.1)(35.46) = 515.4 \text{ kips}$$

EXAMPLE 6.2

Given: Determine the capacity of a W14×90 with an effective length about the x axis of 24 ft and an effective length about the y axis of 12 ft. $F_y = 36$ ksi.

Solution

For the W14×90, $A = 26.5$ in.^{2}, $r_x = 6.14$ in., and $r_y = 3.70$ in. The critical slenderness must first be determined:

$$\frac{KL}{r_x} = \frac{24(12)}{6.14} = 46.9$$

and

$$\frac{KL}{r_y} = \frac{12(12)}{3.70} = 38.9$$

Using the larger slenderness, the x axis, yields

$$\lambda_c = \frac{46.9}{\pi}\sqrt{\frac{36}{29,000}} = 0.53$$

Since $\lambda_c < 1.5$,

$$F_{cr} = (0.658^{0.532})\,36 = 32.0 \text{ ksi}$$

and

$$\phi P_n = 0.85(26.5)(32.0) = 720.8 \text{ kips}$$

EXAMPLE 6.3

Given: Determine the capacity of a W8×28 with an effective length about both axes of 20 ft. $F_y = 50$ ksi.

Solution

For the W8×28, $A = 8.25$ in.^{2}, $r_x = 3.45$ in., and $r_y = 1.62$ in. Since the effective length is the same for both axes, the lower radius of gyration for the y axis will give the greatest slenderness:

$$\frac{KL}{r_y} = \frac{20(12)}{1.62} = 148.1$$

$$\lambda_c = \frac{148.1}{\pi}\sqrt{\frac{50}{29,000}} = 1.96$$

Since $\lambda_c > 1.5$, the critical stress is given as

$$F_{cr} = \left(\frac{0.877}{1.96^2}\right)50.0 = 11.4 \text{ ksi}$$

and

$$\phi P_n = 0.85(8.25)(11.4) = 79.9 \text{ kips}$$

EXAMPLE 6.4

Given: Repeat Example 6.3 using $F_y = 36$ ksi.

Solution

The critical slenderness remains the same; however, with the change in F_y, the nondimensional slenderness will change. Thus,

$$\lambda_c = \frac{148.1}{\pi}\sqrt{\frac{36}{29,000}} = 1.66$$

and

$$F_{cr} = \left(\frac{0.877}{1.66^2}\right)36 = 11.4 \text{ ksi}$$

which yields

$$\phi P_n = 0.85(8.25)(11.4) = 79.9 \text{ kips}$$

Note that these last two examples resulted in the same capacity for the column, even though the material yield stress was different. This is because the columns are very long so as to buckle in the elastic region where material yield is not a controlling factor.

6.8 EFFECTIVE LENGTH OF COLUMNS

The concept of effective length was introduced in Section 6.4.2 to account for the modes of buckling that resulted from the mathematical solution to the buckling equation. The term effective length was again used in Section 6.7 in the design examples. It is appropriate at this point to expand on this important topic. Although effective length resulted from a mathematical solution, it has a physical interpretation that is more useful to the designer. The columns shown in Fig. 6.30 illustrate effective length for situations where the ends of the column remain in their position relative to

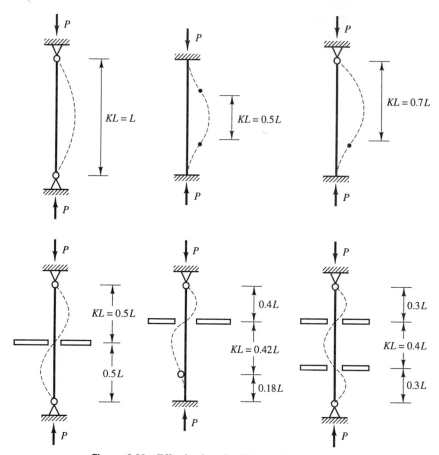

Figure 6.30 Effective length without joint translation.

each other. These columns usually occur in what are referred to as braced frames or frames that do not permit joint translation. The effective length is the length of the longest segment of the column between inflection points and can be seen to vary as a function of column end conditions and intermediate lateral bracing, as shown in the figure. In all cases where translation is not permitted, the effective length of the column will be less than or equal to the column length. Thus, the effective length factor K will be less than or equal to 1.0.

    Unbraced frames with columns whose ends are permitted to move laterally with respect to each other are shown in Fig. 6.31. These columns usually rely on beams framing into their ends, as seen in Fig. 6.31c, to provide stability. Some rotational restraint less than a fixed end but more than a pinned end is provided by the members framing into the column. For all cases in which sway is permitted, the effective length of the

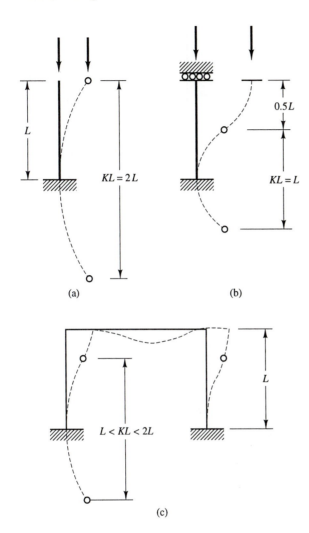

(a)   (b)

(c)

Figure 6.31   Effective length with joint translation.

column will be equal to or greater than the column length, so that the effective length factor K will be equal to or greater than 1.0.

Since the actual end condition for any column will be a function of the connection provided, it is unlikely that true fixed ends or perfect pins can be accomplished. To take this into account, slightly modified effective length factors are generally recommended for use in design. Figure 6.32, taken from the commentary to the LRFD specification, gives values to be used when the cases identified are approximated by the actual conditions.

When columns have different end conditions or bracing arrangements for each principal axis, determination of the critical slenderness ratio must consider both axes. The wide flange column shown in Fig. 6.33

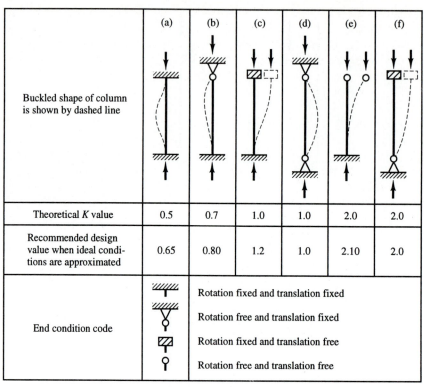

	(a)	(b)	(c)	(d)	(e)	(f)
Buckled shape of column is shown by dashed line						
Theoretical K value	0.5	0.7	1.0	1.0	2.0	2.0
Recommended design value when ideal conditions are approximated	0.65	0.80	1.2	1.0	2.10	2.0
End condition code	Rotation fixed and translation fixed					
	Rotation free and translation fixed					
	Rotation fixed and translation free					
	Rotation free and translation free					

Figure 6.32 Effective length factors for idealized column end conditions. Courtesy the American Institute of Steel Construction, Inc.

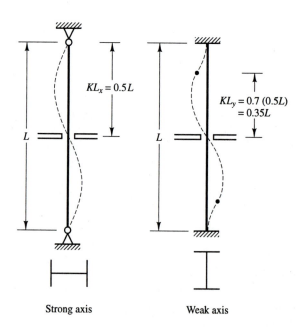

$KL_x = 0.5L$

$KL_y = 0.7\,(0.5L)$
$= 0.35L$

L

L

Strong axis

Weak axis

Figure 6.33 Weak and strong axis buckling.

has different end conditions for each axis. The resulting effective length for each axis is shown. The effective length that leads to the maximum slenderness ratio, however, is not necessarily the longest, since slenderness is also dependent on the radius of gyration for each axis. If the column of Fig. 6.33 is a W8 $\times$ 40, 18 ft in length with r_x = 3.53 in. and r_y = 2.04 in., then the slenderness ratios are KL/r_x = 0.5(18)(12)/3.53 = 30.6 and KL/r_y = 0.35(18)(12)/2.04 = 37.1. As can be seen, the larger slenderness ratio is about the y axis, even though the x axis effective length is the longest.

When columns are connected to other structural elements that do not provide the column with purely pinned or fixed ends, the determination of effective length becomes more complex. For columns in braced, or nonsway, frames, the resulting effective length factor will be somewhere between 1.0 and 0.5. In many cases, the assumption of K = 1.0 is sufficiently accurate for design of columns in braced frames. For those cases where a more accurate value of effective length is required, the use of the nomograph shown in Fig. 6.34a will be helpful. A more detailed discussion of the use of the nomograph will be given in the presentation on unbraced frames.

As was seen earlier, the effective length factor for columns in unbraced frames will be equal to or greater than 1.0. Columns in an unbraced frame rely on the stiffness of the members framing into them to provide their stability. Thus, the stiffer the members framing into a column, the more restraint they provide and the lower will be the effective length factor. In the extreme case of a pin-ended column in an unbraced frame, there is no restraint provided by the members framing into the column. The resulting effective length factor would then be taken as infinity, which implies that the column has no capacity for axial load. This is the case of the "leaning column," which is discussed in Chapter 9.

For normal design situations, it is impractical to carry out a detailed analysis for the determination of buckling strength and effective length. Numerous approaches have been suggested for a more direct determination of the buckling capacity and effective length of columns in a variety of situations [30–34]. The most commonly used approach is the nomograph, already presented in Fig. 6.34, originally developed by Julian and Lawrence and presented in detail by Kavanagh [35]. This approach is also suggested by the commentary to the LRFD specification as one that "affords a fairly rapid method for determining adequate K-values."

The nomographs are based on the assumption of idealized conditions that do not actually occur in real structures. These assumptions are as follows:

1. Behavior is elastic.
2. All members are prismatic.
3. The structure is symmetric.
4. All columns buckle simultaneously.

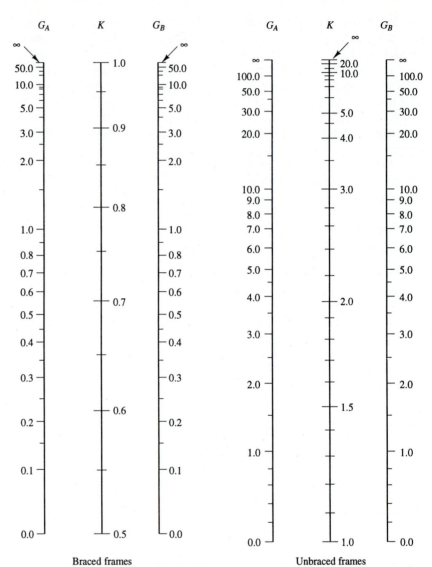

Figure 6.34 Nomograph for effective length factors. Courtesy the American Institute of Steel Construction, Inc.

5. The restraining moment contributed by the girders is distributed to the columns according to their stiffness.

6. Girders carry no axial load.

7. Girders are elastically restrained at their ends.

8. Rotations at the opposite ends of beams are equal in magnitude and produce single curvature in braced frames and double curvature in unbraced frames.

The stability relationship may be solved using either a stiffness or flexibility formulation. The resulting equations are, for a braced frame

$$\frac{G_A G_B}{4}\left(\frac{\pi^2}{K^2}\right) + \left(\frac{G_A + G_B}{2}\right)\left(1 - \frac{\pi/K}{\tan \pi/K}\right) + \frac{2}{\pi/K}\tan\frac{\pi}{2K} = 1 \qquad (6.27)$$

and for unbraced frames

$$\frac{G_A G_B (\pi/K)^2 - 36}{6(G_A + G_B)} = \frac{\pi/K}{\tan(\pi/K)} \qquad (6.28)$$

where $G = \Sigma I_c/L_c \,/\, \Sigma I_g/L_g$, and A and B represent the ends of the column. The graphical solution to these transcendental equations is presented in the alignment charts (nomograph) of Fig. 6.34. Approximate solutions to these equations have been available since 1966 in the French design rules [36]. They provide a reasonably accurate solution to a very complex problem. For braced frames, the approximate solution is given as

$$K = \frac{3G_A G_B + 1.4(G_A + G_B) + 0.64}{3G_A G_B + 2.0(G_A + G_B) + 1.28}$$

and for unbraced frames the solution is

$$K = \sqrt{\frac{1.6G_A G_B + 4.0(G_A + G_B) + 7.5}{G_A + G_B + 7.5}}$$

These approximate equations are said to be accurate within 2 percent of the exact solution, certainly accurate enough to be an aid in design.

EXAMPLE 6.5

Given: Determine the effective length factor for the column AB in an unbraced frame, as shown in Fig. 6.35. The columns are $W10 \times 88$, 14 ft in length. The beams framing into the column are $W16 \times 36$ at the upper end and $W16 \times 77$ at the lower end. All beams span 24 ft.

Solution

At end A:

$$I_c = 534 \text{ in.}^4$$

$$I_g = 448 \text{ in.}^4$$

$$G_A = \frac{2[534/14(12)]}{2[448/24(12)]} = 2.04$$

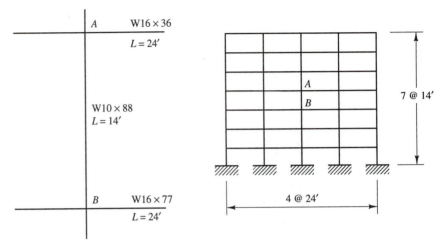

Figure 6.35   Multistory frame for Example 6.5.

At end B:

$$I_c = 534 \text{ in.}^4$$

$$I_g = 1110 \text{ in.}^4$$

$$G_B = \frac{2[534/14(12)]}{2[1110/24(12)]} = 0.83$$

Using the alignment chart as shown in Fig. 6.36, the effective length factor is determined to be $K = 1.41$. It is interesting to note that had this column been a part of a braced frame with the same beams and columns rigidly connected as shown, the effective length factor would have been 0.8.

The first assumption given in the development of the alignment charts was that all members behave elastically. As discussed earlier, the majority of columns in buildings actually do not behave elastically; thus, some modification of the alignment chart solution is in order. With the assumption of elastic members, the modulus of elasticity E drops out when the ratio of column to beam stiffness is determined. For the more common case of elastic beams and inelastic columns, the relative stiffness becomes

$$G_{\text{inelastic}} = \frac{\Sigma E_c I_c / L_c}{\Sigma E_g I_g / L_g}$$

If the additional assumption is made that the modulus of elasticity for all columns at a joint will be the same and equal to the tangent modulus E_T, then

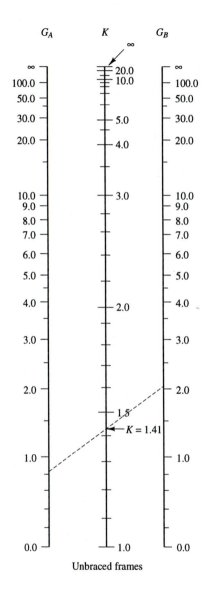

G_A K G_B

Unbraced frames

Figure 6.36 Determination of K for Example 6.5.

$$G_{\text{inelastic}} = \left(\frac{E_T}{E}\right)G_{\text{elastic}} = \beta_s \, G_{\text{elastic}}$$

Numerous approaches have been suggested for the determination of the stiffness reduction factor β_s [37–43], most revolving around the allowable stress specification equations in use at the time. With the development of the LRFD column equations, the determination of β_s became more readily evident. Since the critical stress for an elastic or inelastic column is directly related to E or E_T, respectively, the stiffness reduction factor may be taken as

TABLE 6.1 ADJUSTMENT FACTOR FOR INELASTIC BUCKLING

KL/r	$F_y = 36$ ksi			$F_y = 50$ ksi		
	λ_c	P/A	β_s	λ_c	P/A	β_s
0.00	0.00	36.00	0.00	0.00	50.00	0.00
2.00	0.02	35.99	0.00	0.03	49.99	0.00
4.00	0.04	35.97	0.00	0.05	49.94	0.00
6.00	0.07	35.93	0.01	0.08	49.87	0.01
8.00	0.09	35.88	0.01	0.11	49.77	0.01
10.00	0.11	35.81	0.01	0.13	49.64	0.02
12.00	0.13	35.73	0.02	0.16	49.48	0.03
14.00	0.16	35.63	0.03	0.19	49.29	0.04
16.00	0.18	35.52	0.04	0.21	49.07	0.05
18.00	0.20	35.39	0.05	0.24	48.83	0.06
20.00	0.22	35.25	0.06	0.26	48.56	0.08
22.00	0.25	35.09	0.07	0.29	48.26	0.09
24.00	0.27	34.92	0.08	0.32	47.94	0.11
26.00	0.29	34.74	0.09	0.34	47.59	0.13
28.00	0.31	34.54	0.11	0.37	47.21	0.15
30.00	0.34	34.33	0.12	0.40	46.82	0.17
32.00	0.36	34.11	0.14	0.42	46.39	0.19
34.00	0.38	33.87	0.16	0.45	45.95	0.21
36.00	0.40	33.63	0.17	0.48	45.48	0.23
38.00	0.43	33.36	0.19	0.50	44.99	0.26
40.00	0.45	33.09	0.21	0.53	44.48	0.28
42.00	0.47	32.81	0.23	0.56	43.95	0.31
44.00	0.49	32.51	0.25	0.58	43.40	0.33
46.00	0.52	32.21	0.27	0.61	42.83	0.36
48.00	0.54	31.89	0.29	0.63	42.25	0.39
50.00	0.56	31.56	0.31	0.66	41.65	0.41
52.00	0.58	31.22	0.34	0.69	41.03	0.44
54.00	0.61	30.88	0.36	0.71	40.40	0.47
56.00	0.63	30.52	0.38	0.74	39.75	0.50
58.00	0.65	30.16	0.40	0.77	39.10	0.52
60.00	0.67	29.78	0.43	0.79	38.43	0.55
62.00	0.70	29.40	0.45	0.82	37.75	0.58
64.00	0.72	29.02	0.47	0.85	37.06	0.60

$$\beta_s = \frac{E_T}{E} = \frac{f_{cr\ inelastic}}{f_{cr\ elastic}}$$

Using the critical stress equations from the LRFD specification given in Section 6.7, the stiffness reduction factor becomes

$$\beta_s = \frac{\text{Eq. (6.23)}}{\text{Eq. (6.24)}}$$

Stiffness reduction factors are listed in Table 6.1 for both 36- and 50-ksi steel as a function of KL/r, λ_c, and P/A.

TABLE 6.1 (*Continued*)

KL/r	F_y = 36 ksi			F_y = 50 ksi		
	λ_c	P/A	β_s	λ_c	P/A	β_s
66.00	0.74	28.62	0.50	0.87	36.36	0.63
68.00	0.76	28.22	0.52	0.90	35.66	0.66
70.00	0.79	27.81	0.54	0.93	34.94	0.68
72.00	0.81	27.40	0.57	0.95	34.23	0.71
74.00	0.83	26.98	0.59	0.98	33.50	0.73
76.00	0.85	26.56	0.61	1.00	32.78	0.75
78.00	0.87	26.13	0.63	1.03	32.05	0.78
80.00	0.90	25.70	0.66	1.06	31.31	0.80
82.00	0.92	25.27	0.68	1.08	30.58	0.82
84.00	0.94	24.83	0.70	1.11	29.85	0.84
86.00	0.96	24.39	0.72	1.14	29.11	0.86
88.00	0.99	23.95	0.74	1.16	28.38	0.88
90.00	1.01	23.50	0.76	1.19	27.65	0.89
92.00	1.03	23.06	0.78	1.22	26.93	0.91
94.00	1.05	22.61	0.80	1.24	26.21	0.92
96.00	1.08	22.16	0.81	1.27	25.49	0.94
98.00	1.10	21.71	0.83	1.30	24.77	0.95
100.00	1.12	21.27	0.85	1.32	24.07	0.96
102.00	1.14	20.82	0.86	1.35	23.37	0.97
104.00	1.17	20.37	0.88	1.37	22.67	0.98
106.00	1.19	19.93	0.89	1.40	21.99	0.98
108.00	1.21	19.48	0.91	1.43	21.31	0.99
110.00	1.23	19.04	0.92	1.45	20.64	1.00
112.00	1.26	18.60	0.93			
114.00	1.28	18.16	0.94			
116.00	1.30	17.73	0.95			
118.00	1.32	17.30	0.96			
120.00	1.35	16.87	0.97			
122.00	1.37	16.44	0.98			
124.00	1.39	16.02	0.98			
126.00	1.41	15.61	0.99			
128.00	1.44	15.20	0.99			
130.00	1.46	14.79	1.00			

A quick review of the factors that play a part in the determination of effective length will show that $G_{\text{inelastic}}$ will be less than G_{elastic}. Entry into the alignment chart with lower G values will yield lower K values. Thus, it is seen that the assumption of elastic column buckling in the solution for K is a conservative one—in many instances, an overly conservative one that will waste material.

EXAMPLE 6.6

Given: For the column and framing given in Example 6.5, determine the inelastic column effective length factor if the column carries 510 kips.

Solution

If it is assumed that the column carries the maximum load that would be permitted, the results of using Eq. (6.23) would yield

$$f_a = \frac{P}{A} = \frac{510 \text{ kips}}{25.9 \text{ in.}^2} = 19.7 \text{ ksi}$$

From Table 6.1,

$$\beta_s = 0.90$$

and

$$G_{A \text{ inelastic}} = \beta_s G_{A \text{ elastic}} = 0.90(2.04) = 1.83$$
$$G_{B \text{ inelastic}} = \beta_s G_{B \text{ elastic}} = 0.90(0.83) = 0.75$$

Using the alignment chart yields

$$K = 1.37$$

This compares to an elastic effective length factor $K = 1.41$.

6.9 COLUMNS WITH SLENDER ELEMENTS

6.9.1 Plate Buckling

The discussion of columns presented earlier in this chapter was based on the assumption that the failure limit state would be column buckling, either elastic or inelastic. This is a reasonable assumption since the majority of sections used as columns will behave that way. There is, however, the possibility that for a particular hot-rolled section with a given yield strength or a built-up member composed of specific plates, an individual compression element may have a slenderness that leads to local buckling prior to attaining the full column strength.

Slenderness of individual plate elements controls local buckling in the same way that it controlled column buckling. The slenderness for these plates as elements of a structural shape is defined as the width/thickness ratio b/t, and is shown in Fig. 6.37 for selected shapes. It is evident from the figure that there are two distinct types of element. The plate supported along one edge, such as the flange of the W-shape, is referred to as an unstiffened element, while the plate supported along two edges, such as the web of the W-shape, is called a stiffened element.

Plates under uniform compression can be shown to behave similarly to columns. The critical elastic buckling stress for a plate is given by Bleich [44] and Timoshenko and Gere [45] as

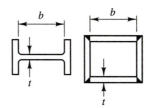

(a) Elements supported along
two edges (stiffened elements)

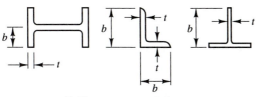

(b) Elements supported along
one edge (unstiffened elements)

Figure 6.37 Stiffened and
unstiffened
compression
elements.

$$F_{cr} = \frac{k\pi^2 E}{12(1 - v^2)(b/t)^2} \qquad (6.29)$$

where k is a constant that reflects edge condition, and v is Poisson's ratio. To insure that local buckling will not occur at a load level below the capacity of the column, Eq. (6.29) could be set equal to the column strength, and the corresponding b/t determined. This would result in a different limiting width/thickness ratio for every column slenderness, which is not a particularly useful situation. To simplify the problem, the plate buckling stress could be limited to F_y, thereby assuring that the column would buckle prior to plate yielding. This too will lead to a fairly conservative limitation on plate slenderness since, as has been shown earlier in this chapter, column buckling takes place well below the yield stress. The result, then, is to establish an arbitrary point at which the plate strength and the column strength would be set equal. The limitations given in the LRFD specification have been established for a plate slenderness that is approximately 0.7 times the slenderness, which will equate plate buckling with yield.

The limiting slenderness ratio, above which local buckling must be considered, is given as λ_r. For unstiffened compression elements it is limited to

$$\text{Flanges} \qquad \frac{95}{\sqrt{F_y}}$$

$$\text{Angle leg} \quad \frac{76}{\sqrt{F_y}}$$

$$\text{T stem} \quad \frac{127}{\sqrt{F_y}}$$

Other conditions are given in Section B5 of the LRFD specification.

6.9.2 Column Strength with Slender Elements

Although it would be preferable that local buckling not control a column design, it must be recognized that this will not always be the case. When local buckling cannot be avoided, the strength of the column must be reduced to account for its influence. This is done by modifying the inelastic buckling equation, Eq. (6.23), and the range over which it is applicable. Using the reduction factor Q, the column equations become

for $\lambda_c \sqrt{Q} \leq 1.5$

$$F_{cr} = Q(0.658^{Q\lambda_c^2}) F_y \tag{6.30}$$

for $\lambda_c \sqrt{Q} > 1.5$

$$F_{cr} = \left(\frac{0.877}{\lambda_c^2}\right) F_y \tag{6.31}$$

For cross sections composed entirely of unstiffened elements, $Q = Q_s$; for cross sections composed entirely of stiffened elements, $Q = Q_a$; and for sections composed of both stiffened and unstiffened elements, $Q = Q_s Q_a$. The value of Q_s is dependent on the type of unstiffened element under consideration. The LRFD specification provides the following:

For single angles:

When $76.0/\sqrt{F_y} < b/t < 155/\sqrt{F_y}$

$$Q_s = 1.340 - 0.00447(b/t) \sqrt{F_y}$$

When $b/t \geq 155/\sqrt{F_y}$

$$Q_s = \frac{15,500}{F_y(b/t)^2}$$

For angles or plates projecting from columns and for projecting elements of compression flanges of girders:

When $95.0/\sqrt{F_y} < b/t < 176/\sqrt{F_y}$

$$Q_s = 1.415 - 0.00437(b/t)\sqrt{F_y}$$

When $b/t \geq 176/\sqrt{F_y}$

$$Q_s = \frac{20,000}{F_y(b/t)^2}$$

For stems of tees:

When $127/\sqrt{F_y} < b/t < 176/\sqrt{F_y}$

$$Q_s = 1.908 - 0.00715(b/t)\sqrt{F_y}$$

When $b/t \geq 176/\sqrt{F_y}$

$$Q_s = \frac{20,000}{F_y(b/t)^2}$$

For stiffened elements, Q_a is taken as the ratio of the effective area to the actual area of the section. The effective area for a W-shape is found by combining the flange areas with the effective web area determined as $b_e t$, where the effective width is given as

$$b_e = \frac{326t}{\sqrt{f}} \left[1 - \frac{57.2}{(b/t)\sqrt{F_y}} \right] \leq b$$

where b is the actual width, t is the actual thickness, and f is the computed elastic compressive stress in the element.

EXAMPLE 6.7

Given: A WT compression member is found in the roof truss of a building. The member is a WT5 × 13, 10 ft in length. Lateral support is provided about the y axis at the center and about the x axis at the ends. Determine the design strength of the compression member.

Solution

(a) Use $F_y = 50$ ksi
    Check the width/thickness ratios:
    For the stem of the WT

$$\lambda = \frac{b}{t} = \frac{5.165}{0.26} = 19.86 > \frac{127}{\sqrt{F_y}} = 17.96$$

For the flange

$$\lambda = \frac{b}{t} = \frac{5.770}{2(0.44)} = 6.56 < \frac{95}{\sqrt{F_y}} = 13.43$$

Thus, the stem is a critical element.
Since

$$\frac{127}{\sqrt{F_y}} < 19.86 < \frac{176}{\sqrt{F_y}}$$

$$Q_s = 1.908 - 0.00715(19.86)\sqrt{50} = 0.904$$

Determining the critical slenderness gives

$$\frac{KL}{r_x} = \frac{10.0(12)}{1.44} = 83.3$$

and

$$\frac{KL}{r_y} = \frac{5.0(12)}{1.36} = 44.12$$

Thus, the x axis controls and

$$\lambda_c = 83.3 \sqrt{\frac{50}{\pi^2(29,000)}} = 1.10$$

and

$$\lambda_c \sqrt{Q} = 1.10\sqrt{0.904} = 1.05 < 1.5$$

Therefore,

$$F_{cr} = 0.658^{Q\lambda^2} QF_y = (0.658^{1.05^2})(0.904)(50)$$
$$= 28.6 \text{ ksi}$$

and

$$\phi F_{cr} = 0.85(28.6) = 24.3 \text{ ksi}$$

and

$$\phi P_n = 3.81 \text{ in.}^2 (24.3 \text{ ksi}) = 92.6 \text{ kips}$$

(b) Use $F_y = 36$ ksi
Check the width/thickness ratios:
For the stem of the WT

$$\lambda = \frac{b}{t} = \frac{5.165}{0.26} = 19.86 < \frac{127}{\sqrt{F_y}} = 21.2$$

For the flange

$$\lambda = \frac{b}{t} = \frac{5.770}{2(0.44)} = 6.56 < \frac{95}{\sqrt{F_y}} = 15.83$$

Thus, local buckling will not be critical.

For the critical x axis slenderness,

$$\frac{KL}{r_x} = \frac{10.0(12)}{1.44} = 83.3$$

$$\lambda_c = 83.3 \sqrt{\frac{36}{\pi^2(29,000)}} = 0.934 < 1.5$$

Therefore,

$$F_{cr} = 0.658^{\lambda_c^2} F_y = (0.658^{0.934^2})(36) = 24.98 \text{ ksi}$$

and

$$\phi F_{cr} = 0.85(24.98) = 21.23 \text{ ksi}$$

and

$$\phi P_n = 3.81 \text{ in.}^2 (21.23 \text{ ksi}) = 80.9 \text{ kips}$$

6.10 TORSIONAL BUCKLING

6.10.1 Limit States

Two limit states have yet to be addressed; torsional buckling and flex-ural–torsional buckling. Doubly symmetric shapes, such as the W-shapes normally used for columns, may fail either through flexural buckling, as discussed earlier in this chapter, or through torsional buckling. Singly symmetric shapes (double angles and T sections) and nonsymmetric shapes (single unequal-leg angles) may fail by flexural, torsional, or flex-ural–torsional buckling. The consideration of either torsional mode of failure for a compression member is a complex undertaking. If at all possible, it is desirable to reduce the possibilities of torsional failure by proper bracing of the column.

When torsion cannot be restricted, calculations must be performed to determine the limiting failure mode. A new property, referred to as the

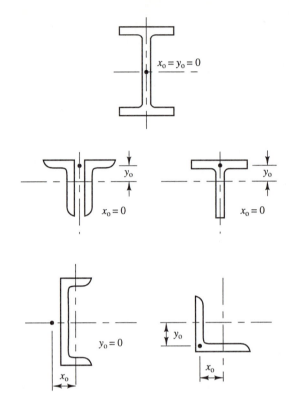

Figure 6.38 Location of shear center.

shear center, for singly symmetric and nonsymmetric sections must be introduced. The shear center is that point through which all lateral loads must pass in order that no torsion be induced in the cross section. For doubly symmetric shapes it coincides with the centroid, for singly symmetric shapes it will fall on the axis of symmetry, and for nonsymmetric shapes it will fall at some distance from each centroidal axis. Figure 6.38 shows the locations for the shear center for selected shapes.

Although the shear center is determined with respect to torsional loads, a column may buckle in a torsional mode, even when it is loaded concentrically. Torsional and flexural–torsional buckling are functions of the cross section rather than the point of application of the axial loads.

6.10.2 Design

To account for the reduction in capacity of a column controlled by one of the torsional modes of failure, equations of the form presented in Section 6.9.2 will be used, with λ_c being replaced by λ_e, where $\lambda_e = \sqrt{F_y/F_e}$. F_e is the critical torsional or flexural–torsional elastic buckling stress as defined by the LRFD specification, Appendix E. For doubly symmetric shapes, the critical torsional elastic buckling stress is

$$F_e = \left[\frac{\pi^2 E C_w}{(K_z L)^2} + GJ\right]\frac{1}{I_x + I_y} \tag{6.32}$$

For singly symmetric shapes with the y axis defined as the axis of symmetry, the critical flexural–torsional elastic buckling stress is

$$F_e = \frac{F_{ey} + F_{ez}}{2H}\left[1 - \sqrt{1 - \frac{4F_{ey}F_{ez}H}{(F_{ey} + F_{ez})^2}}\right] \tag{6.33}$$

where

$$K_z = \text{effective length factor for torsional buckling}$$
$$E = \text{modulus of elasticity, ksi}$$
$$G = \text{shear modulus, ksi}$$
$$C_w = \text{warping constant, in.}^6$$
$$J = \text{torsional constant, in.}^4$$
$$I_x, I_y = \text{moment of inertia about principal axes, in.}^4$$
$$x_o, y_o = \text{coordinates of shear center with respect to centroid, in.}$$

and

$$r_o^2 = x_o^2 + y_o^2 + \frac{I_x + I_y}{A}$$

$$H = 1 - \left(\frac{x_o^2 + y_o^2}{r_o^2}\right)$$

$$F_{ey} = \frac{\pi^2 E}{(K_y L/r_y)^2}$$

$$F_{ez} = \left[\frac{\pi^2 E C_w}{(K_z L)^2} + GJ\right]\frac{1}{A r_o^2}$$

For unsymmetrical shapes, reference should be made to Appendix E of the LRFD specification.

EXAMPLE 6.8

Given: Determine the capacity of the WT5 × 13 from Example 6.7 considering the limit state of flexural–torsional buckling. Assume the torsional length $L_z = 10.0$ ft.

Solution

From Example 6.7

$$Q = 0.904$$
$$KL/r_x = 83.3$$
$$KL/r_y = 44.12$$

From the *LRFD Manual*

$$J = 0.201$$
$$C_w = 0.173$$
$$r_o = 2.15$$
$$H = 0.848$$

Thus,

$$F_{ey} = \pi^2 \frac{(29,000)}{(44.12)^2} = 147.04 \text{ ksi}$$

$$F_{ez} = \left\{ \frac{\pi^2(29,000)(0.173)}{[(10.0)(12)]^2} + 12,000(0.201) \right\} \frac{1}{3.81(2.15)^2}$$

$$F_{ez} = 137.15 \text{ ksi}$$

and

$$F_e = 102.0$$

which yields

$$\lambda_e = \sqrt{50/102.0} = 0.700$$

$$\lambda_e \sqrt{Q} = 0.700\sqrt{(0.904)} = 0.666 < 1.5$$

Therefore,

$$F_{cr} = 0.904(0.658^{(0.904)(0.700)^2})(50) = 37.55 \text{ ksi}$$

and

$$\phi F_{cr} = 0.85(37.55) = 31.9 \text{ ksi}$$

Since $\phi F_{cr} = 31.9 > \phi F_{cr} = 21.23$ from Example 6.7, flexural–torsional buckling is not critical. Thus, $\phi P_n = 92.7$, as found in Example 6.7, is controlled by local buckling ($\phi F_{cr} = 21.23$).

6.11 BUILT-UP COLUMNS

Many situations occur in actual structures where an available rolled shape
is inappropriate for the specific application under consideration. Examples include the lower story columns of a building where loads may be
quite high, requiring the addition of plates to a wide-flange shape, and the
members of a truss where construction details lead to the selection of
multiple shapes combined to form a single member. Cross sections for
some typical built-up compression members are shown in Figure 6.39.

Determination of the critical buckling mode for built-up columns is
essentially the same as it was for single shape columns, with the addition
of a check on slenderness of the individual elements. It is most desirable
that a built-up column function as a single column over its entire length,
rather than as an individual column over the length between connections
of the elements. To assure this type of behavior, the slenderness of the
column must be greater than or equal to the slenderness of the most
slender segment. In addition, to account for shearing deformation in the
connectors between individual shapes in the buckled configuration, a
modified slenderness must be used.

If two channels are placed back-to-back but are not attached, as
shown in Fig. 6.40a, and a concentric compressive load is applied, the
channels can be expected to buckle, as shown in Fig. 6.40b. For this case,
the critical slenderness will be found using the total length and the radius
of gyration of a single channel; the capacity of the two channels is equal to
exactly twice that of a single channel. Note that the two channels slip past
each other along the contact surface, with the greatest slip occurring at
the ends and zero slip at the midpoint of the length. To force these shapes
to work together, they must be attached along their length in such a way
as to prevent this slip.

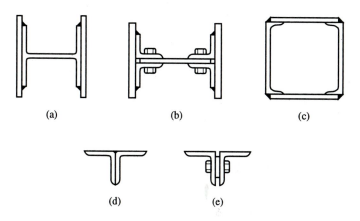

Figure 6.39 Built-up column
cross sections.

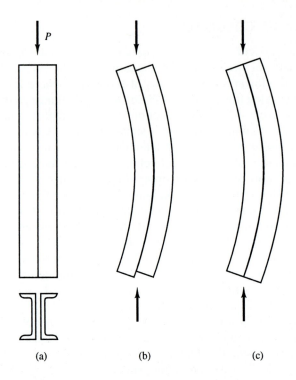

Figure 6.40 Buckling of
built-up section.

        (a)   (b)   (c)

     The resulting buckled shape will be as shown in Fig. 6.40c. Actual
construction practice would have the shapes connected at some interval
along their length and may be such as to permit some slip along the shear
plane. To account for this slip, a modified slenderness, $(KL/r)_m$, must be
used to determine column strength. For snug-tight bolted connections,
that modified slenderness is

$$\left(\frac{KL}{r}\right)_m = \sqrt{\left(\frac{KL}{r}\right)_o^2 + \left(\frac{a}{r_i}\right)^2}$$

For welded connections and for fully tightened bolts, with $a/r_i > 50$

$$\left(\frac{KL}{r}\right)_m = \sqrt{\left(\frac{KL}{r}\right)_o^2 + \left(\frac{a}{r_i} - 50\right)^2}$$

with $a/r_i \leq 50$

$$\left(\frac{KL}{r}\right)_m = \left(\frac{KL}{r}\right)_o^2$$

where

$\left(\dfrac{KL}{r}\right)_o$ = largest column slenderness of built-up member acting as a unit

$\dfrac{a}{r_i}$ = largest column slenderness of individual components

$\left(\dfrac{KL}{r}\right)_m$ = modified column slenderness of built-up member

a = distance between connectors

r_i = minimum radius of gyration of individual component

Once the critical slenderness is determined, column capacity may be determined as a function of the controlling limit states, as previously discussed.

EXAMPLE 6.9

Given: Determine the capacity of the column built-up from two MC10×22 channels, $\frac{3}{8}$ in. back-to-back, as shown in Fig. 6.40. The overall column length is 16 ft. The individual channels are connected with snug-tight bolts at 4-ft intervals.

Solution

$$A = 2(6.45) = 12.9 \text{ in.}^2$$

$$I_x = 2(103) = 206.0 \text{ in.}^4$$

$$r_x = 3.99 \text{ in.}$$

$$I_y = 30.89 \text{ in.}^4$$

$$r_y = 1.55 \text{ in.}$$

For the combined channels

$$\frac{KL}{r_x} = \frac{16.0(12)}{3.99} = 48.12$$

$$\frac{KL}{r_y} = \frac{16.0(12)}{1.55} = 123.87$$

For individual channels

$$r_i = r_y = 1.0$$

$$\frac{a}{r_i} = \frac{4.0(12)}{1.0} = 48.0$$

Since

$$\frac{a}{r_i} = 48.0 < \frac{KL}{r_y} = 123.87$$

the column will behave as a built-up member. The modified slenderness to be used for strength calculations is thus

$$\left(\frac{KL}{r}\right)_m = \sqrt{\left(\frac{KL}{r}\right)_o^2 + \left(\frac{a}{r_i}\right)^2}$$
$$= \sqrt{123.87^2 + 48.0^2}$$
$$= 132.8$$

PROBLEMS

6.1. Determine the axial load capacity for a W14×109, A36 steel with an effective length of 12 ft about the y axis and 24 ft about the x axis.

6.2. A W14×43, A36 steel, column is to be used in a frame where the effective length about each axis is 20 ft. Determine the axial load capacity.

6.3. Determine the axial load capacity for a W12×45 column with $F_y = 50$ ksi. The effective length is 20 ft about the y axis and 40 ft about the x axis. Is this an elastic or inelastic buckling condition?

6.4. A W8×24 column, with $F_y = 50$ ksi, has an effective length of 12.5 ft about the y axis and 28 ft about the x axis. Determine the axial load capacity and indicate if this is due to elastic or inelastic buckling.

6.5. Determine the axial load capacity for a W10×112 column, A36 steel, with an effective length equal to 8 ft about the y axis and equal to 12 ft about the x axis. What maximum effective length could be used about the x axis without altering the column capacity?

6.6. A W14×370, A36, is used as a column in a building with a maximum effective length of 16 ft. Determine if this column will adequately carry an axial live load of 1200 kips and a dead load of 1000 kips.

6.7. For the W10×39 column, with bracing and end conditions shown below, determine the theoretical effective length for each axis and identify the axis that will limit the column capacity.

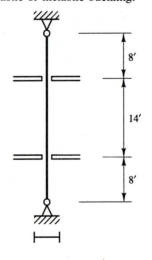
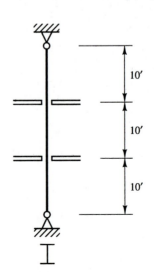

6.8. A W10×45 column with end conditions and bracing is shown. Determine the least theoretical bracing and its location about the y axis, in order that the y axis not control the capacity of the column.

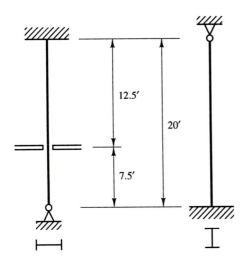

6.9. For design purposes, the commentary to the LRFD specification recommends effective length factors when ideal end conditions are approximated (Fig. 6.32). Determine the effective length for a 20-ft column with fixed ends for buckling about the y axis and pinned and fixed for buckling about the x axis.

6.10. A W10×60 column is called upon to carry an axial live load of 170 kips and an axial dead load of 123 kips. If the effective length is 15 ft, (a) will the column carry the load if $F_y = 36$ ksi? (b) Will the column carry the load if $F_y = 50$ ksi?

6.11. A W8×40 is used as a 12-ft column in a braced frame with W16×26 beams spanning 24 ft, framing in, from both sides, at the top and bottom and providing moment restraint. If the columns above and below are also W8×40, determine the effective length using the nomograph of Fig. 6.34, assuming that the columns are oriented for (a) buckling about the weak axis and (b) buckling about the strong axis. In each case, determine the axial load capacity in the plane being considered. All steel is A36.

6.12. If the structure described in Problem 6.11 is an unbraced frame, determine the effective lengths and axial load capacity as requested in Problem 6.11.

6.13. Select the least weight W12 column to carry a live load of 130 kips and a dead load of 100 kips with an effective length about both axes of 14 ft.

6.14. A W12 column, A36 steel, must carry a live load of 150 kips, a dead load of 130 kips, and a wind load of 175 kips. The effective length is 24 ft about the x axis and 12 ft about the y axis. Determine the least weight W12 column to carry the load.

6.15. A pin-ended column is part of a braced frame. It must carry a design live load of 45 kips and a calculated dead load of 20 kips. The column is 14 ft long. Determine the lightest A36 steel W-shape to carry the load.

6.16. If the column described in Problem 6.15 is 28 ft long, determine the least weight section, with $F_y = 50$ ksi, to carry the load.

6.17. A W12×50 column is an interior column with strong axis buckling in the plane of the frame in an unbraced multistory frame, similar to that shown in Fig. 6.35. The columns above and below are also W12×50. The beams framing in at the top are W16×31 and those at the bottom are W16×36. The columns are 12 ft and the beams span is 22 ft. The column carries a factored load of 300 kips. Determine the inelastic effective length for this condition and the corresponding axial load capacity.

6.18. The plan for a one-story office building is shown. The roof consists of rigid insulation on metal deck supported on open web steel joists with a total dead load of 28 psf. The roof live load is 30 psf. The beams along column lines B and C weigh 25 plf. Columns B2 and C2 are 18 ft in length with pinned ends. Columns B3 and C3 are 24-ft long with pinned ends. Determine the least weight columns for these four locations. Use A36 steel.

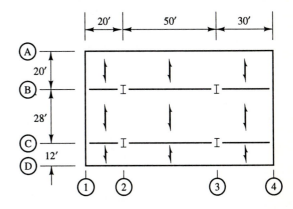

6.19. Determine the design strength of a WT6×11, F_y = 50 ksi, and a length of 12 ft. Assume that the member is pinned and braced at the ends for both axes and braced at the midlength for the *y* axis. For torsion, the length should be taken as 12 ft.

6.20. An A36 compression member in a roof truss is to be a WT4 × 5 with a length of 8 ft. Buckling and torsional restraint are provided at the ends of the member only. Determine the design strength if the ends are considered pinned.

6.21. Two MC8 × 20 channels are used as a 24-ft pin-ended column in a warehouse. They are spaced $\frac{3}{8}$ in. back to back and are connected every 6 ft by snug-tight bolts. Determine the capacity of the built-up column if $F_y = 36$ ksi.

REFERENCES

[1] Tebedge, N., and Tall, L., *Residual Stresses in Structural Steel Shapes—A Summary of Measured Values*, Fritz Engineering Laboratory Report No. 337.34, Lehigh University, Bethlehem, PA, February 1973. (Also published under the title Contraintes Résiduelles dans les Profilés en Acier: Synthèse des Valeurs Mesurées, in *Construction Métallique,* Vol. 11, No. 2, 1974 (pp. 37–47).)

[2] Bjorhovde, R., *Deterministic and Probabilistic Approaches to the Strength of Steel Columns*, Thesis, Lehigh University, Bethlehem, PA, May 1972.

[3] Bjorhovde, R., Brozzetti, J., Alpsten, G. A., and Tall, L., Residual stresses in thick welded plates, *Welding Journal, American Welding Society (AWS),* Vol. 51, No. 8, August 1972 (pp. 392s–405s).

[4] Galambos, T. V., Ed., *Guide to Stability Design Criteria for Metal Structures*, 4th ed., Wiley-Interscience, New York, 1988.

[5] Beedle, L. S., and Tall, L., Basic column strength, *Journal of the Structural Division, ASCE,* Vol. 86, No. ST7, July 1960 (pp. 139-173).

[6] McFalls, R. K., and Tall, L., A study of welded columns manufactured from flame-cut plates, *Welding Journal, American Welding Society (AWS),* Vol. 48, No. 4, April 1970 (pp. 141s–153s).

[7] American Society for Testing and Materials (ASTM), *General Delivery Requirements for Structural Steel*, ASTM Standard No. A6-88, ASTM, Philadelphia, PA, 1988.

[8] Yu, W. W., *Cold-Formed Steel Design*, 2d ed., Wiley-Interscience, New York, 1991.

[9] American Iron and Steel Institute (AISI), *Load and Resistance Factor Design Specification for the Design of Cold-Formed Steel Structural Members*, AISI, Washington, DC, 16 March 1991.

[10] Alpsten, G. A., Residual stresses and mechanical properties of cold straightened H-shapes, *Jernkontorets Annaler,* Swedish Institute of Steel Construction, Stockholm, Sweden, Vol. 154, 1970 (pp. 255–283) (in Swedish; with substantial summary in English).

[11] Bjorhovde, R., Columns: From theory to practice, *Engineering Journal, AISC*, Vol. 25, No. 1, First Quarter, 1988 (pp. 21–34).

[12] Chernenko, D. E., and Kennedy, D. J. Laurie, An analysis of the performance of welded wide-flange columns, *Canadian Journal of Civil Engineering,* Vol. 18, No. 4, August 1991 (pp. 537–555).

[13] Batterman, R. H., and Johnston, B. G., Behavior and maximum strength of metal columns, *Journal of the Structural Division, ASCE,* Vol. 93, No. ST2, February 1967 (pp. 205–230).

[14] Bjorhovde, R., and Birkemoe, P. C., Limit states design of HSS columns, *Canadian Journal of Civil Engineering,* Vol. 6, No. 2, June 1979 (pp. 276–291).

[15] Canadian Standards Association (CSA), *Steel Structures for Buildings—Limit States Design*, CSA Standard No. CAN3-S16.1-M90, CSA, Rexdale, Ontario, Canada, 1990.

[16] Commission of the European Communities (EC), *Eurocode 3: Design of Steel Structures* EC, Brussels, Belgium, June 1992.

[17] Galambos, T. V., *Structural Members and Frames*, Prentice-Hall, Englewood Cliffs, NJ, 1968.

[18] Johnston, B. G., Column buckling theory: Historical highlights, *Journal of the Structural Division, ASCE,* Vol. 107, No. ST4, April 1981 (pp. 649–670).

[19] Johnson, J. B., Bryan, C. W., and Turneaure, F. E., *The Theory and Practice of Modern Framed Structures*, 6th ed., Wiley, New York, 1897.

[20] Chen, W. F., and Lui, E. M., *Structural Stability. Theory and Implementation*, Elsevier Applied Science, London, 1989.

[21] Beedle, L. S., Editor-in-Chief, *Structural Design of Tall Steel Buildings*, Vol. SB, Monograph on Planning and Design of Tall Buildings, American Society of Civil Engineers, New York, 1979.

[22] Shanley, F. R., Inelastic column theory, *Journal of Aeronautical Science,* Vol. 14, No. 5, May 1947 (pp. 261–267).

[23] Johnston, B. G., Buckling behavior above the tangent modulus load, *ASCE Transactions,* Vol. 128, Part I, 1963 (pp. 819–848).

[24] American Institute of Steel Construction, Inc. (AISC), *Specification for the Allowable Stress Design, Fabrication and Erection of Steel Structures for Buildings*, AISC, Chicago, IL, 1 June 1989.

[25] Beer, H., and Schulz, G. W., Theoretical bases of the European column buckling curves, *Construction Métallique, Paris, France,* No. 3, 1970 (pp. 37–55) (in French).

[26] Beaulieu, D., and Adams, P. F., The results of a survey on structural out-of-plumbs, *Canadian Journal of Civil Engineering,* Vol. 5, No. 4, December 1978 (pp. 462–470).

[27] Rondal, J., and Maquoi, R., Single equation for SSRC column strength curves, *Journal of the Structural Division, ASCE,* Vol. 105, No. STl, January 1979 (pp. 247–250).

[28] Rotter, J. M., Multiple column curves by modifying factors, *Journal of the Structural Division, ASCE,* Vol. 108, No. ST7, July 1982 (pp. 1665–1669).

[29] Tide, R. H. R., *Reasonable Column Design Equations*, Annual Technical Session of Structural Stability Research Council, 16–17 April 1985.

[30] Hassan, K., *On the Determination of Buckling Length of Frame Columns*, Publications, International Association for Bridge and Structural Engineering, 28-II, 1968, 91-101 (in German).

[31] Galambos, T. V., Influence of partial base fixity on frame stability, *Journal of the Structural Division, ASCE,* Vol. 86, No. ST5, May 1960 (pp. 85–108).

[32] Gurfinkel, G., and Robinson, A. R., Buckling of elastically restrained columns, *Journal of the Structural Division, ASCE,* Vol. 91, No. ST6, December 1965 (pp. 159–183).

[33] Switsky, H., and Ping Chun Wang, Design and analysis of frames for stability, *Journal of the Structural Division, ASCE,* Vol. 95, No. ST4, April 1969 (pp. 695–713).

[34] Lu, Le-Wu, Effective length of columns in gable frames, *Engineering Journal, AISC,* Vol. 2, No. 1, January 1965 (pp. 6–7).

[35] Kavanagh, T. C., Effective length of framed columns, *Transactions, ASCE,* Vol. 127, 1962, Part II (pp. 81–101).

[36] Dumonteil, P., Simple equations for effective length factors, *Engineering Journal, AISC,* Vol. 29, No. 3, Third Quarter, 1992 (pp. 111–115).

[37] Yura, J. A., The effective length of columns in unbraced frames, *Engineering Journal, AISC,* Vol. 8, No. 2, April 1971 (pp. 37–42); Discussion, Vol. 9, No. 3, October 1972 (pp. 167–168).

[38] Adams, P. F., Discussion of "The effective length of columns in unbraced frames," by Joseph A. Yura, *Engineering Journal, AISC,* Vol. 9, No. 1, January 1972 (pp. 40–41).

[39] Johnston, B. G., Discussion of "The effective length of columns in unbraced frames," by Joseph A. Yura, *Engineering Journal, AISC,* Vol. 9, No. 1, January 1972 (p. 46).

[40] Disque, R. O., Inelastic K-factor for column design, *Engineering Journal, AISC,* Vol. 10, No. 2, Second Quarter, 1973 (pp. 33–35).

[41] Smith, C. V., Jr., On inelastic column buckling, *Engineering Journal, AISC,* Vol. 13, No. 3, Third Quarter, 1976 (pp. 86–88); Discussion, Vol. 14, No. 1, First Quarter, 1977 (pp. 47–48).

[42] Matz, C. A., Discussion of "On inelastic column buckling," by C. V. Smith, Jr., *Engineering Journal, AISC,* Vol. 14, No. 1, First Quarter, 1977 (pp. 47–48).

[43] Stockwell, F. W., Jr., Girder stiffness distribution for unbraced columns, *Engineering Journal, AISC,* Vol. 13, No. 3, Third Quarter, 1976 (pp. 82–85).

[44] Bleich, F., *Buckling Strength of Metal Structures,* McGraw-Hill, New York, 1952.

[45] Timoshenko, S. P., and Gere, J. M., *Theory of Elastic Stability,* 2d. ed., McGraw-Hill, New York, 1961.

CHAPTER SEVEN

Bending Members

7.1 THE BASIC STRUCTURAL ELEMENT

A beam is a bending member that transfers load applied normal to its longitudinal axis, to its support points. In building construction, the most common application of beams is to provide support for floors or roofs. They may be simple span or continuous and normally transfer their load to other structural members, such as columns, girders, or walls. Although the terms beams and girders are often used interchangeably since both are bending members, a beam normally refers to a bending member directly supporting an applied load, while a girder refers to a bending member that supports a beam.

The maximum bending moment that a steel section can resist is defined by its plastic moment. Almost all wide-flange shapes have relatively stocky flanges and webs and will be able to develop this plastic moment.

As load is applied to a bending member with the resulting bending moment, stresses are developed in the cross section. For loads at or below the working load, the entire cross section behaves elastically, with the resulting stresses distributed as shown in Fig. 7.1a.

The relationship between the applied moment and the resulting stresses is given by

$$f_y = \frac{My}{I} \tag{7.1}$$

where

M = any applied moment that will stress the section in the elastic range

y = the distance from the neutral axis to the point where the stress is to be determined

I = moment of inertia (property of the cross section)

f_y = the resulting bending stress at location y

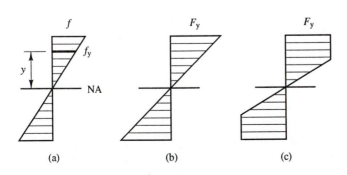

Figure 7.1 Cross-sectional bending stresses: (a) elastic; (b) yield; (c) partial plastic.

Since the normal case of interest is the stress at the extreme fiber, most distant from the neutral axis, the equation may be written as

$$f_b = \frac{M}{S} \tag{7.2}$$

where

  S = section modulus (property of cross section)
  f_b = extreme fiber bending stress

The moment that will cause the extreme fiber to reach the yield stress F_y is called the yield moment M_y; it is shown in Fig. 7.1b. As the load is increased beyond the yield moment, the strain in the extreme fiber increases while the stress remains at F_y, and the stress at some points closer to the neutral axis will also reach the yield stress, as shown in Fig. 7.1c.

If the load continues to increase, the portion of the cross section to reach yield will continue to increase until the entire section is experiencing the yield stress. Equilibrium requires that the total tension force must be equal to the total compression force. For the doubly symmetrical wide-flange shape shown in Fig. 7.2, this occurs when the portion above the neutral axis (NA) is stressed to yield in compression while the area below the NA is stressed to yield in tension. For a nonsymmetric shape, the area above the NA is not necessarily equal to the area below the NA. Thus, a new axis must be defined, which divides the tension and compression zones into two equal areas; this is the plastic neutral axis (PNA).

If moments are taken about this PNA, the moment that corresponds to the given stress distribution will be obtained. This is defined as the plastic moment M_p, and is given as

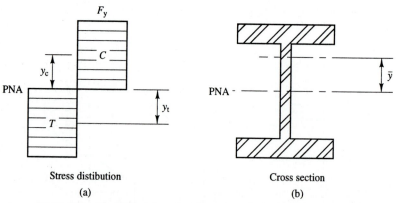

Stress distibution
(a)

Cross section
(b)

Figure 7.2 Equilibrium in a doubly symmetrical wide-flange shape.

$$M_p = F_y (A_c y_c) + F_y (A_t y_t)$$

$$M_p = F_y \left(\frac{A}{2}\right) (y_c + y_t) \tag{7.3}$$

Where A_t and A_c represent the tension and compression area, respectively, while y_t and y_c represent the distance from the PNA to the centroid of the tension and compression area, respectively.

The terms that are multiplied by the yield stress are normally combined, since they refer entirely to the cross-sectional dimensions, and are called the plastic section modulus Z. Thus, the plastic moment, now referred to as the nominal moment, is given as

$$M_n = M_p = ZF_y$$

and the design moment is ϕM_n where $\phi = 0.9$.

For a given beam to attain its full plastic moment capacity, it must satisfy a number of criteria. If these criteria are not met the nominal capacity will be defined as some reduced portion of M_p. The criteria to be addressed are defined by two limit states: local buckling and lateral torsional buckling. Each of these limit states and their impact on beam design are discussed later in this chapter.

7.2 BEHAVIOR OF BEAMS

For the designer to determine if a particular steel section is adequate to carry all the factored loads that are exerted on it, an analysis to determine the maximum applied moment is performed. The factored moment from analysis is compared to the design moment, which is the ability of the section to carry load. If it is less, then the member is satisfactory. The design moment is equal to ϕM_n, where the nominal moment M_n is dependent on the controlling limit state of the particular section: local buckling or lateral torsional buckling.

7.2.1 Local Buckling

Local buckling (LB) is a failure of an element of the cross section so that the section shape has been altered to the point that it can no longer carry load. Figure 7.3a illustrates local buckling of the compression flange of a wide-flange beam, and Fig. 7.3b shows a similar failure of the web. These failures occur when the flange or web are very slender. Since one end of the projecting flange is supported by the web, it is treated as an "unstiffened projecting" element. On the other hand, since the web is connected at both its ends to a flange, it is treated as a "stiffened" element. Similarly, for a square tube loaded as a beam, both its vertical sides (webs) and the horizontal side in compression (flange) would be treated as stiffened elements.

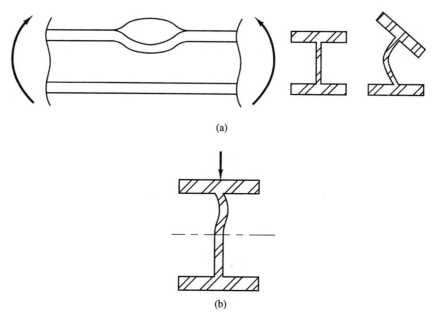

(a)

(b)

Figure 7.3   Local buckling of a wide-flange beam (a) the compression flange and
(b) the web.

There has been considerable research into the capacity of various
steel cross sections [*1–3*] and, as a result, local buckling rules have been
well established. In the LRFD specification, shapes are classified into
three categories: compact, noncompact, and slender. Each of these is
discussed later in this chapter and in Chapter 8.

7.2.2 Lateral–Torsional Buckling

For beams whose compression flange is not sufficiently prevented from
moving laterally or twisting as the load increases, the beam will first
deflect in plane, then twist and finally buckle laterally. From a conceptual
point of view, the beam can be considered as having a constant tendency
to fall over on its weak axis. This failure mode or limit state is called
lateral–torsional buckling (LTB). The three positions of a beam cross
section shown in Fig. 7.4 illustrate the displacement and rotation that
takes place as the midspan of the beam undergoes lateral–torsional buckling.

Resistance of a bending member to lateral–torsional buckling
between points of lateral or torsional support is a function of two torsional
characteristics of the specific cross section. These are described as St.
Venant torsion and warping torsion. St. Venant torsion, sometimes referred to as ''pure'' torsion, is a function of the shear modulus of the
material G and the torsional constant J. Taken together, these two properties provide resistance to shearing in a torsional mode (Fig. 7.5a). Warp-

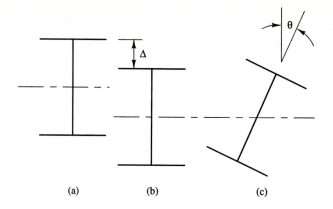

(a)   (b)   (c)

Figure 7.4 The three positions of a beam cross section undergoing lateral–torsional buckling.

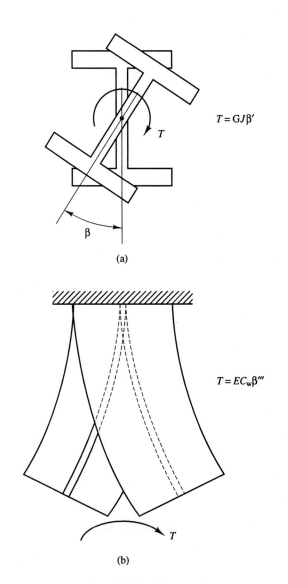

$T = GJ\beta'$

(a)

$T = EC_w\beta'''$

(b)

Figure 7.5 Resistance of a bending member to lateral–torsional buckling.

ing torsional resistance is provided by bending stiffness of the flanges in a horizontal plane (Fig. 7.5b). This resistance is defined by the material modulus of elasticity E and the warping constant C_w. The angle of rotation is taken as β, where β' is the first derivative of the angle of rotation with respect to the applied torque T and β''' is the third derivative of the angle of rotation with respect to the applied torque. Taken together, these two characteristics provide a measure of a section's ability to resist torsion.

The limit state of lateral–torsional buckling is dependent not only on the degree of torsional restraint, but also on the moment gradient along the span. Figure 7.6 illustrates a beam that is called upon to resist the same maximum moment but with three different loading conditions, re-

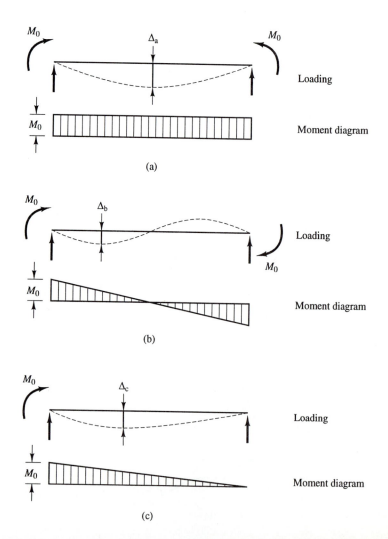

Figure 7.6 Resistance to the maximum moment under three different loading conditions.

sulting in three very different moment diagrams. The maximum elastic extreme fiber bending stress would be the same in all three cases shown.

Even though the maximum moment and the resulting maximum bending stress are the same in all three cases, the maximum in-plane deflection and the deflected shape are quite different because of the difference in the moment gradient. The tendency of the beam to twist, therefore, is also different. Thus, the load-carrying capacity of the beams in Figs. 7.6b and 7.6c can be expected to exceed that of the beam in Fig. 7.6a.

Design rules for LTB, which incorporate the effects of both torsional restraint and moment gradient, are described in Section 7.6.

7.2.3 Shear

Although uncommon with rolled sections, it is possible for a beam to fail by excessive shear yielding on the gross area of the web (Fig. 7.7a) or by shear fracture on the net area of the web (Fig. 7.7b).

The nominal shear yielding strength is based on the von Mises criterion, which states that for an unreinforced beam web that is stocky enough not to fail by buckling, the shear strength may be taken as $F_y/\sqrt{3} = 0.58F_y$. On this basis, the LRFD specification stipulates that shear yielding on the gross area can occur if

$$\frac{h}{t_w} \leq 187\sqrt{\frac{k}{F_y}}$$

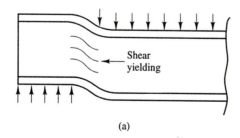

(a)

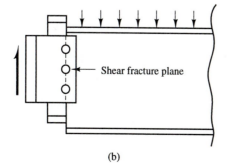

(b)

Figure 7.7 (a) Shear yielding of the gross area of the web; (b) shear fracture on the net area of the web.

where $k = 5$ for rolled beams when stiffeners are not required, which results in

$$\frac{h}{t_\text{w}} \leq \frac{418}{\sqrt{F_\text{y}}} \tag{7.4}$$

and the nominal shear yield strength V_n is taken as

$$V_\text{n} = 0.6F_\text{y}dt_\text{w}$$

Fracture of the beam web occurs through the net area of the web and is taken through the line of connection bolts (Fig. 7.7b). In this case, the nominal strength is a function of the ultimate strength F_u of the beam web. Thus,

$$V_\text{n} = 0.6F_\text{u}\, h_\text{net}t_\text{w}$$

With the specified $\phi = 0.9$, the shear yield design strength is

$$\phi V_\text{n} = 0.9V_\text{n} = 0.54F_\text{y}dt_\text{w} \tag{7.5}$$

and with the specified $\phi = 0.75$, the design shear fracture strength is

$$\phi V_\text{n} = 0.75V_\text{n} = 0.45F_\text{u}\, h_\text{net}t_\text{w} \tag{7.6}$$

Which of the two shear limit states governs depends on the ratios of the fracture to yield stress of the material and the gross to net section. Shear yield is almost never a limit state with a rolled beam. However, because of the relatively thin webs of most plate girders, it is important in the design of these members; shear yield is discussed in detail in Chapter 8. Shear fracture may be a possible limit state for a rolled beam with a thin web connected by high strength bolts; shear fracture is discussed in Chapter 11.

7.3 LOCAL BUCKLING

Local buckling characteristics of bending members can be defined as a function of the slenderness λ of the webs and flanges. The dimensions of the section used to define λ are the projected length of the beam flange width b, the flange thickness t_f, the web depth h_c, and the web thickness t_w. Figure 7.8 illustrates each of these dimensions for commonly used sections.

The slenderness of the flange of an I-shaped section is

$$\lambda = \frac{b}{t_\text{f}} = \frac{b_\text{f}}{2t_\text{f}} \tag{7.7}$$

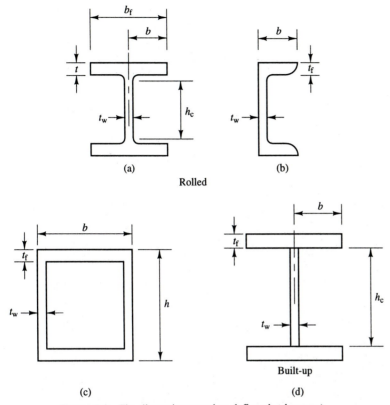

Figure 7.8   The dimensions used to define slenderness λ.

while for the web of the same section,

$$\lambda = \frac{h_c}{t_w} \tag{7.8}$$

7.3.1 Flange Local Buckling

The resisting moment M_n of a cross section can be expressed as a function of λ. Figure 7.9 shows this relationship for flange local buckling. There are three distinct zones: plastic, inelastic, and elastic. Shapes that conform to the definition of each of these zones are termed compact, noncompact, and slender, respectively. For I-shaped sections, the dividing line between compact and noncompact beams is given by $\lambda_p = 65/\sqrt{F_y}$, and the division between noncompact and slender beams occurs at $\lambda_r = 141/\sqrt{(F_y - F_r)}$, as shown in Fig. 7.9. To provide some additional control on noncompact sections in seismic areas, it is recommended that λ_p be reduced to $\lambda_p = 52/\sqrt{F_y}$.

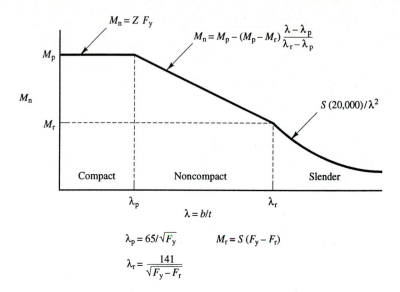

Figure 7.9 Flange local buckling flexure.

Within the compact or plastic zone, the nominal moment is

$$M_n = M_p = ZF_y \tag{7.9}$$

At the junction of the noncompact and slender zone, the moment is defined as

$$M_r = S(F_y - F_r) \tag{7.10}$$

where F_r is the assumed "average" residual stress and is taken as 10 ksi for rolled shapes and 16.5 ksi for welded shapes; and S is the section modulus (in.^{3}).

7.3.2 Web Local Buckling

Figure 7.10 illustrates the relationship between M_n and slenderness for web local buckling. In this case, for an I-shaped section, the limit of the plastic (or compact section) region is defined by $\lambda_p = 640/\sqrt{F_y}$; the limit for the inelastic (or noncompact section) region is given by $\lambda_r = 970/\sqrt{F_y}$. At the division between inelastic and elastic behavior, the nominal moment is given by

$$M_n = M_r = SF_y \tag{7.11}$$

If the web is slender and behaves elastically, the member is designed by plate girder rules, as described in Chapter 8. Table 7.1 summarizes the local buckling rules for both I-shaped members described here and some other common sections.

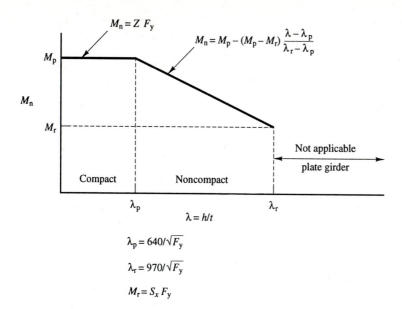

$M_n = Z\,F_y$

$M_n = M_p - (M_p - M_r)\dfrac{\lambda - \lambda_p}{\lambda_r - \lambda_p}$

$\lambda = h/t$

$\lambda_p = 640/\sqrt{F_y}$

$\lambda_r = 970/\sqrt{F_y}$

$M_r = S_x\,F_y$

Figure 7.10 Web local buckling flexure.

TABLE 7.1 LOCAL BUCKLING RULES FOR COMMON SHAPES

Element	λ	λ_p	λ_r
Rolled	$\dfrac{b}{t_f}$	$\dfrac{65}{\sqrt{F_y}}$	$\dfrac{141}{\sqrt{F_y}-10}$
Welded or hybrid	$\dfrac{b}{t_f}$	$\dfrac{65}{\sqrt{F_{yf}}}$	$\dfrac{106}{\sqrt{F_y}-16.5}$
Box	$\dfrac{b}{t_f}$	$\dfrac{190}{\sqrt{F_y}}$	$\dfrac{238}{\sqrt{F_y}}$
Webs	$\dfrac{h}{t_w}$	$\dfrac{640}{\sqrt{F_{yw}}}$	$\dfrac{970}{\sqrt{F_{yw}}}$

261

For both flange local buckling (Fig. 7.9) and web local buckling (Fig. 7.10), the relationship between λ and M_n in the inelastic range is considered linear. Therefore, M_n can be easily calculated through similar triangles. Thus, when

$$\lambda_p < \lambda \le \lambda_r \tag{7.12}$$

$$M_n = M_p - (M_p - M_r)\frac{\lambda - \lambda_p}{\lambda_r - \lambda_p} \tag{7.13}$$

7.4 DESIGN REQUIREMENTS FOR LATERALLY SUPPORTED BEAMS

Section 7.3 illustrated how the nominal strength of a member is determined when the limit state is local buckling of either the web or flange. The overall bending strength of a beam is also a limit state and is dependent on the distance between lateral braces, as measured by

$$\lambda = \frac{L_b}{r_y} \tag{7.14}$$

where L_b is the distance between bracing, which provides lateral or torsional restraint to the beam (in.), and r_y is the radius of gyration of the section (in.)

Figure 7.11 is a plot of the nominal bending strength M_n of a beam under uniform moment as a function of λ, as defined in Eq. (7.14). In this section we discuss the region of the curve from where the beam would be considered fully braced against lateral–torsional buckling, λ = 0, to a bracing defined by λ_p; in Section 7.5 we discuss the case where the beam has reduced lateral support.

When the unbraced length of the compression flange is less than L_{pd}, as defined in the LRFD specification, then the nominal moment can be taken as M_p and plastic analysis is permitted. When the unbraced length is between L_{pd} and L_p, the nominal moment may still be taken as M_p, but the analysis must be elastic. The difference is due to the fact that plastic design requires that the first "hinges" to form be capable of undergoing sufficient rotation to permit a second hinge to form without the first hinge becoming unstable. A discussion of plastic design follows in Section 7.6.

For an I-shaped section, L_{pd} and L_p are defined as

$$L_{pd} = \frac{3600 + 2200(M_1/M_2)}{F_y}r_y \tag{7.15}$$

and

$$L_p = \frac{300}{\sqrt{F_{yf}}}r_y \tag{7.16}$$

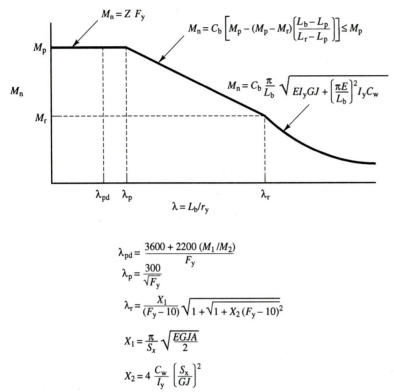

$$M_n = Z\,F_y$$

$$M_n = C_b\left[M_p - (M_p - M_r)\left[\frac{L_b - L_p}{L_r - L_p}\right]\right] \le M_p$$

$$M_n = C_b\frac{\pi}{L_b}\sqrt{EI_yGJ + \left[\frac{\pi E}{L_b}\right]^2 I_yC_w}$$

$$\lambda_{pd} = \frac{3600 + 2200\,(M_1/M_2)}{F_y}$$

$$\lambda_p = \frac{300}{\sqrt{F_y}}$$

$$\lambda_r = \frac{X_1}{(F_y - 10)}\sqrt{1 + \sqrt{1 + X_2\,(F_y - 10)^2}}$$

$$X_1 = \frac{\pi}{S_x}\sqrt{\frac{EGJA}{2}}$$

$$X_2 = 4\frac{C_w}{I_y}\left[\frac{S_x}{GJ}\right]^2$$

Figure 7.11   Lateral–torsional buckling flexure.

where M_1 is the smaller and M_2 is the larger moment at the end of each unbraced segment of the beam. M_1/M_2 is positive when the moments cause reverse curvature and negative when they cause single curvature, as seen in Fig. 7.12.

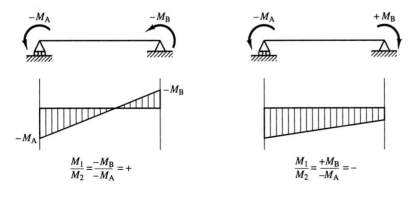

$$\frac{M_1}{M_2} = \frac{-M_B}{-M_A} = +$$

$$\frac{M_1}{M_2} = \frac{+M_B}{-M_A} = -$$

Reverse curvature   Single curvature

Figure 7.12   Moment diagrams showing the relationships between M_1 and M_2.

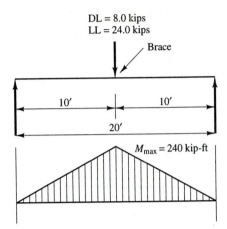

DL = 8.0 kips
LL = 24.0 kips

Brace

10'   10'

20'

M_{max} = 240 kip-ft

Figure 7.13 Beam used in
Example 7.1.

EXAMPLE 7.1

Given: A beam (F_y = 36 ksi), simply supported at both ends, spans 20 ft and is
loaded at its center line with a dead load of 8.0 kips and a live load of 24.0 kips.
Assume full lateral support and a compact section. See Fig. 7.13.

Solution

$$P_u = 1.2(8.0) + 1.6(24.0) = 48.0 \text{ kips}$$

$$M_u = \frac{48(20)}{4} = 240 \text{ kip-ft}$$

For a compact, fully braced section,

$$\phi M_n = \phi M_p = \phi Z F_y = M_u$$

Thus,

$$Z_{req} = \frac{240(12)}{0.9(36)} = 88.9 \text{ in.}^3$$

From the plastic section modulus economy table (*LRFD Manual*, p. 3–15),

$$\text{try W21} \times 44, \; (Z = 95.4 \text{ in.}^3)$$

Checking for shear,

$$V_u = 24 \text{ kips}$$

Conservatively using the beam depth,

$$\frac{d}{t_w} = \frac{20.66}{0.35} = 59.0 < \frac{418}{\sqrt{36}} = 69.7$$

Thus,

$$\phi V_n = 0.9(0.6)(36)(20.66)(0.35) = 140.6 \text{ kips} > 24 \text{ kips}$$

OK for shear

Use W21×44

7.5 DESIGN REQUIREMENTS FOR LATERALLY UNSUPPORTED BEAMS

When the unbraced length of a beam is between L_p and L_r, the beam will fail in an inelastic buckling mode. In this range, the nominal moment M_n under uniform moment is represented by a straight line, as was shown in Fig. 7.11. By using actual lengths (L_b, L_p, and L_r) instead of the equivalent λ values, the moment capacity equation may be written as

$$M_n = M_p - (M_p - M_r)\frac{L_b - L_p}{L_r - L_p} \tag{7.17}$$

where L_b is the distance between points bracing the compression flange against lateral displacement or the cross section against twist, L_p is from Table 7.2, L_r is from Table 7.3, and M_r is from Table 7.4.

The design moment may be determined by multiplying Eq. (7.17) by ϕ and combining terms to yield

TABLE 7.2 LATERAL-TORSIONAL BUCKLING RULES, L_p.

Shape	L_p
	$\dfrac{300r_y}{\sqrt{F_{yf}}}$
	$\dfrac{3750r_y}{M_p}\sqrt{JA}$

TABLE 7.3 LATERAL-TORSIONAL BUCKLING RULES, L_r.

Shape	L_r
	$\dfrac{r_y X_1}{(F_{yw} - F_r)} \sqrt{1 + \sqrt{1 + X_2 (F_{yw} - F_r)^2}}$ $X_1 = \dfrac{\pi}{S_x} \sqrt{\dfrac{EGJA}{2}}$ $X_2 = 4 \dfrac{C_w}{I_y} \left(\dfrac{S_x}{GJ}\right)^2$
	$\dfrac{57{,}000 r_y \sqrt{JA}}{M_r}$

TABLE 7.4 LATERAL-TORSIONAL BUCKLING RULES, M_r.

Shape	M_r
	$[F_{yw} - F_r]S_x$
	$F_y S_x$

$$\phi M_n = \phi M_p - \frac{\phi(M_p - M_r)}{L_r - L_p}(L_b - L_p) \tag{7.18}$$

Since the ratio $\phi(M_p - M_r)/(L_r - L_p)$ is a constant for each beam shape, it is tabulated as BF in the *LRFD Manual*. Thus, Eq. (7.18) can be rewritten as

$$\phi M_n = \phi M_p - BF(L_b - L_p) \tag{7.19}$$

where

$$BF = \frac{\phi(M_p - M_r)}{L_r - L_p} \tag{7.20}$$

and

$$\phi = 0.9$$

The nominal strength of a beam as defined in Eqs. (7.17) and (7.19) assumes that the moment is applied uniformly, as shown in Fig. 7.6a. This is the most severe loading case. For any other loading pattern and resulting moment gradient, the nominal strength of the beam is greater because the tendency for LTB is reduced under a moment gradient with an associated reduced in-plane deflection.

For the usual case where the moment diagram is not uniform, the design moment, as calculated in Eq. (7.17) or (7.19), may be increased by a factor to account for that difference. The factor is defined as C_b [4], so that the ϕM_n is

$$\phi M_n = C_b \phi M_n \le \phi M_p \qquad (7.21)$$

C_b is a function of the moment gradient and is specified by AISC as

$$C_b = 1.75 + 1.05 \left(\frac{M_1}{M_2}\right) + 0.3 \left(\frac{M_1}{M_2}\right)^2 \le 2.3 \qquad (7.22)$$

where M_1 is the smaller and M_2 is the larger moment at the end of each unbraced segment of the beam. M_1/M_2 is positive when the moments cause reverse curvature and negative when they cause single curvature. C_b is 1.0 under uniform moment, which is the basic case, and may be conservatively taken as 1.0 for other cases. In doing so, however, the designer may be sacrificing significant economy.

The effect of the moment gradient factor C_b is to alter the moment-unbraced length relationship by a constant, as shown in Fig. 7.14. The

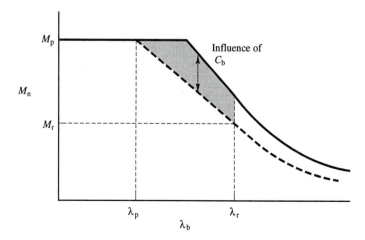

Figure 7.14 Effect of moment gradient.

shaded area shows the increase in moment capacity as a result of the use of C_b.

EXAMPLE 7.2

Given: Figure 7.15 shows a beam fixed at one support and pinned at the other with a 48-kip factored concentrated load at the center. Assume A36 steel and a lateral brace at the supports and the load point. Check to determine if a W16 × 40 will carry the load assuming (a) $C_b = 1.0$ and (b) C_b from Eq. (7.22). (c) Determine the lightest weight section to carry the load using the correct C_b.

Solution

The maximum moment from an elastic analysis is determined to be 180 kip-ft at the fixed end, as shown in Fig. 7.15.
(a) Assume $C_b = 1.0$

$$W16 \times 40, \ Z = 72.9 \text{ in.}^3, \ L_r = 19.3 \text{ ft}, \ L_p = 6.5 \text{ ft}, \ M_p = 218.7 \text{ kip-ft},$$
$$BF = 5.53, \ L_b = 10 \text{ ft}$$

Since $L_b = 10 \text{ ft} > L_p$

$$\phi M_n = 0.9(218.7) - 5.53(10 - 6.5) = 177.5 \text{ kip-ft} < 180 \text{ kip-ft}$$

Therefore, the beam will not work without considering C_b.

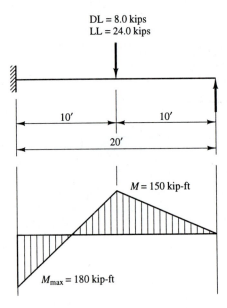

Figure 7.15 Beam used in Example 7.2.

(b) Use the calculated value of C_b

$$C_b = 1.75 + 1.05\left(\frac{150}{180}\right) + 0.3\left(\frac{150}{180}\right)^2 = 2.83 > 2.3$$

Calculate ϕM_n using $C_b = 2.3$

$$\phi M_n = 2.3(177.5) = 408.3 \text{ kip-ft} > \phi M_p = 196.8 \text{ kip-ft}$$

Since

$$\phi M_n = 196.8 \text{ kip-ft} > 180 \text{ kip-ft}$$

the W16×40 is adequate for bending.

(c) Considering that $C_b = 2.3$, a smaller section may be tried.
Try W18×35, $M_p = 199.5$ kip-ft, $L_r = 14.8$ ft, $L_p = 5.1$ ft, BF = 6.93

Since $L_b = 10$ ft $> L_p = 5.1$ ft,

$$\phi M_n = 0.9(199.5) - 6.93(10.0 - 5.1) = 145.6 \text{ kip-ft}$$

Considering $C_b = 2.3$,

$$\phi M_n = 2.3(145.6) = 334.9 \text{ kip-ft} > \phi M_p = 180.0 \text{ kip-ft}$$

Thus,

$$\phi M_n = 180.0 \text{ kip-ft} = 180 \text{ kip-ft}$$

So the W18×35 will just work in bending.

The beams shown in Fig. 7.16 illustrate the determination of C_b for four different moment gradients. Figure 7.16a shows a simply supported, uniformly loaded beam with lateral supports at the ends and at midspan. Thus, $M_1 = 0$ for both segments of unbraced length and $M_1/M_2 = 0$, which yields

$$C_b = 1.75 + 1.05(0) + 0.3(0)^2 = 1.75 \leq 2.3$$

Figure 7.16b shows a continuous, uniformly loaded beam with braces at the supports and at the midspan. The moment diagram between braces causes the beam to bend in reverse curvature resulting in the ratio of M_1/M_2 being positive. In this case,

$$C_b = 1.75 + 1.05\left(\frac{M_1}{M_2}\right) + 0.3\left(\frac{M_1}{M_2}\right)^2 \leq 2.3$$

Figure 7.16 Four moment
gradients.

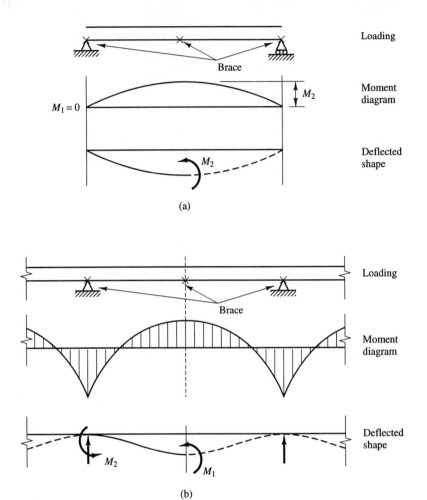

(a)

(b)

The beam of Fig. 7.16c is similar to that in Fig. 7.16b. The moments cause
reverse curvature for each unbraced length, producing positive moment
ratios, which results in C_b between 1.75 and 2.3. Figure 7.16d illustrates
the case where the moments are opposite in rotational direction and
M_1/M_2 is negative:

$$C_b = 1.75 - 1.05\left(\frac{M_1}{M_2}\right) + 0.3\left(\frac{M_1}{M_2}\right)^2 \leq 2.3$$

EXAMPLE 7.3

Given: A simply supported W6×15 spans 10 ft. It is braced at the ends and at
the midspan ($L_b = 5$ ft). Determine the maximum design moment. Steel grade is
$F_y = 50$ ksi.

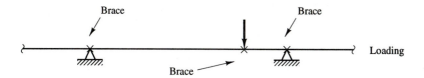

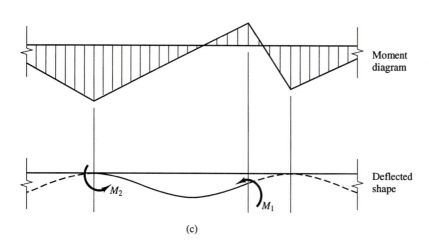

(c)

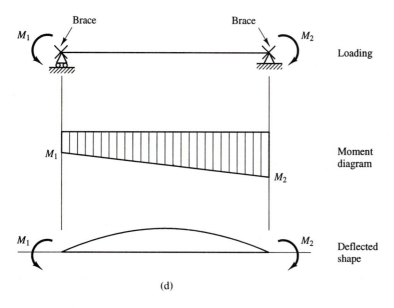

(d)

Solution

Determine the nominal moment for local buckling (Section 7.3):

Flange: $\lambda_p = \dfrac{65}{\sqrt{F_y}} = \dfrac{65}{\sqrt{50}} = 9.19$

$$\lambda = \frac{b_f}{2t_f} = \frac{5.99}{2(0.26)} = 11.5 > 9.19$$

Thus, the flange is noncompact.

$$\lambda_r = \frac{141}{\sqrt{F_y - 10}} = \frac{141}{\sqrt{40}} = 22.3$$

Web: $\lambda_p = \dfrac{640}{\sqrt{F_y}} = \dfrac{640}{\sqrt{50}} = 90.5$

$$\lambda = \frac{h_c}{t} = 21.6 < 90.5$$

Thus, the web is compact.

Since the shape is noncompact (flange), the nominal moment is calculated by Eq. (7.13) with

$$M_r = (F_y - F_r)S_x = (50 - 10)9.72 = 389 \text{ kip-in.}$$
$$M_p = Z_x F_y = (10.8)50 = 540 \text{ kip-in.}$$
$$M_n = 540 - (540 - 389)(11.5 - 9.19)/(22.3 - 9.19)$$
$$= 513 \text{ kip-in.} = 42.8 \text{ kip-ft}$$

Design moment for flange local buckling $= 0.9(42.8)$

$$= 38.5 \text{ kip-ft}$$

Determine the design moment for lateral–torsional buckling (Section 7.4):
For this noncompact shape,

$$L_p = \frac{300r_y}{\sqrt{F_y}} = 5.2 \text{ ft}$$

Since the unbraced length, $L_b = 5$ ft $< L_p = 5.2$ ft, the bracing is adequate for the beam to resist M_p. Thus,

$$\phi M_n = 0.9(540) = 486 \text{ kip-in.} = 40.5 \text{ kip-ft}$$

Since the moment determined through the noncompact calculations, 38.5 kip-ft, is less that the moment based on the bracing length, 40.5 kip-ft, local buckling controls and there is no value in calculating C_b. The design moment is 38.5 kip-ft.

When the unbraced length of a beam exceeds L_r, lateral–torsional buckling is an elastic phenomenon. The general behavior of these members was described in Section 7.2.2.

For I-shaped members, doubly symmetric and singly symmetric members with the compression flange larger than the tension flange, and channels loaded in the plane of the web, the design moment is given by

$$\phi M_n = \phi C_b M_r \qquad (7.23)$$

where

$$M_r = \frac{\pi}{L_b}\sqrt{EI_y GJ + \left(\frac{\pi E}{L_b}\right)^2 I_y C_w} \qquad (7.24)$$

The first term under the radical relates the capacity of the shape to resist St. Venant torsion, and the second term reflects its warping torsion resistance [1].

7.6 PLASTIC ANALYSIS

Up to this point, it has been assumed that the plastic moment capacity of a bending member could be equated to the maximum elastic moment occurring on a beam in order to determine member capacity. This is appropriate for determinate members where the occurrence of the plastic moment results in the development of a single plastic hinge which leads to member failure. For indeterminate structures, such as continuous beams, more than one plastic hinge must form before the beam will fail. It is the formation of the necessary number of plastic hinges in the appropriate locations that develop what is called a failure mechanism.

The formation of a beam failure mechanism may best be understood by following the load history of a fixed ended beam with a uniformly distributed load. The beam is shown in Fig. 7.17a along with the moment diagram that results from an elastic indeterminate analysis. It can be seen that the largest moments occur at the fixed ends and are given by $wL^2/12$. If the load on the beam is increased, the beam will behave elastically until the moments on the ends are equal to the plastic capacity of the member, as shown in Fig. 7.17b. Since the application of additional load will cause the member to rotate at its ends while maintaining the plastic moment, these points will behave as pins. These pins are normally referred to as plastic hinges. In this case, the load will be designated as w_1. From this load on, the member may continue to accept load, functioning as a simple

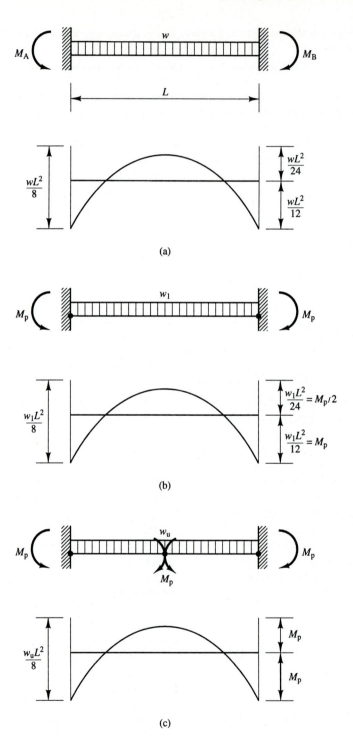

Figure 7.17 Beam and moment diagrams for the development of a plastic mechanism.

beam, until a third plastic hinge forms at the beam centerline. The formation of this third hinge makes the beam unstable, thus forming what is referred to as the collapse (or failure) mechanism. The mechanism and corresponding moment diagram are given in Fig. 7.17c.

For the collapse mechanism just described, equilibrium requires that the simple beam moment $wL^2/8$ be equal to twice the plastic moment; thus,

$$M_{\text{p}} = \frac{wL^2}{16} \qquad (7.25)$$

It can be seen that, through plastic analysis, a smaller plastic moment capacity must be provided for in the design of the beam than would have been the case for an elastic analysis where the plastic moment was equated to $wL^2/12$.

An additional advantage to the use of plastic analysis for indeterminate beams is the simplicity of the analysis. By observation, regardless of the overall geometry of the continuous beam, each segment between supports may be evaluated independently of each other segment. This means that any beam segment, continuous at each end and loaded with a uniformly distributed load, will exhibit the same collapse mechanism. Thus, the relation between the applied load and the plastic moment will be as given in Eq. (7.25). Plastic analysis results for additional loading and beam configurations are given in Fig. 7.18. Additional examples and the development of these relations through application of energy principles may be found in numerous textbooks, including Ref. [5].

To insure that a given beam cross section is capable of undergoing the necessary rotation at each plastic hinge, the LRFD specification requires that the compression flange of a wide flange shape be braced such that the unbraced length in the area of the hinge is less than that given as L_{pd} in Eq. (7.15). If this limit is not satisfied, the member design must be based on an elastic analysis.

EXAMPLE 7.4

Given: A beam, simply supported at one end and fixed at the other as shown in Fig. 7.15 (Example 7.2), spans 20 ft and is loaded at its centerline with a dead load of 8.0 kips and a live load of 24.0 kips. Lateral support is provided at the ends and at the concentrated load.

Design the beam using plastic design and A36 steel. Assume that the final section will be compact.

Solution

$$P_{\text{d}} = 1.2(8.0) + 1.6(24.0) = 48.0 \text{ kips}$$

$$\phi M_{\text{p req'd}} = \frac{Pab}{a + 2b} = 160 \text{ kip-ft}$$

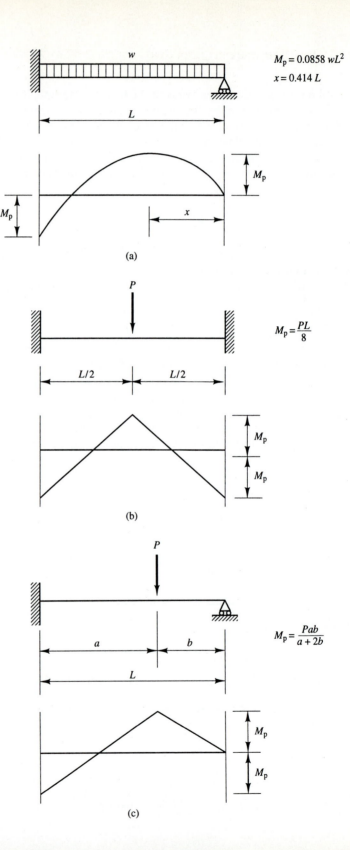

$M_p = 0.0858\, wL^2$
$x = 0.414\, L$

M_p

L

M_p

x

(a)

P

$M_p = \dfrac{PL}{8}$

$L/2$

$L/2$

M_p

M_p

(b)

P

$M_p = \dfrac{Pab}{a + 2b}$

a

b

L

M_p

M_p

(c)

Figure 7.18 Loading and beam
configurations
resulting from
276   plastic analysis.

Try W18×35, r_y = 1.22, ϕM_p = 180 kip-ft

Calculate the required minimum unbraced length through Eq. (7.15).

$$L_{pd} = [3600 + 2200\,(0/180)]\,(1.22)/36 = 122 \text{ in.}$$

Since the provided unbraced length equals 10 ft, or 120 in., which is less than the maximum permitted of 122 in., the W18×35 is acceptable under plastic design.

7.7 SPECIAL REQUIREMENTS FOR DOUBLE-ANGLE AND TEE MEMBERS

The development of equations to represent the theoretical bending resistance of a structural tee, or a member built up from double angles, is extremely complex and is beyond the scope of this book. The reader is referred to Galambos [6] for an in-depth treatment of this development. The limit state for bending in these shapes comes from the interaction of local and lateral torsional buckling.

The LRFD specification stipulates that the nominal moment is

$$M_n = \frac{\pi\sqrt{EI_yGJ}}{L_b}\left(B + \sqrt{1 + B^2}\right) \leq M_y \qquad (7.26)$$

where

$$B = \pm\,2.3\left(\frac{d}{L_b}\right)\sqrt{\frac{I_y}{J}} \qquad (7.27)$$

With ϕ = 0.9, the design moment = $0.9M_n$.

Equation (7.26) is applicable only when local buckling of the elements of the member is prevented by limiting λ_r as specified in Table 7.5.

The sign of B in Eq. (7.27) depends on whether the stem of the member is in tension or compression, since the tendency for lateral torsional buckling is greater when the stem is in compression than when it is in tension. The two cases can be defined as follows:

Case 1: Stem in tension, B is positive.
Case 2: Stem in compression, B is negative.

EXAMPLE 7.5

Given: A WT9×17.5 has support provided at 5-ft (60-in.) intervals. Determine the nominal moment for the member, where (a) the flange is at the top and (b) the flange is at the bottom.

TABLE 7.5   LATERAL-TORSIONAL BUCKLING RULES, λ_r.

Shape	λ_n
	$\dfrac{b}{t} = \dfrac{76}{\sqrt{F_y}}$
	$\dfrac{b}{t_f} = \dfrac{76}{\sqrt{F_y}}$ $\dfrac{d}{t_w} = \dfrac{127}{\sqrt{F_y}}$

Solution

$$M_n = \frac{3.14\sqrt{30,000(7.67)(11,200)(0,252)}}{60.0}\,(B + \sqrt{1 + B^2})$$

$$M_n = 1334(B + \sqrt{1 + B^2})$$

with

$$B = 2.3\left(\frac{9}{60}\right)\sqrt{\frac{7.67}{0.252}} = 1.90$$

Case 1 (*B* is positive, stem in tension):

$$M_n = 1334(+1.91 + \sqrt{1 + 1.91^2}) = 5424 \text{ kip-in.}$$

Case 2: (*B* is negative, stem in compression):

$$M_n = 1334(-1.91 + \sqrt{1 + 1.91^2}) = 328 \text{ kip-in.}$$

The great difference between case 1 and case 2 reflects the greater tendency of case 2 toward lateral–torsional buckling.

7.8 SERVICEABILITY CRITERIA FOR BEAMS

There are several serviceability considerations that the designer should review. These should be discussed in detail with the client so the quality of the building is consistent with the desire of the client. For instance, experience may indicate that a certain amount of vibration in a floor system may be annoying at first but as time goes on, occupants become used to it. The client may be unwilling to pay for the cost of a stiffer floor and should have the right to make that decision, as long as the client is warned by the structural engineer and there is no public safety aspect involved.

For beams there are three potential serviceability problems:

1. *Excessive live load deflection.* The most serious problem is cracking of plastered ceilings and distress in nonstructural elements attached to the structure. Experience has demonstrated that this is not a problem if the live load deflection is limited to $\frac{1}{360}$ of the span [7–9].

2. *Vibration.* Vibration of floor systems is not a safety consideration, but it can be annoying. The most common problem is with wide open spaces with very little damping, such as the jewelry department in a department store. To reduce the risk of annoyance, a general rule is to space the beams or joists sufficiently far apart that the slab thickness will be large enough to provide the needed stiffness and damping [10–12].

3. *Drift.* Under wind loading, a building has the potential to move laterally. This lateral displacement is referred to as drift. As with vibration, drift is usually not a safety consideration, but it can be annoying. Beams and girders are important in reducing the drift, and their final size might actually be determined by drift considerations. Since drift is a function of the stiffness of beams, columns, and connection, it is treated in Chapter 9.

Since beam deflection is a serviceability consideration, calculations are carried out using unfactored loads, usually only the unfactored live load. Numerous elastic analysis techniques are available for the determination of the maximum deflection of a given beam and loading. Some common loading conditions with their corresponding maximum deflection are shown in Fig. 7.19. These and many others are also given in the *LRFD Manual.*

EXAMPLE 7.6a

Given: Using the problem statement given in Example 7.1, check that the live load deflection does not exceed $\frac{1}{360}$ of the span.

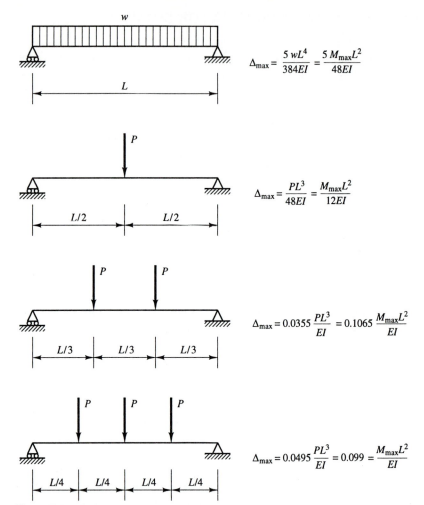

$$\Delta_{max} = \frac{5\,wL^4}{384EI} = \frac{5\,M_{max}L^2}{48EI}$$

$$\Delta_{max} = \frac{PL^3}{48EI} = \frac{M_{max}L^2}{12EI}$$

$$\Delta_{max} = 0.0355\frac{PL^3}{EI} = 0.1065\frac{M_{max}L^2}{EI}$$

$$\Delta_{max} = 0.0495\frac{PL^3}{EI} = 0.099 = \frac{M_{max}L^2}{EI}$$

Figure 7.19   Some common loading conditions with their corresponding maximum deglection.

Solution

The section selected previously was a W21 × 44, $I = 843$ in.^{4}. The service live load is 24 kips applied at the center of a 20-ft span.

The deflection equation found in Fig. 7.19 gives

$$\text{Deflection} = \frac{24(20)^3}{48(29,000)(843)}\,(12)^3 = 0.28 \text{ in.}$$

$$\text{Deflection limit} = \frac{20(12)}{360} = 0.67 \text{ in.} > 0.28 \text{ in.} \quad \text{OK}$$

EXAMPLE 7.6b

Given: Same as Example 7.6a except that the deflection limit is set to the more severe level of span/1000.

Solution

$$\text{Deflection limit} = \frac{20(12)}{1000} = 0.24 \text{ in.} < 0.28 \text{ in.}$$

Since this beam deflects too much, the minimum acceptable moment of inertia I is determined as

$$I = \frac{24(20)^3}{48(29,000)(0.24)} (12)^3 = 993 \text{ in.}^4$$

Thus, a beam must be found with $I \geq 993$ in.^{4} and from Example 7.1, $Z \geq 88.9$ in.^{3}.

Select a W24 × 55, $I = 1350$ in.^{4} and $Z = 134$ in.^{3}

PROBLEMS

7.1–7.4. Determine the plastic neutral axis and plastic section modulus for the cross sections shown.

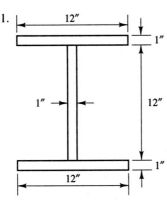

1.

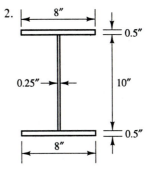

2.

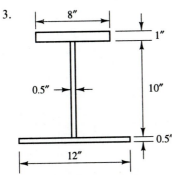

3.

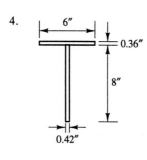

4.

7.5–7.8. Assume that the wide flange beams specified below are made from rectangles with the dimensions given for the beam in the *LRFD Manual*. Determine the plastic section modulus and compare it to the value given in the manual.

5. W14×90   6. W18×71   7. W10×77   8. W8×67

7.9. A simply supported beam spans 20 ft and carries a uniformly distributed dead load of 0.8 k/ft including the beam weight and a live load of 2.3 k/ft. Determine the minimum required plastic section modulus and select the lightest weight wide flange shape to carry the moment. Assume full lateral support and A36 steel.

7.10. Considering only bending, determine the lightest weight W section to carry a uniform dead load of 1.2 k/ft including the weight of the beam and a live load of 3.2 k/ft on a simple span of 24 ft. Assume full lateral support and A36 steel.

7.11. A beam is required to carry a uniform dead load of 0.85 k/ft including its own weight and a concentrated live load of 12 kips at the center of the 30-ft span. Considering only bending, determine the least weight W-shape to carry the load. Assume full lateral support and A36 steel.

7.12. Considering both shear and bending, determine the lightest weight W-shape to carry the following load: a uniform dead load of 0.6 k/ft plus the self-weight, a concentrated dead load of 2.1 kips, and a concentrated live load of 6.4 kips, located at the center of a 16-ft span. Use A36 steel and assume full lateral support.

7.13. Considering both shear and bending, determine the lightest W-shape to carry a uniform dead load of 4.0 k/ft plus the self-weight and a uniform live load of 2.3 k/ft on a simple span of 10 ft. Assume full lateral support and A36 steel.

7.14. A 24-ft simple span laterally supported beam is required to carry a total uniformly distributed service load of 8.0 k/ft. Determine the lightest, A36, W section to carry this load if it is broken down as follows:
(a) Live load = 6.0 k/ft; dead load = 2.0 k/ft
(b) Live load = 1.0 k/ft; dead load = 7.0 k/ft
(c) Live load = 6.7 k/ft; dead load = 1.3 k/ft
(d) Live load = 7.0 k/ft; dead load = 1.0 k/ft

7.15. Repeat the design specified in Problem 7.9 using F_y = 50 ksi.

7.16. Repeat the design specified in Problem 7.10 using F_y = 50 ksi.

7.17. Repeat the design specified in Problem 7.12 using F_y = 50 ksi.

7.18. Repeat the design specified in Problem 7.14 using F_y = 50 ksi.

7.19. A 30-ft simply supported beam is loaded at the third points of the span with dead loads of 4.0 kips and live loads of 6.0 kips. Lateral supports are provided at the load points and the supports. The self-weight of the beam may be ignored. Determine the least weight W section to carry the load. Use A36 steel and assume C_b = 1.0.

7.20. An 18-ft simple span beam is loaded with a uniform dead load of 1.4 k/ft, including the beam weight and a uniform live load of 2.3 k/ft. The lateral supports are located at 6.0-ft intervals. Determine the least weight W section to carry the load. Use A36 steel.

7.21. A W18×60 is used on a 36-ft simple span to carry a uniformly distributed load. Determine the locations of the lateral supports in order to provide just enough strength to carry a design moment of 280 kip-ft. Use A36 steel.

7.22. A girder that carries a uniformly distributed dead load of 1.7 k/ft plus its self-weight and three 15-kip concentrated live loads at the quarter points of the 36-ft span is to be sized. Using F_y = 50 ksi, determine the lightest W section to carry the load with lateral supports provided at the supports and the load points.

7.23. A 32-ft simple span beam carries a uniform dead load of 2.3 k/ft plus its self-weight and a uniform live load of 3.1 k/ft. The beam is laterally supported at the supports only. Determine the minimum weight W-shape to carry the load using F_y = 50 ksi.

7.24. A fixed ended beam on a 28-ft span is required to carry a total factored uniformly distributed load of 22.5 kips. Using plastic analysis and A36 steel, determine the design moment and select the lightest weight W section assuming (a) full lateral support and (b) lateral support at the ends and the center line.

7.25. A beam is fixed at one support and simply supported at the other. A concentrated factored load of 32.0 kips is applied at the center of the 40-ft span. Using plastic analysis and A36 steel, determine the lightest weight W-shape to carry the load when the nominal depth is limited to 18 in. Assume (a) full lateral support and (b) lateral supports at the ends and at the load.

7.26. A 36-ft simple span beam carries a uniformly distributed dead load of 3.4 k/ft plus its self-weight and a uniformly distributed live load of 4.2 k/ft. Determine the lightest W-shape to carry the load while limiting the service live load deflection to $\frac{1}{360}$ of the span. Use $F_y = 50$ ksi and assume full lateral support.

7.27. A simple span beam with a uniformly distributed dead load of 1.1 k/ft including the self-weight and concentrated dead loads of 3.4 kips and live loads of 6.0 kips at the third points of a 24-ft span is to be designed with lateral supports at the third points and service live load deflection limited to $\frac{1}{360}$ of the span. Determine the least weight W section to carry the loads. $F_y = 36$ ksi.

REFERENCES

[1] Johnson, B. G., Ed., *Guide to Stability Design Criteria for Metal Structures*, 3d ed., Structural Stability Research Council, Wiley, New York, 1976.

[2] Trahair, N. S., *The Behavior and Design of Steel Structures*, Chapman & Hall, London, 1977.

[3] Nethercot, D. A., and Trahair, N. S., Design of laterally unsupported beams, in *Beams and Columns: Stability and Strength*, R. Narayanan, Ed., Applied Science, Barking, Essex, UK, 1983.

[4] Salvadori, M. G., Lateral buckling of I-beams, *Transactions, ASCE,* Vol. 120, 1955 (p. 1165).

[5] Disque, R. O., *Applied Plastic Design in Steel*, Van Nostrand Reinhold, New York, 1971.

[6] Galambos, T. G., Ed., *Guide to Stability Design Criteria for Metal Structures*, 4th ed., Wiley Interscience, New York, 1988.

[7] Galambos, T. G., and Ellingwood, B., Serviceability limit states: Deflection, *Journal of Structural Engineering, ASCE,* Vol. 110, No. ST5, May 1985 (pp. 1158–1161).

[8] Ellingwood, B., Serviceability guidelines for steel structures, *Engineering Journal, AISC,* Vol. 26, First Quarter, 1989 (pp. 1–8).

[9] Ad Hoc Committee on Serviceability Research, Committee on Research of the Structural Division, Structural serviceability: A critical appraisal and research needs, *Journal of Structural Engineering, ASCE,* Vol. 112, No. ST12, December 1986 (pp. 2646–2664).

[10] Ellingwood, B., and Tallin, A., Structural serviceability: Floor vibrations, *Journal of Structural Engineering, ASCE,* Vol. 110, No. ST2, February 1984 (pp. 1158–1161).

[11] Murray, T., Design to prevent floor vibrations, *Engineering Journal, AISC,* Vol. 12, Third Quarter, 1975 (pp. 82–87).

[12] Murray, T., Acceptability criterion for occupant-induced floor vibrations, *Engineering Journal, AISC,* Vol. 18, Second Quarter, 1981 (pp. 62–70).

CHAPTER EIGHT

Plate Girders

8.1 BACKGROUND

In normal practice, a plate girder is thought of as a bending member made up of individual steel plates. Although they are normally the member of choice for situations where the available rolled shapes are not large enough to carry the intended load, there is no requirement that they always be at the large end of the scale of member size. Beams fabricated from individual steel plates to meet a specific requirement are generally identified in the field as plate girders.

Thus, plate girders used in building structures are generally for special situations, very large spans, or very large loads. Their most common application is as a transfer girder, a bending member that supports a structure above and permits the column spacing to be changed below. They are also common in industrial structures, as crane girders and as support for large pieces of equipment. In commercial buildings, they are often designed to span large open areas as required by the particular architectural constraints, as seen in Fig. 8.1. Because of their normally greater depth and resulting stiffness, plate girders tend to deflect less than other potential long-span members.

The cross section of a typical plate girder is shown in Fig. 8.2. Although it is possible to combine steel plates into numerous geometries, the plate girders addressed here will be those formed from three plates, one for the web and two for the flanges, as noted in the figure. Since the web and flanges of the plate girder are fabricated from individual plates, they may be designed with web and flanges from the same grade of steel or from different grades of steel; those from different grades of steel are known as hybrid girders. For hybrid girders, the flanges are usually fabricated with a higher grade of steel than that used in the web. This takes advantage of the higher stresses developed in the flange, which is

Figure 8.1   Application of a plate girder.

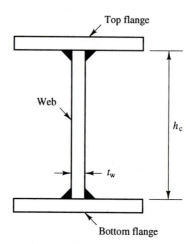

Figure 8.2 Plate girder definitions.

located at a greater distance from the plastic neutral axis than the web, resulting in a higher plastic moment capacity. Although hybrid girders are relatively common in bridge construction, they are rarely used in buildings and thus are not discussed in this book.

Plate girder research performed at Lehigh University by Konrad Basler [1–3], has formed the basis of the AISC rules for plate girder design. The AISC definition of a plate girder is somewhat different from that given above [4]. With regard to the LRFD specification, two shapes are to be treated as plate girders. The first is any I-shape with stiffeners and the second is any beam with a depth/web thickness ratio h_c/t_w that exceeds $970/\sqrt{F_y}$. The terms h_c and t_w are shown in Fig. 8.2. The first definition, in the opinion of the authors, is not really significant, since rolled shapes are rarely, if ever, designed with stiffeners. Rolled shapes are produced with relatively thick webs, primarily to ease the problems of handling during rolling and transporting. As a result, shear in the beam web rarely controls the selection of the shape and, since the only function of a stiffener is to increase shear strength, stiffeners are rarely needed. The authors suggest that the definition of a plate girder be ''an H-shaped girder built up of three plates.''

For plate girders that meet the compact web and flange criteria presented in Chapter 7, the rules as presented for rolled beams shall apply. Therefore, the discussion of plate girders presented in this chapter will address built-up H-shapes that are noncompact or slender as defined later in this chapter [5]. The behavior of a plate girder may best be understood by considering flexure and shear separately.

8.2 HOMOGENEOUS PLATE GIRDERS: FLEXURE

In flexure, a plate girder is considered to be either noncompact or slender according to the proportions of the web. Thus, it is possible for a noncompact web plate girder to have a slender flange, potentially controlling the capacity of the member. The design rules for each type of girder are considered separately.

8.2.1 Noncompact Web Plate Girders

A noncompact web plate girder could have its flexural strength governed by flange local buckling (FLB), web local buckling (WLB), or lateral torsional buckling (LTB). The failure modes for these limit states were discussed in Chapter 7 and the rules are specified in the LRFD specification, Appendix F1. The limit state that results in the lowest nominal moment is, of course, the one that governs.

A noncompact plate girder is one with the slenderness parameters λ defined as follows:

Flange local buckling: $\lambda = b_f/2t_f$

$$\frac{65}{\sqrt{F_y}} < \lambda < \frac{106}{\sqrt{F_y - 16.5}} \tag{8.1}$$

Web local buckling: $\lambda = h_c/t_w$

$$\frac{640}{\sqrt{F_y}} < \lambda < \frac{970}{\sqrt{F_y}} \tag{8.2}$$

The moment capacity for a noncompact plate girder is found through the linear relation shown in Figs. 8.3 and 8.4 and is given as follows:

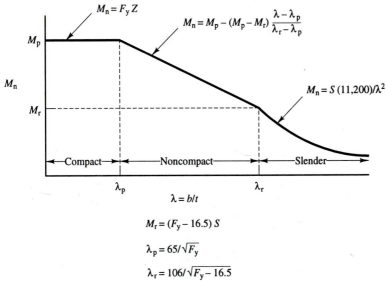

Figure 8.3 Noncompact plate girder flexure: flange local buckling.

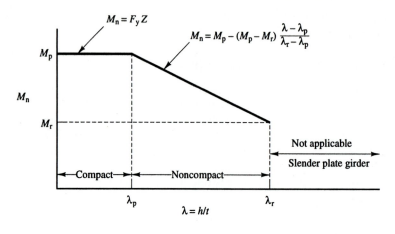

$$M_r = F_y S_x$$

$$\lambda_p = 640/\sqrt{F_y}$$

$$\lambda_r = 970/\sqrt{F_y}$$

Figure 8.4 Noncompact plate girder flexure: web local buckling.

$$M_n = C_b \left[M_p - (M_p - M_r) \frac{\lambda - \lambda_p}{\lambda_r - \lambda_p} \right] \qquad (8.3)$$

where

$$M_p = F_y Z$$

and for FLB

$$\lambda = \frac{b_f}{2 t_f}$$

$$\lambda_p = \frac{65}{\sqrt{F_y}}$$

$$\lambda_r = \frac{106}{\sqrt{F_y}}$$

$$M_r = (F_y - 16.5)S$$

and for WLB

$$\lambda = \frac{h_c}{t_w}$$

$$\lambda_p = \frac{640}{\sqrt{F_y}}$$

$$\lambda_r = \frac{970}{\sqrt{F_y}}$$

$$M_r = F_y S$$

For the case where the flange of the plate girder is slender, that is, $b_f/2t_f > 106/\sqrt{F_y}$, the moment capacity is given by

$$M_n = \left(\frac{11,200}{\lambda^2}\right) S \qquad (8.4)$$

Lateral–torsional buckling for noncompact plate girders is the same as for rolled beams, as shown in Fig. 8.5, except for the change in the residual stress, which is to be taken as 16.5 ksi for welded members. Thus,

$$M_p = F_y Z$$

$$M_r = (F_y - 16.5)S$$

$$\lambda = \frac{L_b}{r_y}$$

$$\lambda_p = \frac{300}{\sqrt{F_y}}$$

$$\lambda_r = \frac{X_1}{F_y - 16.5} \sqrt{1 + \sqrt{1 + X_2(F_y - 16.5)^2}}$$

with

$$X_1 = \frac{\pi}{S_x} \sqrt{\frac{EGJA}{2}}$$

and

$$X_2 = 4\left(\frac{C_w}{I_y}\right)\left(\frac{S_x}{GJ}\right)^2$$

For unbraced lengths that fall between λ_p and λ_r, Eq. (8.3) will yield the moment capacity, as discussed in Chapter 7. For unbraced lengths beyond λ_r, moment capacity is given by

$$M_n = C_b S \left(\frac{X_1\sqrt{2}}{\lambda} \sqrt{1 + \frac{X_1^2 X_2}{2\lambda^2}}\right)$$

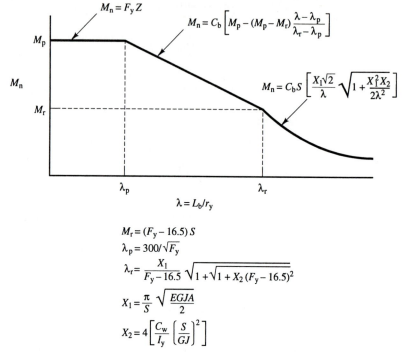

$$M_n = F_y Z$$

$$M_n = C_b \left[M_p - (M_p - M_r) \frac{\lambda - \lambda_p}{\lambda_r - \lambda_p} \right]$$

$$M_n = C_b S \left[\frac{X_1 \sqrt{2}}{\lambda} \sqrt{1 + \frac{X_1^2 X_2}{2\lambda^2}} \right]$$

$$\lambda = L_b / r_y$$

$$M_r = (F_y - 16.5) S$$

$$\lambda_p = 300/\sqrt{F_y}$$

$$\lambda_r = \frac{X_1}{F_y - 16.5} \sqrt{1 + \sqrt{1 + X_2 (F_y - 16.5)^2}}$$

$$X_1 = \frac{\pi}{S} \sqrt{\frac{EGJA}{2}}$$

$$X_2 = 4 \left[\frac{C_w}{I_y} \left[\frac{S}{GJ} \right]^2 \right]$$

Figure 8.5 Noncompact plate girder flexure: lateral–torsional buckling.

8.2.2 Slender Web Plate Girders

A slender web plate girder is defined as a plate girder with

$$\frac{h_c}{t_w} > \frac{970}{\sqrt{F_y}}$$

There are two possible limit states and both are elastic: tension flange yield and compression flange buckling. Both must be checked to determine the moment capacity of the member. The two equations, as given in the LRFD specification, Appendix G2, are

Tension flange yield:

$$M_{n\text{-ten}} = S_t R_{PG} F_y \tag{8.5}$$

Compression flange buckling:

$$M_{n\text{-comp}} = S_c R_{PG} F_y \tag{8.6}$$

where

R_{PG} = Plate girder bending reduction factor due to web buckling

$$= 1 - 0.0005a_r\left(\frac{h_c}{t_w} - \frac{970}{\sqrt{F_{cr}}}\right) \le 1.0$$

a_r = ratio of web area to compression flange area

F_{cr} = the critical buckling stress, which depends on λ

For $\lambda \le \lambda_p$

$$F_{cr} = F_y$$

For $\lambda_p < \lambda \le \lambda_r$

$$F_{cr} = C_b F_y \left[1 - \frac{\lambda - \lambda_p}{2(\lambda_r - \lambda_p)}\right] \le F_y \qquad (8.7)$$

For $\lambda > \lambda_r$

$$F_{cr} = \frac{C_{PG}}{\lambda^2}$$

For the limit state of lateral–torsional buckling, the following parameters are to be used:

$$\lambda = \frac{L_b}{r_T}$$

$$\lambda_p = \frac{300}{\sqrt{F_y}}$$

$$\lambda_r = \frac{756}{\sqrt{F_y}}$$

$$C_{PG} = 286{,}000C_b$$

For flange local buckling, the parameters are

$$\lambda = \frac{b_f}{2t_f}$$

$$\lambda_p = \frac{65}{\sqrt{F_y}}$$

$$\lambda_r = \frac{150}{\sqrt{F_y}}$$

$$C_{PG} = 11{,}200 \quad \text{and} \quad C_b = 1$$

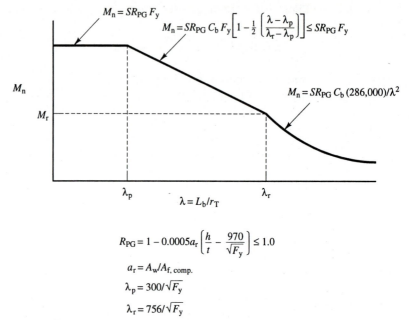

$$R_{PG} = 1 - 0.0005a_r \left[\frac{h}{t} - \frac{970}{\sqrt{F_y}} \right] \le 1.0$$

$$a_r = A_w/A_{f, comp.}$$

$$\lambda_p = 300/\sqrt{F_y}$$

$$\lambda_r = 756/\sqrt{F_y}$$

Figure 8.6 Slender plate girder flexure: lateral–torsional buckling.

The capacity of slender web plate girders is given in Figs. 8.6 and 8.7 for LTB and FLB, respectively. The limit state of WLB is not applicable to slender web plate girders since the slender web has been included in the R_{PG} parameter, which reduces the bending capacity of the section due to

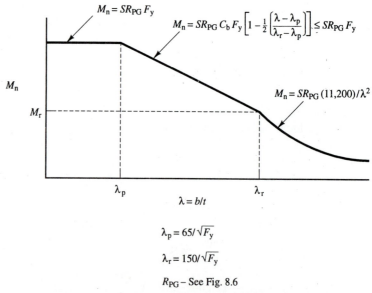

$$\lambda_p = 65/\sqrt{F_y}$$

$$\lambda_r = 150/\sqrt{F_y}$$

R_{PG} – See Fig. 8.6

Figure 8.7 Slender plate girder flexure: flange local buckling.

the buckling of the web. As will be seen in the discussion of shear to follow, transverse stiffeners are always required with slender web plate girders.

8.3 HOMOGENEOUS PLATE GIRDERS: SHEAR

Unlike in rolled shapes, shear is an important factor in the behavior and design of plate girders, since the webs are relatively thin. There are two recognized design procedures: nontension field action and tension field action. Transverse stiffeners may or may not be used when designing a nontension field girder, but are always required with a tension field girder.

In a tension field girder, the post buckling strength of the girder is recognized. Research has demonstrated that a plate girder with transverse stiffeners and a thin web will act as a Pratt truss after the web buckles, thus providing some additional strength [1]. The web acts as the diagonal tension member between the stiffeners, while the stiffeners act as the compression member as shown in Fig. 8.8. The designer must decide whether to utilize this tension field action or to design a conventional, nontension field girder.

8.3.1 Nontension Field Action: Shear

The design shear strength for a nontension field plate girder is specified in the LRFD specification, Section F2. With $\phi = 0.9$, these design shear strength equations are

For $h/t_w \leq 187\sqrt{k/F_y}$

$$\phi V_n = \phi 0.6 F_y A_w \qquad (8.8)$$

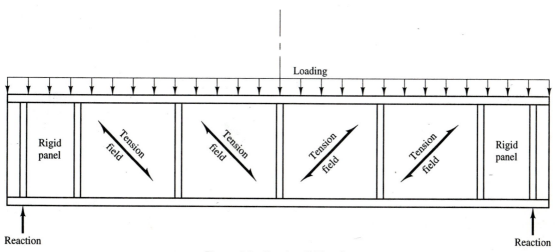

Figure 8.8   Tension field action.

For $187\sqrt{k/F_y} < h/t_w \leq 234\sqrt{k/F_y}$

$$\phi V_n = \phi 0.6 F_y A_w \frac{187\sqrt{k/F_y}}{h/t_w} \qquad (8.9)$$

For $h/t_w > 234\sqrt{k/F_y}$

$$\phi V_n = \phi A_w \frac{26{,}400k}{(h/t_w)^2} \qquad (8.10)$$

If stiffeners are not used, then k, the web buckling coefficient in Eqs. (8.8)–(8.10), is taken as 5.0. When stiffeners are used, the design shear may be increased as discussed in Section 8.3.3.

8.3.2 Tension Field Action: Shear

A plate girder with transverse stiffeners may take advantage of the tension field action, with the design shear strength calculated as a function of the stiffener spacing. The rules are specified in the LRFD specification and are a function of both the web slenderness h/t_w and the stiffener aspect ratio a/h. With $\phi = 0.9$, the shear capacity is given by

For $h/t_w \leq 187\sqrt{k/F_y}$

$$\phi V_n = \phi\, 0.6 F_y A_w \qquad (8.11)$$

For $h/t_w > 187\sqrt{k/F_y}$

$$\phi V_n = \phi 0.6 A_w F_y \left[C_v + \frac{1 - C_v}{1.15\sqrt{1 + (a/h)^2}} \right] \qquad (8.12)$$

In Eq. (8.12), C_v is defined as follows:

For $187\sqrt{k/F_y} \leq h/t_w \leq 234\sqrt{k/F_y}$

$$C_v = \frac{187\sqrt{k/F_y}}{h/t_w}$$

For $h/t_w > 234\sqrt{k/F_y}$

$$C_v = \frac{44{,}000k}{(h/t_w)^2\, F_y}$$

The stiffener spacing for the panel next to the support must be less than that within the span. The end panel must be especially rigid for the remainder of the web to properly function as a Pratt truss. Shear in the end panel must conform to the rules for a nontension field girder.

8.3.3 Intermediate Stiffeners: Tension and Nontension Field

When stiffeners are utilized, whether required as for the tension field girders or not required, the design shear for the girder is to be determined using k, the web plate buckling coefficient, as given by

$$k = 5 + \frac{5}{(a/h)^2} \qquad (8.13)$$

where a/h is the aspect ratio of the stiffener panel and a is the stiffener clear spacing. When a/h exceeds 3 or $[260/(h/t_w)]^2$, k is taken as 5. For practical purposes, there would be no reason to have stiffeners in this case.

The LRFD specification, Appendix G1, gives maximum (h/t_w) limits as

For $a/h \leq 1.5$

$$\left(\frac{h}{t_2}\right)_{\max} = \frac{2000}{\sqrt{F_y}} \qquad (8.14)$$

For $a/h > 1.5$

$$\left(\frac{h}{t_w}\right)_{\max} = \frac{14,000}{\sqrt{F_y (F_y - 16.5)}} \qquad (8.15)$$

A summary of the design shear for an A36 plate girder as a function of h/t_w is given in Fig. 8.9. The solid line represents the case of nontension field action, and the dashed lines represent the tension field case. It can be seen that the design shear strength for a plate girder increases with the addition of stiffeners and with a decrease in the aspect ratio of the stiffened panels.

EXAMPLE 8.1

Given: The cross section of a homogeneous, A36, plate girder is shown in Fig. 8.10. The span is 120 ft and the unbraced length is 20 ft. Assume $C_b = 1.0$. Determine the flexural and shear capacity. Assume a web thickness of $\frac{3}{8}$ in. (compact) and $\frac{1}{4}$ in. (slender).

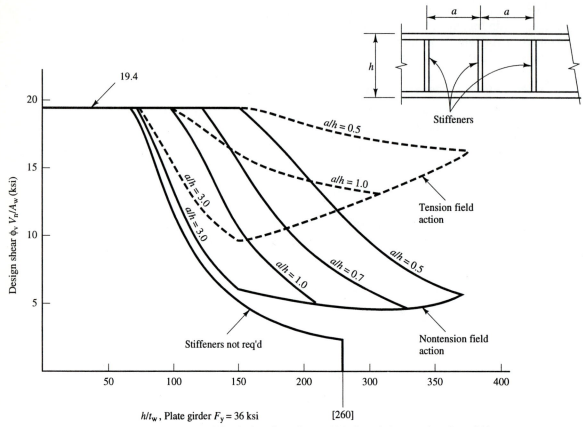

Figure 8.9 A summary of the design shear for an A36 plate girder as a function of h/t_w:
——, nontension field action; ----, tension field.

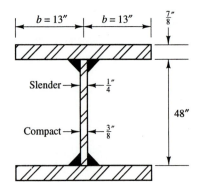

Figure 8.10 Cross-section of a
homogeneous A36
plate girder
(Example 8.1).

Solution

Determine the section properties for both plate girders.

(a) $t_w = 0.375$ in.   (b) $t_w = 0.25$ in.

$A = 63.5$ in.^{2} $A = 57.5$ in.^{2}
$I_x = 17039$ in.^{4} $I_x = 15887$ in.^{4}
$I_y = 2563$ in.^{4} $I_y = 2563$ in.^{4}
$S_x = 685$ in.^{3} $S_x = 639$ in.^{3}
$Z_x = 1328$ in.^{3} $r_T = 7.4$ in.
$r_y = 6.35$ in. $r_y = 6.68$ in.

For plate girder (a):

Check the web

$$\frac{h_c}{t_w} = \frac{48}{0.375} = 128 > \frac{640}{\sqrt{36}} = 106.7$$

$$< \frac{970}{\sqrt{36}} = 161.7$$

Thus, this is a noncompact web girder.

Check FLB

$$\frac{b_f}{2t_f} = \frac{26}{2(0.875)} = 14.86 > \frac{65}{\sqrt{36}} = 10.8$$

$$< \frac{106}{\sqrt{(36 - 16.5)}} = 24.0$$

Thus, the flange is noncompact, so that

$$M_p = \frac{1328(36)}{12} = 3984 \text{ kip-ft}$$

$$M_r = \frac{(36 - 16.5)(685)}{12} = 1113 \text{ kip-ft}$$

and

$$M_n = 3984 - (3984 - 1113)\frac{14.86 - 10.8}{24.0 - 10.8} = 3101 \text{ kip-ft}$$

Check WLB (noncompact)

$$M_r = \frac{36(685)}{12} = 2055 \text{ kip-ft}$$

and

$$M_n = 3984 - (3984 - 2055)\frac{128.0 - 106.7}{161.7 - 106.7} = 3237 \text{ kip-ft}$$

Check LTB

$$L_b/r_y = \frac{20(12)}{6.35} = 37.8 < \frac{300}{\sqrt{36}} = 50$$

Thus, the girder is adequately braced; therefore,

$$M_n = M_p = 3984 \text{ kip-ft}$$

Selecting the lowest M_n value, computed in this case according to FLB, yields

$$M_n = 3101 \text{ kip-ft}$$

For nontension field shear, with

$$\frac{h_c}{t_w} = 128 > 234\sqrt{\frac{k}{F_y}} = 87.2$$

$$V_n = 18\left[\frac{26,400(5)}{128^2}\right] = 145 \text{ kips}$$

For plate girder (b):

$$\frac{h_c}{t_w} = \frac{48}{0.25} = 192 > \frac{970}{\sqrt{36}} = 161.7$$

Thus, this is a slender web.

For LTB

$$\frac{L_b}{r_T} = \frac{20(12)}{7.2} = 33.3 < \frac{300}{\sqrt{36}} = 50$$

Therefore,

$$F_{cr} = F_y = 36 \text{ ksi}$$

$$R_{PG} = 1.0 - 0.0005\left[\frac{48(0.25)}{26(0.875)}\right](192 - 161.7) = 0.992$$

$$M_n = S_x R_{PG} F_y = \frac{639(0.992)(36)}{12} = 1902 \text{ kip-ft}$$

For FLB

$$\frac{b_f}{2t_f} = 14.86 > \frac{65}{\sqrt{36}} = 10.8$$

$$< \frac{150}{\sqrt{36}} = 25$$

Therefore,

$$F_{cr} = (1)(36)\left[1 - \frac{(14.86 - 10.8)}{2(25 - 10.8)}\right] = 30.85$$

$$R_{PG} = 1.0 - 0.0005\left[\frac{48(0.25)}{26(0.875)}\right]\left(192 - \frac{970}{\sqrt{30.85}}\right) = 0.995$$

$$M_n = \frac{639(0.995)(36)}{12} = 1907 \text{ kip-ft}$$

Selecting the lowest value of M_n, resulting from LTB, gives

$$M_n = 1902 \text{ kip-ft}$$

Nontension field shear,

$$\frac{h_c}{t_w} = 192 > 234\sqrt{\frac{k}{F_y}} = 87.0$$

Therefore,

$$V_n = 48(0.25)\left[\frac{26,400(5)}{192^2}\right] = 43.0 \text{ kips}$$

PROBLEMS

For Problems 8.1–8.6 assume girders have full lateral support.

8.1. Determine the moment capacity of an A36 plate girder with a web plate of $75 \times \frac{3}{8}$ in. and equal flange plates of $14 \times 1\frac{1}{4}$ in.

8.2. Determine the moment capacity of an A36 plate girder with a web plate of $50 \times \frac{1}{2}$ in. and equal flange plates of 12×1 in.

8.3. Using steel with $F_y = 50$ ksi, determine the moment capacity of a plate girder with a web plate of $80 \times \frac{1}{2}$ in. and equal flange plates of $16 \times 1\frac{1}{2}$ in.

8.4. Determine the shear strength for an A36 plate girder without transverse stiffeners. The web plate is $100 \times \frac{3}{4}$ in. and the flange plates are $15 \times 1\frac{1}{2}$ in.

8.5. Determine the shear strength for an A36 plate girder with transverse stiffeners located at each end and 20 in. from each end along with stiffeners every 40 in. within the span. The web plate is $40 \times \frac{1}{4}$ in. and the flange plates are $8 \times \frac{3}{4}$ in.

8.6. Determine the moment capacity for the plate girder from Problem 8.5 and the factored uniform load that the girder may carry on an 80-ft span.

REFERENCES

[1] Basler, K., Strength of plate girders in shear, *Journal of the Structural Division, ASCE,* Vol. 87, No. ST10, October 1961 (pp. 151–180).

[2] Basler, K., Strength of plate girders under combined bending and shear, *Journal of the Structural Division, ASCE,* Vol. 87, No. ST10, October 1961 (pp. 181–195).

[3] Basler, K., and Thürlimann, B., Strength of plate girders in bending, *Journal of the Structural Division, ASCE,* Vol. 87, No. ST8, August 1961 (pp. 153–181).

[4] Zahn, C. J., Plate girder design using LRFD, *Engineering Journal, AISC,* Vol. 24, 1st Quarter, 1987 (pp. 11–20).

[5] Galambos, T. V., Cooper, P. B., and Ravindra, M. K., LRFD criteria for plate girders, *Journal of the Structural Division, ASCE,* Vol. 104, No. ST9, September 1978 (pp. 1389–1407).

CHAPTER NINE

Beam-Columns and Frame Behavior

9.1 INTRODUCTION

Beam-columns are members that are subjected to axial forces and bending moments simultaneously. Thus, their behavior falls somewhere between that of a pure, axially loaded column and a beam with only moments applied. Practical applications of beam-columns are numerous. They occur as chord members of trusses, as elements of rigidly connected frameworks, and as members of pin-connected structures with eccentric loads.

The manner in which the combined loads reach any particular beam-column will have a significant impact on the ability of the member to resist those loads. Starting with the axially loaded column, bending moments may occur from various sources. Lateral load may be applied directly to the member, as would be the case for a truss member or a column supporting the lateral load from a wall. The axial force may actually be applied at some eccentricity from the centroid of the column as a result of the specific connections. Or, the member may receive end moments from its connection to other members of the structure, such as in a rigid frame. In all cases, the relation of the beam-column to the other elements of the structure is important in determining both the applied forces and the resistance of the member.

To understand the behavior of beam-columns, it is common practice to look at the response as predicted through an interaction equation. The

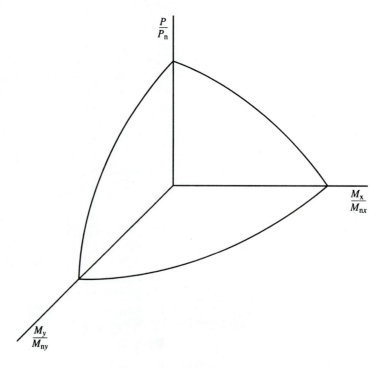

Figure 9.1   Ultimate interaction surface for a stocky beam-column.

response of a beam-column to a given load P, major axis moment M_x, and minor axis moment M_y is presented on the three-dimensional graph shown in Fig. 9.1. Each axis represents the capacity of the member when it is subjected to loading of one type only, while the curves represent the combination of two types of loading. The surface formed by connecting the three curves represents the interaction of axial load and biaxial bending. It is this interaction surface that is of interest to the designer.

The end points of the curves shown in Fig. 9.1 are dependent on the capacities of the members as described for beams (Chapter 7) and columns (Chapter 6). The shape of the curves between these end points will depend on several of the same properties as for the beams and columns, as well as those properties for additional members of the structure.

9.2 SECOND-ORDER EFFECTS

The single most complicating factor in the analysis and design of a beam-column is what is known as second-order effects. Second-order effects are the direct result of structural deformations. Since the common elastic methods of structural analysis assume that all deformations are small, they are not able to account for these additional effects. The results of these elastic analyses are referred to as primary or first-order forces. Since the effect of the deformations requires an additional analysis, they are referred to as second-order effects.

Second-order effects may be included in an analysis by either of two approaches. A complete second-order inelastic analysis would take into account the actual deformation of the structure and the resulting forces as well as the sequence of loading. This approach to analysis is generally more complex than is necessary for normal design. An approach that is consistent with normal design office practice uses an elastic analysis to approximate the inelastic conditions. In this approach, a classical first-order analysis is carried out. The resulting forces are used to calculate the elastic deformations of the structure. The loads are then applied to the deformed structure, and secondary moments and forces are determined. This process is repeated, incrementing the load until the analysis shows that the structure has become unstable. Thus, the load capacity of the frame is determined.

Two different deflection components will influence the moments in the beam-column. The first, illustrated in Fig. 9.2a, is the deflection along the length of the member that results from the moment gradient along the member. In this case the member ends must remain in their original position; thus, no sway is considered. The moment created by the load P acting at an eccentricity δ from the deformed member will be superimposed on the moment gradient resulting from the applied end moments. The magnitude of this additional, second-order, moment is dependent on the properties of the column itself. Thus, this is referred to as the member effect.

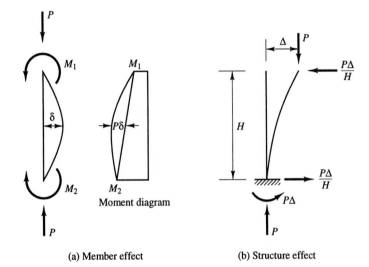

Figure 9.2 Column displacements for second-order effects.

(a) Member effect   (b) Structure effect

When the beam-column is part of a structure, the displacements of the structure will also influence the moments in the member. For a beam-column that is permitted to sway an amount Δ, as shown in Fig. 9.2b, the additional moment will be given by $P\Delta$. Since the lateral displacement of a given member is a function of the properties of all members in the given story, this moment will be referred to as the structure effect.

9.3 INTERACTION PRINCIPLES

The interaction of axial load and bending within the elastic response range of a beam-column may be investigated through the straightforward techniques of superposition. This is the approach normally considered in elementary strength of materials where the normal stress due to an axial force is added to the normal stress due to a bending moment.

Although the superposition of individual stress effects is both simple and correct for elastic stresses, there are significant limitations when applying this approach to the limit states of real structures:

1. Superposition of stress is correct only for behavior in the elastic range and only for similar stress types.
2. Superposition of strain can be extended into the inelastic range only when deformations are small.
3. Superposition cannot account for member deformations or stability effects such as local buckling.
4. Superposition cannot account for structural deflections and system stability.

With these limitations in mind, it is desirable to develop interaction equations that will reflect the true limit states behavior of beam-columns.

Any limit state interaction equation must reflect the following characteristics:

AXIAL LOAD

1. Maximum column strength
2. Individual column slenderness

BENDING MOMENT

1. Lateral support conditions
2. Sidesway conditions
3. Member second-order effects
4. Structure second-order effects
5. Moment gradient along member

It is also important that the resulting equations provide a close correlation with test results and theoretical analyses for beam-columns, including the two limiting cases of pure bending and pure compression.

Application of the resulting interaction equations may be regarded as a process of determining available axial load capacity in the presence of a given bending moment or determining the available moment capacity in the presence of a given axial load. An applied bending moment consumes a portion of the column strength, leaving a reduced axial load capacity. When the two actions are added together, they must not exceed the maximum column strength. Conversely, the axial load can be regarded as consuming a fraction of the moment capacity of the member. This fraction, plus the applied moments, which include all second-order effects, must not exceed the maximum beam strength.

9.4 INTERACTION EQUATIONS

The basic form of the three-dimensional interaction equation is

$$\frac{P}{P_n} + \frac{M_x}{M_{nx}} + \frac{M_y}{M_{ny}} \leq 1.0 \tag{9.1}$$

It can be seen from Fig. 9.3 that this results in a straight-line representation of the interaction between any two of the components. It should also be apparent that the three-dimensional aspect is represented by a plane with intercepts given by the straight lines on the three coordinate planes.

The LRFD interaction equations result from fitting interaction equations of the form of Eq. (9.1) to a set of data. Using the data of Kanchanalai [1], two equations were fit to the data developed from the analysis of 82 different sidesway permitted frames. The resulting equations, given as Eqs. (H1-1a) and (H1-1b) in the LRFD specification, are given here as

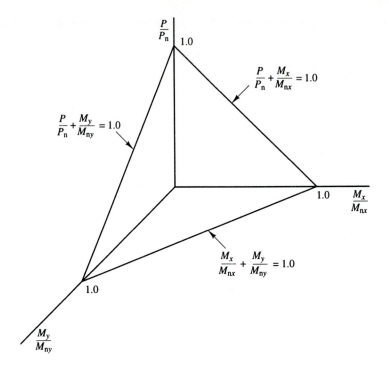

Figure 9.3 Simplified interaction surface.

Eqs. (9.2) and (9.3). The equations shown here and plotted in Fig. 9.4 consider bending about only one principal axis.

For $P_u/\phi P_n \geq 0.2$

$$\frac{P_u}{\phi P_n} + \frac{8M_u}{9\phi_b M_n} = 1.0 \tag{9.2}$$

and for $P_u/\phi P_n < 0.2$

$$\frac{P_u}{2\phi P_n} + \frac{M_u}{\phi_b M_n} = 1.0 \tag{9.3}$$

In both Eqs. (9.2) and (9.3) it is important to note the following:

1. The nominal column strength P_n is based on the axis of the column with the largest slenderness ratio. This is not necessarily the axis about which bending takes place.
2. The nominal bending strength M_n is based on the bending strength of the beam without axial load, including the influence of all of the beam limit states.
3. The applied factored load P_u is the load on the column found through an elastic analysis.

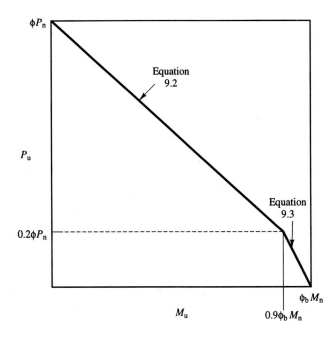

Figure 9.4 LRFD interaction equations.

4. The moment M_u is the second-order bending moment on the member. It may be determined through a second-order analysis, either elastic or inelastic, or by a modification of the first-order elastic moments using the amplification factors provided in the specification. These amplification factors are discussed as they relate to braced frames (Section 9.5) and unbraced frames (Section 9.6).

9.5 BRACED FRAMES

A frame is considered braced if a positive system, such as a shear wall (masonry or other material) or diagonal steel members, as illustrated in Fig. 9.5, serves to resist the lateral loads and to stabilize the frame under gravity loads. In both cases, the columns are considered braced against lateral translation and the in-plane K factor is taken as 1.0 or less. This is the type of column that was presented in Chapter 6.

   If the column in a braced frame is rigidly connected to a girder, bending moments will result from the application of the gravity loads. These moments may be determined through a first-order elastic analysis. The additional moments resulting from the displacement along the column length will be determined through the application of an amplification factor.

   The full derivation of the amplification factor has been presented by various authors [2, 3]. A somewhat simplified derivation is presented here to help establish the background. An axially loaded column with equal

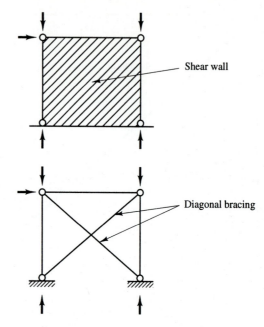

Figure 9.5   Braced frame.

and opposite end moments is presented in Fig. 9.6a. The resulting moment diagram is shown in Fig. 9.6b where the moments resulting from both the end moments and the secondary effects are given. The maximum moment occurring at the midheight of the column M_u is seen to be

$$M_u = M + P\delta$$

The amplification factor is defined as

$$AF = \frac{M_u}{M} = \frac{M + P\delta}{M}$$

Rearranging terms yields

$$AF = \frac{1}{1 - P\delta/(M + P\delta)}$$

Two simplifying assumptions will be made [3]. The first is based on the assumption that δ is sufficiently small that

$$\frac{\delta}{M + P\delta} \approx \frac{\delta}{M}$$

and the second assumes that

$$\frac{M}{\delta} = \frac{8EI}{L^2} \approx \frac{\pi^2 EI}{L^2} = P_e$$

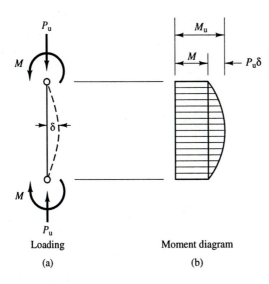

Loading

(a)

Moment diagram

(b)

Figure 9.6 An axially loaded column with equal and opposite end moments.

Since these simplifying assumptions are in error in opposite directions, they tend to be offsetting. This results in a fairly accurate indication of the amplification. Thus,

$$AF = \frac{1}{1 - P_u/P_e} \qquad (9.4)$$

A comparison between the actual amplification and that given by Eq. (9.4) is shown in Fig. 9.7.

The discussion so far has assumed that the moments at each end of the column are equal and opposite and that the resulting moment diagram is uniform. This is the most severe loading case for a column. If the moment is not uniformly distributed, the displacement δ will be less and the resulting amplified moment will be less than indicated. It has been customary to use the case of uniform moment as a base and to provide for other moment gradients by converting them to an equivalent uniform moment through the use of an additional factor, C_m.

Numerous studies [4, 5] have shown that a reasonably accurate correction results, for beam-columns braced against translation and not loaded in bending between their supports, if the moment is reduced by multiplying by C_m, where

$$C_m = 0.6 - 0.4\left(\frac{M_1}{M_2}\right) \qquad (9.5)$$

M_1/M_2 is the ratio of the smaller to larger moments at the ends of the member unbraced length in the plane of bending. M_1/M_2 is positive when the member is bent in reverse curvature and negative when the member is bent in single curvature.

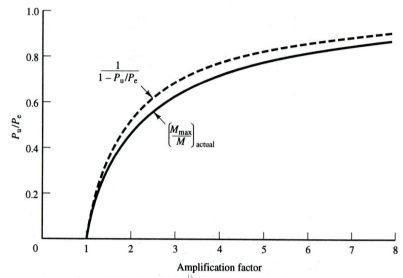

Figure 9.7 Amplified moment: exact and approximate.

For beam-columns in braced frames where the member is subjected to transverse loading between their supports, C_m may be taken from Table 9.1 or approximated as follows:

For members whose ends are rotationally restrained

$$C_m = 0.85$$

For members whose ends are rotationally unrestrained

$$C_m = 1.0$$

Although these values are approximate, they have been compared to accurate analyses [6] and have been considered sufficiently accurate to be adopted for design.

The combination of the amplification factor AF and the equivalent moment factor C_m accounts for the total member secondary effects. This combined factor is given as B_1 in the LRFD specification and is shown here as

$$B_1 = \frac{C_m}{1 - P_u/P_e} \geq 1.0 \tag{9.6}$$

Thus, the value of M_u in Eqs. (9.2) and (9.3) is taken as

$$M_u = B_1 M$$

TABLE 9.1 AMPLIFICATION FACTORS ψ AND C_m, $C_m = 1 + \psi P_u / P_e$

Case	ψ	C_m
	0	1.0
	-0.4	$1 - 0.4\,\dfrac{P_u}{P_e}$
	-0.4	$1 - 0.4\,\dfrac{P_u}{P_e}$
	-0.2	$1 - 0.2\,\dfrac{P_u}{P_e}$
	-0.3	$1 - 0.3\,\dfrac{P_u}{P_e}$
	-0.2	$1 - 0.2\,\dfrac{P_u}{P_e}$

Source: Reprinted by permission of the American Institute of Steel Construction, Inc, from the 1st edition of the *AISC-LRFD Manual*.

where M is the maximum moment on the beam-column. It is possible for C_m to be less than 1.0 and for Eq. (9.6) to give an amplification factor less than 1.0. This is an indication that the combination of the $P\delta$ effects and the nonuniform moment gradient working together result in a moment less than the maximum moment on the beam-column from first-order effects. In those cases, the amplification factor B_1 will be taken as 1.0.

EXAMPLE 9.1

Given: The three-dimensional braced frame for a single-story structure is given in Fig. 9.8. Rigid connections are provided at the roof level for columns $A1$, $B1$, $A4$, and $B4$. All other column connections are pinned. Design column $A1$ for the given loads using the LRFD specification provided amplification factor.

Dead load = 50 psf, snow load = 20 psf, roof live load = 10 psf, wind load = 20 psf horizontal. Use A36 steel. Assume that the X-bracing is so much

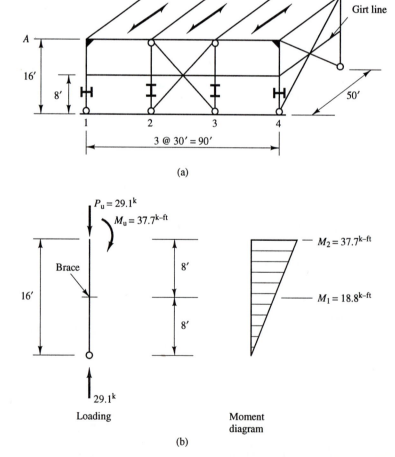

(a)

(b)

Figure 9.8 Three-dimensional braced frame for a single-story structure.

stiffer than the rigid frames that it resists all lateral load and that the column bracing at midheight acts to resist buckling only; that is, it does not participate in resisting bending.

Solution

Step 1. Determine the appropriate load factor combinations. From Section A4.1 of the LRFD specification, two combinations must be checked:

$$1.2D + 1.6(L_r \text{ or } S \text{ or } R) + (0.5L \text{ or } 0.8W) \qquad \text{(A4-3)}$$

$$1.2D + 1.3W + 0.5L + 0.5(L_r \text{ or } S \text{ or } R) \qquad \text{(A4-4)}$$

Step 2. Determine the factored loads based on these load combinations.

$$1.2(50) + 1.6(20) = 92 \text{ psf} \qquad \text{(A4-3)}$$

$$1.2(50) + 0.5(20) = 70 \text{ psf} \qquad \text{(A4-4)}$$

Therefore, use a uniformly distributed load of 92 psf.

Step 3. Carry out a preliminary first-order analysis. Since the structure is indeterminate, a number of approaches may be taken. If a 6 to 1 ratio of moment of inertia for beams to columns is assumed, a moment distribution analysis will yield the moment and force given in Fig. 9.8b. Thus, the column will be designed to carry $P_u = 29.1$ kips and $M_u = 37.7$ kip-ft.

Step 4. Select a trial size for column A1.

Try W8 × 24.

$$A = 7.08 \text{ in.}^2, \frac{r_x}{r_y} = 2.12, r_x = 3.42 \text{ in.}, I_x = 82.8 \text{ in.}^4$$

The column is oriented so that bending is about the x axis of the section. It is braced against sidesway, pinned at the bottom, and rigidly connected at the top. Lateral support for buckling is provided about both axes at midheight; thus, $KL_x = KL_y = 8.0$ ft.
    From the *LRFD Manual*, $\phi P_n = 180$ kips > 29.1 kips

$$\phi M_n = 61.0 \text{ kip-ft for } L_b = 8.0 \text{ ft}$$

Step 5. Check W8 × 24 for combined axial load and bending in-plane. For an unbraced length of 8 ft, the Euler load is

$$P_e = \frac{\pi^2 EI}{KL^2} = \frac{\pi^2 (29,000)(82.8)}{[8(12)]^2} = 2571 \text{ kips}$$

The column is bent in single curvature between bracing points so that $M_1/M_2 = -0.5$. Thus,

$$C_m = 0.6 - 0.4(-0.5) = 0.8$$

Therefore, the amplification factor becomes

$$B_1 = \frac{0.8}{1 - (29.1/2571)} = 0.81 < 1.0$$

The specification requires that B_1 not be less than 1.0. Therefore, taking $B_1 = 1.0$ yields

$$M_{ux} = B_1 M_x = 1.0(37.7) = 37.7 \text{ kip-ft}$$

Determine which equation to use from

$$P_u/\phi P_n = \frac{29.1}{180} = 0.16 < 0.2$$

Therefore, use Eq. (9.3) (H1-1b)

$$0.5(0.16) + \frac{37.7}{61.0} = 0.7 < 1.0$$

Thus, the W8 × 24 will carry the given loads. The solution to Eq. (H1-1b) indicates that there is a fairly wide extra margin of safety. It would be appropriate to consider a slightly smaller column for a more economical design.

9.6 UNBRACED FRAMES

An unbraced frame depends on the stiffness of the beams and columns that make up the frame for stability under gravity loads and under combined gravity and lateral loads. Unlike braced frames, there is no external structure to lean against for stability. Columns in unbraced frames are subjected to both axial load and moment and will experience lateral translation.

The same interaction equations, Eqs. (9.2) and (9.3), are used to design beam-columns in unbraced frames. However, in addition to the member second-order effects discussed in Section 9.5, there is the additional second-order effect that results from the sway or lateral displacement of the frame.

Figure 9.9 shows a cantilever column under the action of an axial load and a lateral load. Figure 9.9a is the column as viewed for a first-order elastic analysis where equilibrium requires a moment at the bottom, $M_{lt} = HL$. The deflection that results at the top of the column Δ_1 is an elastic deflection of a cantilever beam; thus,

$$\Delta_1 = \frac{HL^3}{3EI} \tag{9.7}$$

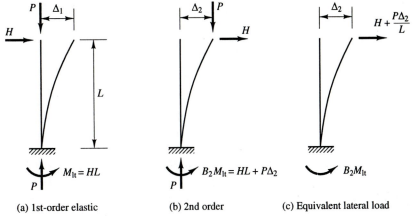

(a) 1st-order elastic   (b) 2nd order   (c) Equivalent lateral load

Figure 9.9   Structure second-order effect: sway.

A second-order analysis would yield the forces and displacements as seen in Fig. 9.9b. The displacement Δ_2 is the total displacement including second-order effects, and the moment including second-order effects is

$$B_2 M_{lt} = HL + P\Delta_2 \qquad (9.8)$$

An equivalent lateral load may be determined that will result in the same moment at the bottom of the column as the second-order analysis. This load is $H + P\Delta_2/L$ and is shown in Fig. 9.9c.

It may be assumed, with only slight error, that the displacement at the top of the column for the cases in Figs. 9.9b and c are the same. Thus, from the equivalent load

$$\Delta_2 = \left(\frac{H + P\Delta_2}{L}\right)\frac{L^3}{3EI} \qquad (9.9)$$

Equation (9.9) may now be solved for Δ_2 and the result substituted into Eq. (9.8). Solving the resulting equation for the amplification factor yields

$$B_2 = \frac{1}{1 - P\Delta_1/HL} \qquad (9.10)$$

Considering that the beam-column will be part of a total structure, this equation is modified to include the total load on the columns in a story and the total lateral load in a story. The amplification factor given by the LRFD specification as Eq. (H1-5) is

$$B_2 = \frac{1}{1 - \Sigma P_u \, \Delta_{oh}/\Sigma HL} \qquad (9.11)$$

where Δ_{oh} is the story drift from a first-order analysis using factored loads.

Often it is desirable to limit the lateral displacement, or drift, of a structure during design. This limit may be defined using a drift index which is the story drift divided by the story height, Δ_{oh}/L. The design then proceeds with sections selected so that the final structure performs as desired. This is similar to beam design where deflection is a serviceability criteria. Since the drift index can be established without knowing member sizes, it may be used in Eq. (9.11). Thus, an analysis with assumed member sizes is unnecessary.

In lieu of the more accurate value of B_2 calculated from Eq. (9.11), a conservative alternative is given in the specification as Eq. (H1-6) and here as

$$B_2 = \frac{1}{1 - \Sigma P_u / \Sigma P_e} \tag{9.12}$$

where

ΣP_u = total gravity load on the story

ΣP_e = Euler load total for all columns in story where the K factor is determined in the plane of bending and is greater than 1.0

The moment M_u to be used in Eqs. (9.2) and (9.3) must include both the member and structure second-order effects. Thus, a first-order analysis without sidesway must be carried out, yielding moments M_{nt}, that is, no translation, to be amplified by B_1. Then, a first-order analysis including lateral loads and permitting translation must be carried out. This will yield moments M_{lt} with translation, to be amplified by B_2. Thus, the second-order moment is

$$M_u = B_1 M_{nt} + B_2 M_{lt} \tag{9.13}$$

where

B_1 is given by Eq. (9.6)

B_2 is given by Eq. (9.11) or Eq. (9.12)

M_{nt} = first-order moments with no translation

M_{lt} = first-order moments that result from lateral translation

It should be noted that M_{lt} could include moments that result from unsymmetrical frame properties or loading as well as from lateral loads. In most real structures, however, moments resulting from this lack of symmetry are usually small and may be ignored.

For situations where there is no lateral load on the structure and significant lateral translation does not result from the structure geometry or load placement, a first-order analysis should be used and the amplification factor B_2 set to 0.0. This would be the case for loading combinations that do not include lateral loads.

EXAMPLE 9.2

Given: An exterior column from an intermediate level of a multistory rigid frame is shown in Fig. 9.10. The column is part of a braced frame out of the plane of the figure. Figure 9.10a shows the members to be checked. The same column section will be used for the level above and below the column AB. A first-order analysis of the frame for gravity loads results in the forces shown in Fig. 9.10b, while the results for gravity plus wind are shown in Fig. 9.10c.

Determine whether the W14 × 109, A36 column is adequate to carry the imposed loading. Assume that the frame drift under factored loads will be limited to height/300.

Solution

Using the effective length nomograph introduced in Chapter 6 and given in the *LRFD Manual*, determine the effective length for buckling in the plane of the frame:

$$G_t = G_b = \frac{2(1240/12.5)}{(2100/30)} = 2.83$$

Thus, $K = 1.79$ (nomograph)

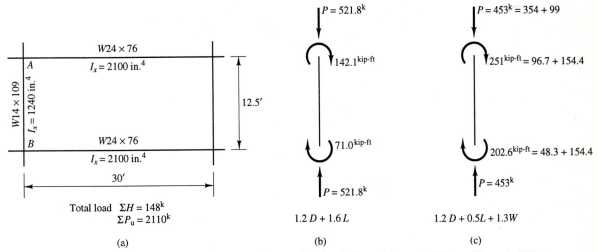

Figure 9.10   An exterior column from an intermediate level of a multistory rigid frame (Example 9.2).

Determine the controlling effective length. With $r_x/r_y = 1.67$, an equivalent effective length for the x axis is determined.

$$(KL_x)_{\text{equiv}} = \frac{1.79(12.5)}{1.67} = 13.4 \text{ ft}$$

$$KL_y = 1.0(12.5) = 12.5 \text{ ft}$$

From the column tables, for $KL = 13.4$ ft, $\phi P_n = 887.8$ kips.

First, consider the loading case including wind. The column end moments given in Fig. 9.10c are a combination of moments resulting from a nonsway gravity load analysis and a wind analysis. These moments are

$$M_{\text{nt}} = 96.7 \text{ kip-ft}$$

$$M_{\text{lt}} = 154.4 \text{ kip-ft}$$

The nontranslation moments must be amplified by B_1 as follows:

$$C_m = 0.6 - 0.4\frac{48.3}{96.7} = 0.4$$

$$P_e = \frac{\pi^2(29,000.0)(1240)}{[1.0(12.5)(12)]^2} = 15,773.8 \text{ kips}$$

$$B_1 = \frac{0.4}{1.0 - 354.0/15,773.8} = 0.41 < 1.0$$

Thus,

$$B_1 = 1.0$$

The translation moment must be amplified by B_2. There are two options for calculation of B_2. Since the complete design is not known and the design drift limit is known, LRFD specification equation (H1-5) will be used here.

The total lateral load on this story is

$$\Sigma H = 148 \text{ kips}$$

The total gravity load for this load combination is

$$\Sigma P_u = 2110 \text{ kips}$$

The drift limit is

$$\Delta_{\text{oh}} = \frac{L}{300}$$

Thus,

$$B_2 = \frac{1.0}{1.0 - (2110/148)(1/300)} = 1.05$$

The amplified moment is given as

$$M_u = 1.0(96.7) + 1.05(154.4) = 258.8 \text{ kip-ft}$$

The unbraced length of the compression flange for pure bending is 12.5 ft, which is less than $L_p = 15.5$ ft for this section. Thus, the moment capacity of the section is $\phi M_p = 518$ kip-ft.

Determine the appropriate interaction equation:

$$\frac{P_u}{\phi P_n} = \frac{453.0}{887.8} = 0.51 > 0.2$$

Thus, use Eq. (H1-1a), which yields:

$$0.51 + \frac{8}{9}\left(\frac{258.8}{518}\right) = 0.95 < 1.0 \qquad \text{OK}$$

Check the section for $1.2D + 1.6L$. For $C_m = 0.4$, $P_e = 15{,}773$ kips as before and $P_u = 521.8$,

$$B_1 = 1.0$$

Since there is no lateral load, $M_{lt} = 0.0$ and B_2 is unnecessary. Again using Eq. (H1-1a),

$$\frac{521.8}{887.8} + \frac{8}{9}\left(\frac{142.1}{518.0}\right) = 0.83 < 1.0 \qquad \text{OK}$$

Thus, the $W14 \times 109$ is adequate for both loading conditions considered.

The moments in the beams and the beam-to-column connections must also be amplified for the lateral translation case. This is done by considering equilibrium of the beam-to-column joint. The amplified moments in the column above and below the joint are added together and this sum is distributed to the beams which frame into the joint according to their stiffness. These moments are then added to the beam moments determined from the no translation case to determine the beam and connection design moments.

9.7 INITIAL BEAM-COLUMN SELECTION

Beam-column design is a trial and error process that requires the column section to be known before any of the critical parameters may be determined for use in the appropriate interaction equations. There are numerous approaches to determining a preliminary column section. Each incorporates its own level of sophistication and results with its own level of accuracy. Regardless of the approach used to select the trial section, one factor remains: The trial section must satisfy the appropriate interaction equation.

To establish a simple yet useful approach to selecting a trial section, LRFD specification Eq. (H1-1a) will be rewritten. This equation was written as Eq. (9.2) for bending about one axis. Using the complete equation and multiplying each term by ϕP_n yields

$$P_u + \frac{8\phi P_n}{9\phi_b M_{nx}} M_{ux} + \frac{8\phi P_n}{9\phi_b M_{ny}} M_{uy} = \phi P_n \qquad (9.14)$$

Multiplication of the third term by M_{nx}/M_{nx}, letting

$$m = \frac{8\phi P_n}{9\phi_b M_{nx}} \qquad (9.15a)$$

and

$$U = \frac{M_{nx}}{M_{ny}} \qquad (9.15b)$$

and substituting into Eq. 9.14 yields

$$P_u + mM_{ux} + mUM_{uy} = \phi P_n = P_{eff} \qquad (9.16)$$

The accuracy used in the evaluation of m and U will dictate the accuracy with which Eq. (9.16) will represent the column being selected. Since at this point in a design, the actual column section is not known, exact values of m and U cannot be determined.

Numerous approaches to the evaluation of these multipliers have been presented [7, 8]. If the influence of length on P_n and M_{nx} is neglected, the ratio P_n/M_{nx} becomes A/Z_x. Evaluation of this m for all W6 to W14 shapes results in the average m values given in Table 9.2. If the relationship between the area A and the plastic section modulus Z_x is established using an approximate internal moment arm of $0.84d$, then m would reduce to $24/d$. This value is also presented in Table 9.2. It is easily

TABLE 9.2   LRFD BENDING FACTORS

Shape	$m = (A/Z_x)_{avg}$	$m = 24/d$	U
W6	4.12	4.0	2.89
W8	3.07	3.0	3.17
W10	2.47	2.4	3.62
W12	1.96	2.0	3.47
W14	1.63	1.71	2.86

TABLE 9.3 BENDING FACTOR m

F_y	36 ksi							50 ksi						
	Values of m													
KL (ft):	10	12	14	16	18	20	22 and over	10	12	14	16	18	20	22 and over
1st Approximation														
All shapes	2.0	1.9	1.8	1.7	1.6	1.5	1.3	1.9	1.8	1.7	1.6	1.4	1.3	1.2
Subsequent Approximation														
S4,5,6	1.3	1.0	0.8	0.7	0.6	0.5	0.5	1.1	0.9	0.8	0.7	0.6	0.5	0.5
W,M4	3.1	2.3	1.7	1.4	1.1	1.0	0.8	2.4	1.8	1.4	1.1	1.0	0.9	0.8
W,M5	3.2	2.7	2.1	1.7	1.4	1.2	1.0	2.8	2.2	1.7	1.4	1.1	1.0	0.9
W,M6	2.8	2.5	2.1	1.8	1.5	1.3	1.1	2.5	2.2	1.8	1.5	1.3	1.2	1.1
W8	2.5	2.3	2.2	2.0	1.8	1.6	1.4	2.4	2.2	2.0	1.7	1.5	1.3	1.2
W10	2.1	2.0	1.9	1.8	1.7	1.6	1.4	2.0	1.9	1.8	1.7	1.5	1.4	1.3
W12	1.7	1.7	1.6	1.5	1.5	1.4	1.3	1.7	1.6	1.5	1.5	1.4	1.3	1.2
W14	1.5	1.5	1.4	1.4	1.3	1.3	1.2	1.5	1.4	1.4	1.3	1.3	1.2	1.2

Source: Uang et al. [7]. Reprinted by permission of the American Institute of Steel Construction, Inc.

seen that this new m is close enough to the average m that it may be readily used.

When bending occurs about the y axis, U must be evaluated. A review of the same W6 to W14 shapes results in the average U values given in Table 9.2. However, an in-depth review of the U values for these sections shows that only the smallest sections for each depth have U values appreciably larger than 3. Thus, a reasonable value of $U = 3.0$ may be used for the first trial.

A more accurate evaluation of m, including length effects, has been conducted by Uang et al. [7]. The resulting values are presented in Table 9.3. These values have proven to be more accurate than the values presented in Table B, Section 2 of the *LRFD Manual*.

EXAMPLE 9.3

Given: Determine an initial trial section for the column and loadings of Fig. 9.10c using A36 steel and the simplified values of Table 9.2.

Solution

From Fig. 9.10c, $P_u = 453$ kips, $M_u = 251$ kip-ft

Assuming a W14, $m = 1.71$. Thus,

$$P_{\text{eff}} = 453 + 1.71(251) = 882 \text{ kips}$$

Using an effective length $KL = 12.5$ ft,

$$\text{try W14} \times 109, \quad \phi P_n = 899 \text{ kips}$$

From Example 9.2 it is seen that this column will adequately carry the imposed load.

It must be remembered that every column section selected must be checked through the appropriate interaction equations. Thus, for this initial selection, the approach should be quick and reasonable. The designer will soon learn to rely on experience rather than these simplified approaches.

9.8 COMBINED SIMPLE AND RIGID FRAMES

The practical design of steel structures often results in frames that combine segments of rigidly connected elements with segments that are pin connected. If these structures rely on the unbraced rigid frame to resist lateral load and to provide the overall stability of the structure, the rigidly connected columns may be called upon to carry more load than they had been originally designed to resist. In these combined simple and rigid frames, the simple columns "lean" on the rigid frames to maintain their stability. Thus, they are called *leaning columns*. Leaning columns may be designed with an effective length factor K of 1.0. But, since leaning columns have no stability of their own, the rigid frame columns must be designed to carry a load that is sufficient to provide stability for the full frame. In addition, the effective length factor, greater than 1.0, must be appropriate for the rigid framing conditions. Although this combination of framing types may add to the considerations for design, it may also be economically advantageous since a reduced number of moment connections may result.

A number of design approaches have been proposed for consideration of the leaning column and the associated rigid frame. Yura [9] proposes to design the columns that provide stability for the total load on the frame at the story in question, and LeMessurier [10] presents a modified effective length factor that accounts for the full frame stability. The approach presented here is based on the presentation by Yura, although the effective length factor equation suggested by LeMessurier gives a more accurate design.

The two-column frame shown in Fig. 9.11a is an unbraced frame with pinned base columns and a rigidly connected beam. The column sizes are selected so that under the loads shown they will buckle in a sidesway mode simultaneously. Equilibrium in the displaced position is

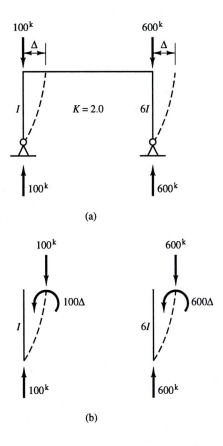

(a)

(b)

Figure 9.11 Pinned base
unbraced frame.

shown in Fig. 9.11b. The lateral displacement of the frame Δ results in a moment at the top of each column equal to the column applied load times the displacement. These are the second-order effects that were discussed in Section 9.6. The total load on the frame is 700 kips and the total $P\Delta$ moment is 700Δ, divided between the two columns according to the load each carries.

If the load on the right-hand column is reduced to 500 kips, the column will not buckle sideways, since the moment at the top would be less than 600Δ. To reach the buckling condition, a horizontal force must be applied at the top of the column, as shown in Fig. 9.12b. This force can result only from action on the left column as transmitted through the beam. Equilibrium of the left column, shown in Fig. 9.12a, requires that an additional column load of 100 kips must be applied to that column in order for the frame to be in equilibrium in this displaced position. It can be seen that the total frame capacity is still 700 kips and the total second-order moment is still 700Δ.

The maximum load that an individual column can resist is limited to that permitted for the column in a braced frame for which $K = 1.0$. In the example case, the left column could resist 400 kips and the right column

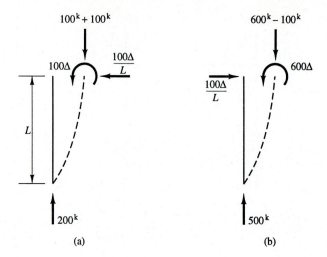

Figure 9.12 Columns from unbraced frame with revised loading.

could resist 2400 kips, both as a result of the effective length factor changing from 2.0 to 1.0. It should also be noted that the additional capacity of the left column is only with respect to the bending axis. The column would have the same capacity about the other axis as it did prior to reducing the load on the right column.

The ability of one column to carry increased load when another column in the frame is called upon to carry less than its lateral buckling capacity is an important characteristic. It will allow a pin-ended column to lean on an unbraced rigid frame column, provided that the total gravity load on the frame can be carried by the rigid frame.

EXAMPLE 9.4

Given: The frame shown in Fig. 9.13 is similar to that in Example 9.1 except that the in-plane stability and lateral load resistance is provided by the rigid frame action at the four corners. Out-of-plane stability and lateral load resistance is

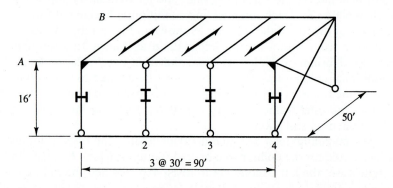

Figure 9.13 Frame used in Example 9.4.

provided by X-bracing along column lines 1 and 4. The frame is to be checked for both stability and strength.

The loading will be the same as that for Example 9.1: dead load = 50 psf, snow load = 20 psf, roof live load = 10 psf, wind load = 20 psf horizontal. Use A36 steel.

Solution

Step 1. The analysis of the frame for gravity loads as found for Example 9.1 will be used. Since different load combinations may be critical, the analysis results for unfactored snow load and unfactored dead load are given in Fig. 9.14b. The analysis results for unfactored wind load is given in Fig. 9.14a.

Step 2. Check column *A1* for strength, considering gravity plus wind acting toward the left:

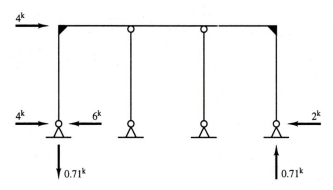

(a) Unfactored wind load

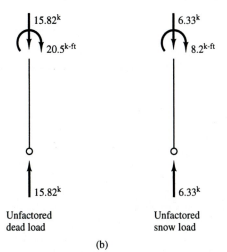

Unfactored   Unfactored
dead load   snow load

(b)

Figure 9.14 Unfactored wind load, unfactored snow load, and unfactored dead load (Example 9.4).

For load case A4-3

$$P_u = 1.2(15.82) + 1.6(6.33) + 0.8(0.71) = 29.7 \text{ kips}$$
$$M_u = 1.2(20.5) + 1.6(8.2) + 0.8(32.0)$$
$$= 37.7 + 25.6 = 63.3 \text{ kip-ft}$$

For load case A4-4

$$P_u = 1.2(15.82) + 0.5(6.33) + 1.3(0.71) = 23.1 \text{ kips}$$
$$M_u = 1.2(20.5) + 0.5(8.2) + 1.3(32.0)$$
$$= 28.7 + 41.6 = 70.3 \text{ kip-ft}$$

Total story gravity load acting on one frame:

$$\text{Dead} = \frac{0.05 \text{ ksf (90 ft)(50 ft)}}{2 \text{ frames}} = 112.5 \text{ kips}$$

$$\text{Snow} = \frac{0.02 \text{ ksf (90 ft)(50 ft)}}{2 \text{ frames}} = 45.0 \text{ kips}$$

Step 3. Check loading case A4-3.

$$P_u = 29.7 \text{ kips}, M_{nt} = 37.7 \text{ kip-ft}, M_{lt} = 25.6 \text{ kip-ft}$$
$$\Sigma P_u = 1.2(112.5) + 1.6(45.0) = 207.0 \text{ kips}$$
$$\Sigma H = 0.8(4.0) = 3.2 \text{ kips}$$

A drift index of 0.003 is chosen for amplification of moments. Note that this index is related to wind speed associated with ultimate load, not service load. Assume values for $K_x = 1.5$ and $K_y = 1.0$.

Try a W8×35:

$$A = 10.3 \text{ in.}^2, r_x = 3.51 \text{ in.}, \phi P_{ny} = 197 \text{ kips for } KL = 16 \text{ ft}$$

$$\frac{r_x}{r_y} = 1.73 \text{ in.}, \phi M_{nx} = 84.4 \text{ kip-ft for } L_b = 16 \text{ ft}$$

$$P_{ex} = 1042 \text{ kips for } KL = 16 \text{ ft}$$

With r_x/r_y greater than K_x, the y axis will control the axial capacity.

Checking for appropriate interaction equation,

$$\frac{P_u}{\phi P_{ny}} = \frac{29.7}{197} = 0.15 < 0.20$$

Therefore, use Eq. (H1-1b).

$$C_m = 0.6 - 0.4\left(\frac{0}{37.7}\right) = 0.6$$

$$B_1 = \frac{C_m}{1 - P_u/P_e} = \frac{0.6}{1 - 29.7/1042} = 0.62 < 1.0$$

Use $B_1 = 1.0$.

$$B_2 = \frac{1}{1 - (\Sigma P_u/\Sigma H)(\Delta/L)} = \frac{1}{1 - (207/3.2)(0.003)} = 1.24$$

$$M_u = B_1 M_{nt} + B_2 M_{lt} = 1.0(37.7) + 1.24(25.6) = 69.4 \text{ kip-ft}$$

Equation (H1-1b) yields

$$\frac{29.7}{2(197)} + \frac{69.4}{84.4} = 0.90 < 1.0$$

Thus the column is adequate for this case.
Step 4. Check loading case A4-4:

$$P_u = 23.1 \text{ kips}, M_{nt} = 28.7 \text{ kip-ft}, M_{lt} = 41.6 \text{ kip-ft}$$

$$\Sigma P_{ux} = 1.2(112.5) + 0.5(45.0) = 157.5 \text{ kips}$$

$$\Sigma H = 1.3(4.0) = 5.2 \text{ kips}$$

Checking for appropriate interaction equation,

$$\frac{P_u}{\phi P_{ny}} = \frac{23.1}{197} = 0.12 < 0.20$$

Therefore, use Eq. (H1-1b).

$$C_m = 0.6 - 0.4(0) = 0.6$$

$$B_1 = \frac{C_m}{1 - P_u/P_e} = \frac{0.6}{1 - 23.1/1042} = 0.61 < 1.0$$

Use $B_1 = 1.0$.

$$B_2 = \frac{1}{1 - (\Sigma P_u/\Sigma H)(\Delta/L)} = \frac{1}{1 - (157.5/5.2)(0.003)} = 1.10$$

$$M_u = B_1 M_{nt} + B_2 M_{lt} = 1.0(28.7) + 1.10(41.6) = 74.5 \text{ kip-ft}$$

Equation (H1-1b) yields

$$\frac{23.1}{2(197)} + \frac{74.5}{84.4} = 0.94 < 1.0$$

Thus the column is adequate for this case also. The W8×35 is shown to be adequate for gravity and wind loads in combination.

Step 5. Check to see that these columns have sufficient capacity to brace the interior pinned columns for gravity load only.

The total load on the structure is to be resisted by the four corner columns; thus,

$$\text{Dead load} = \frac{0.05 \text{ ksf (50 ft)(90 ft)}}{4 \text{ columns}} = 56.3 \text{ kips}$$

$$\text{Snow load} = \frac{0.02 \text{ ksf (50 ft)(90 ft)}}{4 \text{ columns}} = 22.5 \text{ kips}$$

For load combination A4-3, with no lateral translation

$$P_u = 1.2(56.3) + 1.6(22.5) = 103.6 \text{ kips}$$

$$M_u = 1.2(20.5) + 1.6(8.2) = 37.7 \text{ kip-ft}$$

The effective length factor was assumed as 1.5 for the column about the x axis. Thus, $(KL_x)_{\text{equiv}} = 1.5(16.0)/1.73 = 13.95$ ft and $\phi P_{nx} = 221$ kips. As determined earlier, for an unbraced flange length of 16 ft, $\phi M_{nx} = 84.4$ kip-ft.

Checking for the appropriate interaction equation, we obtain

$$\frac{P_u}{\phi P_{nx}} = \frac{103.6}{221} = 0.47 > 0.2$$

Thus, use Eq. (H1-1a).

For an effective length $KL_x = 1.0(16.0) = 16.0$ ft, $P_{ex} = 985$ kips. As before, $C_m = 0.6$. Thus,

$$B_1 = \frac{0.6}{1 - 103.6/985} = 0.67 < 1.0$$

Therefore use $B_1 = 1.0$.

$$\frac{P_u}{\phi P_{nx}} + \frac{8}{9}\frac{M_{ux}}{\phi M_{nx}} = \frac{103.6}{221} + \frac{8}{9}\frac{37.7}{84.4} = 0.87 < 1.0$$

Step 6. Check the stability under combined gravity plus wind described by load combination A4-4.

$$P_u = 1.2(56.3) + 0.5(22.5) + 1.3(0.71) = 79.7 \text{ kips}$$

$$M_u = 1.2(20.5) + 0.5(8.2) + 1.3(32.0)$$

$$= 28.7 + 41.6 = 70.3 \text{ kip-ft}$$

Checking for the appropriate interaction equation, we obtain

$$\frac{P_u}{\phi P_{nx}} = \frac{79.7}{221} = 0.36 > 0.2$$

Thus, use Eq. (H1-1a).

For the determination of B_1, use $K = 1.0$ and $P_{ex} = 985$ kips. Thus,

$$B_1 = \frac{0.6}{1 - 79.7/985} = 0.65 < 1.0$$

Therefore use $B_1 = 1.0$.

As before $B_2 = 1.1$ so that $M_u = 1.0(28.7) + 1.1(41.6) = 74.5$ kip-ft.

$$\frac{P_u}{\phi P_n} + \frac{8}{9}\frac{M_{ux}}{\phi M_{nx}} = \frac{79.7}{221} + \frac{8}{9}\frac{74.5}{84.4} = 1.15 > 1.0$$

Therefore the W8×35 will not be adequate for stability of the frame by this method even though it was adequate for strength. The next larger column should be checked.

9.9 PARTIALLY RESTRAINED FRAMES

The beams and columns in the frames considered to this point have all been connected with either moment-resisting, fully rigid (FR) connections, or simple, pinned connections. Pinned connections are actually a special case of the more general, partially restrained (PR) connections provided for in the LRFD specification. Partially restrained connections have historically been referred to as semirigid connections. When these PR connections are included as the connecting elements in a structural frame, they influence both the strength and stability of the structure.

Prior to consideration of the partially restrained frame, it will be helpful to look at the partially restrained beam [11]. The relationship between end moment and end rotation for a symmetric, uniformly loaded prismatic beam may be obtained from the well-known slope deflection equation as

$$M = \frac{-2EI\,\theta}{L} + \frac{WL}{12} \qquad (9.17)$$

This equation is plotted in Fig. 9.15a and labeled as the beam line.

All PR connections will exhibit some rotation as a result of an applied moment. The moment–rotation characteristics of these connections are the key to determining the type of a connection and thus the behavior of the structure. Moment–rotation curves for three generic connections are shown in Fig. 9.15b and are labeled rigid, simple, and PR. A great deal of research has been conducted in an effort to identify the

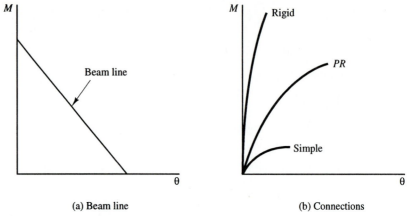

(a) Beam line   (b) Connections

Figure 9.15   Moment rotation curves for uniformly loaded beam and typical connections.

moment–rotation curves for real connections. Two compilations of these curves have been published [*12, 13*].

The relationship between the moment–rotation characteristics of a connection and a beam may be seen by plotting the beam line and the connection curve, as in Fig. 9.16. Normal engineering practice treats connections capable of resisting at least 90% of the fixed end moment as rigid and those capable of resisting no more than 20% of the fixed end moment as simple. All connections that exhibit an ability to resist moment between these limits must be treated as partially restrained connections, accounting for their true moment–rotation characteristics.

The influence of the PR connection on the maximum positive and negative moments on the beam is seen in Fig. 9.17. Here, the ratio of positive or negative moment to the fixed end moment is plotted against the ratio of beam stiffness EI/L to connection stiffness M/θ. It is seen that the maximum moment for which the beam must be designed ranges from 0.75 times the fixed end moment to 1.5 times the fixed end moment, depending on the stiffness of the connection.

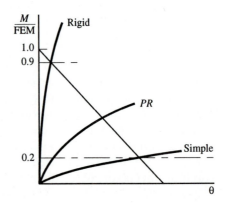

Figure 9.16   Beam line and connection curves.

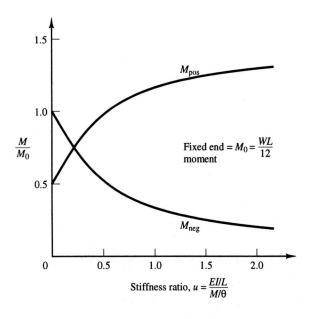

Figure 9.17 Influence of the PR connection on the maximum positive and negative moments of a beam.

When PR connections are used to connect beams and columns to form PR frames, the analysis becomes much more complex. Numerous studies dealing with this issue have been reported. Recent examples include the work of Gerstle and Ackroyd [*14*], Deierlein et al. [*15*], and Lui and Chen [*16*]. Although some practical designs have been carried out, widespread practical design of PR frames is still some time off. In addition to the problems associated with modeling a particular connection, the question of loading sequence arises. Since real, partially restrained connections behave nonlinearly, the sequence of applied loads will influence the structural response. In fact, the approach to load application may have more significance than the accuracy of the connection model used in the analysis.

Although a complete, theoretical analysis of a partially restrained frame may be beyond the scope of normal engineering practice, there is a simplified approach that is not only well within the scope of practice, but is commonly carried out in everyday design. This approach may be referred to as *flexible wind connections*. It has also been referred to as *Type 2 with wind* when applied to allowable stress design. The flexible wind connection approach relies heavily on the nonlinear moment–rotation characteristics of the PR connection, although the actual curve is not considered. In addition, it relies on a phenomenon called *shakedown*, which shows that the connection, although exhibiting a nonlinear behavior initially, behaves linearly after a limited number of applications of wind load [*17*].

The moment–rotation curve for a typical PR connection is shown in Fig. 9.18a along with the beam line for a uniformly loaded beam. The

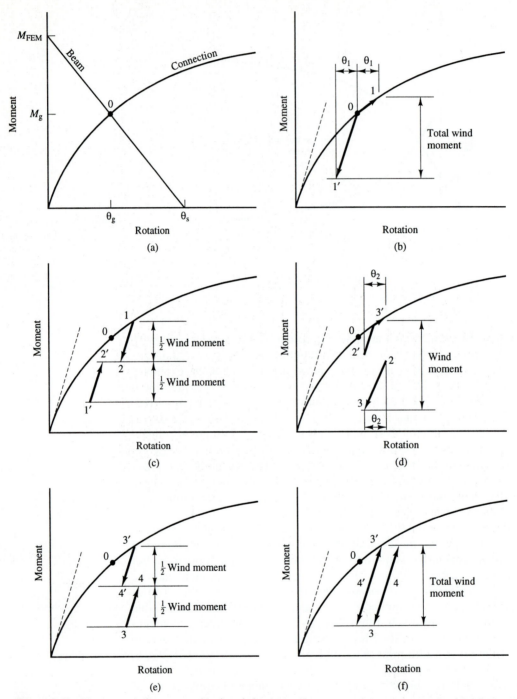

Figure 9.18 Moment–rotation curves showing shake down. Courtesy American Iron and Steel Institute.

point labeled O represents equilibrium for the applied gravity loads. The application of wind load produces moments at the beam ends which add to the gravity moment at the leeward end of the beam and subtract at the windward end. Since moment at the windward end is being removed, the connection behaves elastically with a stiffness close to the original connection stiffness; at the leeward end, the connection continues to move along the nonlinear connection curve. Points labeled 1 and 1′ in Fig. 9.18b represent equilibrium under the first application of wind to the frame.

When the wind load is removed, the connection moves from points 1 and 1′ to points 2 and 2′, as shown in Fig. 9.18c. The next application of a wind load that is larger than the first will see the connection behavior move to points 3 and 3′. Note that on the windward side, the magnitude of this applied wind moment will dictate whether the connection behaves linearly or follows the nonlinear curve, as shown in Fig. 9.18d. Removal of this wind load causes the connection on one end to unload and on the other end to load, both linearly. Any further application of wind load less than the maximum already applied will see the connection behave linearly. In addition, the maximum moment on the connection is still close to that applied originally from the gravity load. Thus, the condition described in Fig. 9.18f shows that shakedown has taken place and the connection now behaves linearly for both loading and unloading.

The design procedure employed to account for shakedown is straightforward. All beams are designed as simple beams using the appropriate load factors. This assures that the beams will be adequate regardless of the actual connection stiffness, as was seen in Fig. 9.17. Wind load moments may be determined through an indeterminate analysis or a simplified analysis, such as the portal method. Connections will be sized to resist these moments, again with the appropriate load factor. In addition, it is particularly important to provide connections that have sufficient ductility to accommodate the large rotations that will occur, without overloading the bolts or welds under combined gravity and wind.

Columns must be designed to provide frame stability under gravity loads as well as gravity plus wind. The columns may be designed using the approach that was presented for columns in unbraced frames, but with two essential differences from the conventional rigid frame design:

1. Since the gravity load is likely to load the connection to its plastic moment capacity, the column can be restrained only by a girder on one side and this girder will act as if it is pinned at its far end. Therefore, in computing the girder stiffness rotation factor I_g/L_g for use in the effective length nomograph, the girder stiffness factor should be reduced by 0.5.
2. One of the external columns, the leeward column for the wind loading case, will not be able to participate in frame stability since it will be attached to a connection that is at its plastic moment capacity. The stability of the frame may be assured,

however, by designing the remaining columns to support the total frame load.

For the exterior column, the moment in the beam-to-column joint is equal to the capacity of the connection. It is sufficiently accurate to assume that this moment is distributed one-half to the upper column and one-half to the lower column. For interior columns, the greatest realistically possible difference in moments resulting from the girders framing into the column should be distributed equally to the columns above and below the joint.

EXAMPLE 9.5

Given: An intermediate story of a five-story building is given in Fig. 9.19. Story height is 12 ft. The frame is braced in the direction normal to that shown. Design the girders and columns using flexible wind connections and determine the moments for which the connections must be designed. Use A36 steel.

Solution

The loads shown in Fig. 9.19 are service loads. The factored loads are calculated using tributary areas as follows.

Gravity loads on exterior columns:

$$DL = 1.2[25 \text{ k} + 0.75 \text{ k/ft}(15 \text{ ft})] = 43.5 \text{ kips}$$

$$LL = 1.6[75 \text{ k} + 2.25 \text{ k/ft}(15 \text{ ft})] = \underline{174.0 \text{ kips}}$$

$$\text{Total} \quad 217.5 \text{ kips}$$

Gravity loads on interior columns:

$$DL = 1.2[\ 50 \text{ k} + 0.75 \text{ k/ft}(30 \text{ ft})] = 87.0 \text{ kips}$$

$$LL = 1.6[150 \text{ k} + 2.25 \text{ k/ft}(30 \text{ ft})] = \underline{348.0 \text{ kips}}$$

$$\text{Total} \quad 435.0 \text{ kips}$$

Gravity load on girders:

$$DL = 1.2[0.75 \text{ k/ft}(30 \text{ ft})] = 27.0 \text{ kips}$$

$$LL = 1.6[2.25 \text{ k/ft}(30 \text{ ft})] = \underline{108.0 \text{ kips}}$$

$$\text{Total} \quad 135.0 \text{ kips}$$

Girder design

$$M_u = \frac{135.0(30)}{8} = 506 \text{ kip-ft}$$

Use W24×62 ($\phi M_p = 574$ kip-ft, $I_x = 1550$ in.[4])

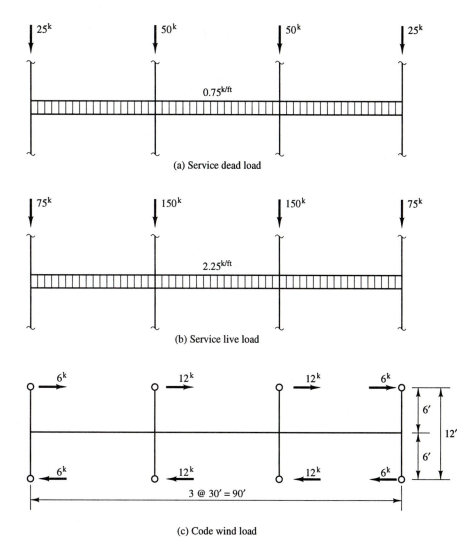

(a) Service dead load

(b) Service live load

(c) Code wind load

Figure 9.19   An intermediate story of a five-story building (Example 9.5).

Interior column design

y axis gravity load
P_u = 435 kips
Try W14×61 (ϕP_n = 457 kips)

x axis gravity load (frame stability check)
Check the W14×61

Determine the effective length factor from the nomograph, Fig. 6.34, with

$$G_t = G_b \text{ (elastic)} = \frac{2(640/12)}{0.5 \ (1550/30)} = 4.13$$

Note that only one beam will be capable of restraining the column and that beam is pinned at its far end. By considering the stress in the column under load, the stiffness reduction factor can be determined.

$$f_a = \frac{435 \text{ kips}}{17.9 \text{ in.}^2} = 24.3 \text{ ksi}$$

Thus, from Table 6.1, β_s, the stiffness reduction factor (SRF) = 0.724. Giving

$$G_t = G_b \text{ (inelastic)} = (0.724)(4.13) = 2.99$$

which yields, from the nomograph

$$K = 1.8$$

The equivalent effective length in the plane of bending is

$$\frac{KL}{r_x/r_y} = \frac{1.8(12.0)}{2.44} = 8.9 \text{ ft}$$

and

$$\phi P_n = 494 \text{ kips}$$

The two interior columns can support a total of 2(494) = 988 kips out of the total gravity load of 2(217.5 + 435.0) = 1305 kips. The remaining load, 317 kips, must be supported by the windward exterior column.

x axis combined loads

Factored wind moment per column

$$M_u = (1.3)(12.0 \text{ kips})(6.0 \text{ ft}) = 93.6 \text{ kip-ft}$$

Factored gravity load per column

$$P_u = 1.2[50 + 0.75(30)] \\ + 0.5[150 + 2.25(30)] = 195.8 \text{ kips}$$

Using the approximate interaction equations discussed in Section 9.7, $m = 1.71$ for a W14; thus,

$$P_{\text{eff}} = 195.8 + 1.71(93.6) = 356.0 \text{ kips} < 494 \text{ kips}$$

Therefore, the W14×61 is adequate for both strength and stability.

Exterior Column

x axis gravity load (frame stability)

Each exterior column must be designed to support that portion of the total frame load not supported by the interior columns. Thus, $P_u = 317$ kips.

Try W12×45

$$G_t = G_b \text{ (elastic)} = \frac{2(350/12)}{0.5(1550/30)} = 2.26$$

$$f_a = \text{actual load/area} = \frac{217.5}{13.2} = 16.5 \text{ ksi}$$

Thus, from Table 6.1, $\beta_s = \text{SRF} = 0.979$ and

$$G_t = G_b \text{ (inelastic)} = (0.979)(2.26) = 2.21$$

From the nomograph, Fig. 6.34, $K = 1.63$

The equivalent effective length in the plane of bending is

$$\frac{KL}{r_x/r_y} = \frac{1.63(12)}{2.65} = 7.4 \text{ ft}$$

which yields $\phi P_n = 362 \text{ kips} > 317 \text{ kips}$

x axis combined loads

$$M_u = 1.3(6 \text{ kips})(6 \text{ ft}) = 46.8 \text{ kip-ft}$$
$$P_u = 1.2[25 + 0.75(15)]$$
$$+ 0.5[75 + 2.25(15)] = 97.9 \text{ kips}$$

Using $m = 2.0$,

$$P_{\text{eff}} = 46.8 + 2.0(97.9) = 242.6 < 362 \text{ kips}$$

y axis gravity load

$$P_u = 217 \text{ kips}, K = 1.0, KL = 12.0$$
$$\phi P_n = 302 \text{ kips} < 217 \text{ kips}$$

Therefore, use the W12×45 for each exterior column.
    The beam connections must be designed to resist the wind moments as determined from the forces in Fig. 9.19c as

Exterior connection

$$M_u = 1.3(6 \text{ kips})(6 \text{ ft}) (2 \text{ columns}) = 93.6 \text{ kip-ft}$$

Interior connection

$$M_u = \frac{1.3(12 \text{ kips})(6 \text{ ft}) (2 \text{ columns})}{(2 \text{ girders})}$$

$$= 93.6 \text{ kip-ft}$$

9.10 BRACING DESIGN

Braces in steel structures are used for two distinctly different functions:
(1) to force a particular mode in a buckling member such as a column or
the compression flange of a beam, and (2) to restrain a frame from
excessive lateral movement resulting in serviceability problems or instability under factored loads. The first case may be called a point brace and
the second a relative brace [18]. The following treatment of bracing design
follows procedures developed by Yura and widely used by the profession.

9.10.1 Bracing: Member Support

Figure 9.20a illustrates an elevation of a column with a brace at midheight, and Fig. 9.20b shows a plan of a loaded beam supported at the
ends and braced at midspan. In each case, the brace serves to reduce the
buckling length of the member by one-half, greatly increasing the load-
carrying capacity of the member. The question is, What strength and
stiffness are required for the brace to adequately serve its function?

Although both the beam and the column are similar, for illustration
purposes only the column will be discussed.

Figure 9.21a shows a column with a brace at midheight. Braces are,
in reality, elastic springs, as shown in Fig. 9.21b. If the spring is stiff
enough, the column will buckle, as shown in Fig. 9.21c. However, if the
spring is too soft, the column will buckle, as shown in Fig. 9.21d under a
load approaching one-quarter of that for the column of Fig. 9.21c. To be
effective, therefore, the spring must have a stiffness $\beta = F/\delta$ sufficiently
large to prevent the buckling shown in Fig. 9.21d.

It is possible for the stiffness β to be sufficient while the strength of
the brace F is small. Thus, the brace could fail even though the critical
deflection δ has not been reached. Therefore, there is also a minimum
force that the brace must be able to develop.

There is no rigorous method available to derive the minimum values
for β and F. A long-standing rule of thumb states that both stiffness and

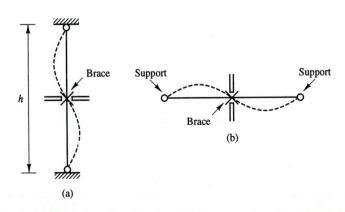

Figure 9.20 (a) Elevation of a
column with a
brace at midheight;
(b) plan of a
loaded beam
supported at the
ends and braced at
midspan.

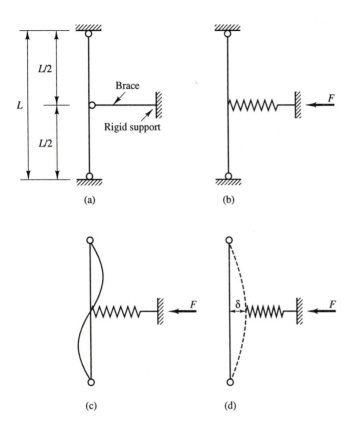

Figure 9.21 A column with a brace at midheight.

strength will be satisfied if the force F equals 2 percent of the force in the compression member to be braced: the capacity of the column or the maximum force in the compression flange of a beam. This approach is generally considered to be conservative [*19*].

Another approach may be devised using the model shown in Fig. 9.22. The column with an initial lateral displacement of δ_o resulting from erection tolerances is loaded with a factored load P. A moment of $P\delta_o$ results, which causes an additional displacement δ. If δ_o is kept within the specified tolerance of a slope of 1:500, then $P\delta_o = 0.002PL$. If we assume that the additional displacement is equal to the tolerance, $\delta = \delta_o$, then $P(\delta_o + \delta) = 0.004PL$. From the free-body diagram shown in Fig. 9.22b, it can be seen that $(F/2)(L/2) = 0.004PL$. Solving for the required brace force,

$$F = 0.016P \qquad (9.18)$$

The required brace stiffness is

$$\beta = \frac{F}{\delta} = \frac{0.016P}{0.002L} = 8PL \qquad (9.19)$$

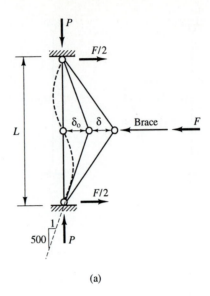

(a)

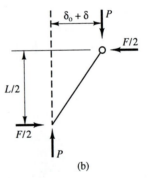

Figure 9.22 Model for design of column point brace.

(b)

9.10.2 Bracing: Frame Support

The second function of a brace is to control the overturning moment caused by the relative end displacement of δ_o and a displacement under load of $\delta + \delta_o$. For the column shown in Fig. 9.23, a brace is applied at the top and bottom of the column. Each has a horizontal force component F and a spring constant β. The column is axially loaded by P.

The overturning moment OM on the column is

$$OM = P(\delta + \delta_o)$$

and the resisting moment RM is

$$RM = F(L) = \beta\delta(L)$$

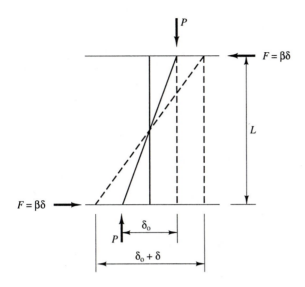

Figure 9.23 Model for design of a relative brace.

By equating the overturning moment to the resisting moment, the required brace force F and stiffness β can be determined as

$$F = \frac{P(\delta + \delta_o)}{L} \qquad (9.20)$$

and

$$\beta = \frac{P(1 + \delta_o/\delta)}{L} \qquad (9.21)$$

Equations (9.20) and (9.21), like Eqs. (9.18) and (9.19), are useful only for a specific value of δ and δ_o. If the same assumptions for the displacements are used here that were used for the buckling case, the same equations will result. It is important that designers decide for themselves whether or not these assumptions are valid for a particular design.

EXAMPLE 9.6

Given: The bracing shown in Fig. 9.8a will be selected to resist the service wind load of 4 kips and provide stability for a gravity live load of 112.5 kips and dead load of 45 kips.

Solution

For the wind load:

$$H = 1.3(4.0) = 5.2 \text{ kips}$$

Force in single tension diagonal

$$F = 5.2 \frac{(34 \text{ ft})}{30 \text{ ft}} = 5.9 \text{ kips}$$

Area required to resist force

$$A_{req} = \frac{5.9}{(0.9)(36)} = 0.182 \text{ in.}^2$$

For stability under gravity load:

$$P = 1.2(112.5) + 1.6(45) = 207 \text{ kips}$$

Force in brace by Eq. 9.18

$$F = 0.016(207) = 3.3 \text{ kips}$$

$$A_{req} = \frac{3.3}{(0.9)(36)} = 0.10 \text{ in.}^2$$

Required stiffness by Eq. 9.19

$$\beta = \frac{8(207)}{16(12)} = 8.625 \text{ kip/in.}$$

Required area

$$\frac{P}{\delta} = \frac{AE}{L} = 8.625 \text{ kip/in.}$$

Thus,

$$A_{req} = \frac{8.625(34 \text{ ft})(12)}{29,000} = 0.12 \text{ in.}^2$$

Select a tension member with a minimum area, $A = 0.182$ in.², which will be required to resist the applied load and results in a member of adequate stiffness to brace the frame.

PROBLEMS

Unless noted otherwise, all columns should be considered pinned in a braced frame out of the plane being considered in the problem.

9.1. Will a W14×90 (A36 steel) with a length of 12.5 ft be adequate in a braced frame to carry a factored axial load of 450 kips and factored moments, about the strong axis, of 100 kip-ft at the top and 50 kip-ft at the bottom, bending the member in reverse curvature?

9.2. A W12×58 (A36 steel) is used as a 14-ft column in a braced frame to carry an axial factored load of 200 kips. Will it be adequate to carry moments about the strong axis of 80 kip-ft at each end, bending the column in single curvature?

9.3. Given a W14×120 (A36 steel) in a braced frame with an axial factored load of 400 kips on a 16-ft column, determine the maximum moment that may be applied about the strong axis on the upper end when the lower end is a pin support.

9.4. Reconsider the column and loading of Problem 9.1 if the column is bent in single curvature.

9.5. Reconsider the column and loading of Problem 9.2 if the column is bent in reverse curvature.

9.6. A pin-ended column in a braced frame must carry a factored axial load of 200 kips along with a transverse factored load of 1.6 k/ft on an 18-ft length. Will a W14×74, A36 member be adequate if the transverse load is applied to put bending about the strong axis?

For Problems 9.7–9.10, assume that the ratio of total gravity load on the story to total Euler load for the story is the same as the ratio for the specific column being considered.

9.7. An unbraced frame contains a 12-ft column which is called upon to carry a factored axial load of 425 kips, a no-translation moment $M_{nt} = 110$ kip-ft, and a translation-permitted moment $M_{lt} = 154$ kip-ft, at the top with half of these moments at the bottom causing reverse curvature. Will a W14 × 109 (A36 steel) be adequate to carry this loading? Assume that the effective length factor in the plane is $K_x = 1.66$.

9.8. A W14 × 176 (A36 steel) is proposed for use as a 12.5-ft column in an unbraced frame. Will this member be adequate to carry a factored axial load of 700 kips with moments $M_{nt} = 50$ kip-ft and $M_{lt} = 300$ kip-ft at the top and a pin support at the bottom? Assume $K_x = 1.5$.

9.9. Will a W14 × 48 be adequate as a 13-ft column in an unbraced frame with a factored axial load of 112 kips along with moments $M_{nt} = 100$ kip-ft and $M_{lt} = 40$ kip-ft at the top and bottom, with the column bent in reverse curvature? Use $F_y = 50$ ksi. Assume $K_x = 1.3$.

9.10. Check a W14 × 43 (A36 steel) for its ability to carry an axial load of 125 kips and moments $M_{nt} = 43$ kip-ft and $M_{lt} = 48$ kip-ft. One-half of these moments are applied at the other end of the member, bending it in single curvature. The braced frame column is 10 ft in length.

9.11. A two-story, single-bay frame is shown below. The uniform live and dead loads are indicated along with the wind load. A first-order elastic analysis has yielded the results shown in the figure for the given service loads. Assuming that the structure drift will be limited to height/300, determine whether the first- and second-story columns are adequate. The structure is A36 steel.

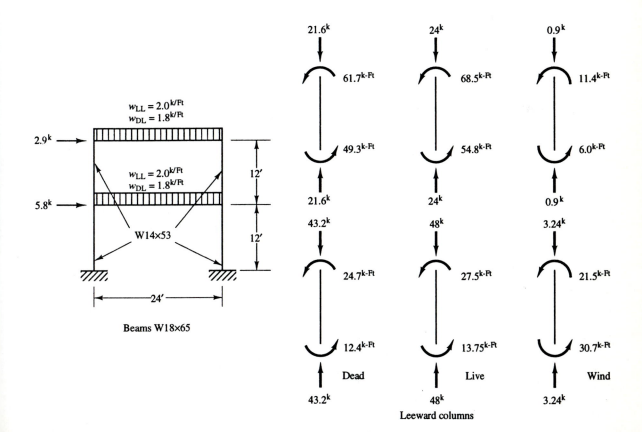

Leeward columns

9.12. The columns of the two-bay unbraced frame shown below are to be checked to determine if they are adequate to carry the given loading. Results for the first-order elastic analysis are provided. Since the structure drift is unknown, use the ratio of total applied load to total Euler load in calculating the amplification factors. Use steel with $F_y = 50$ ksi.

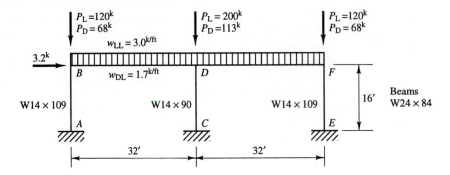

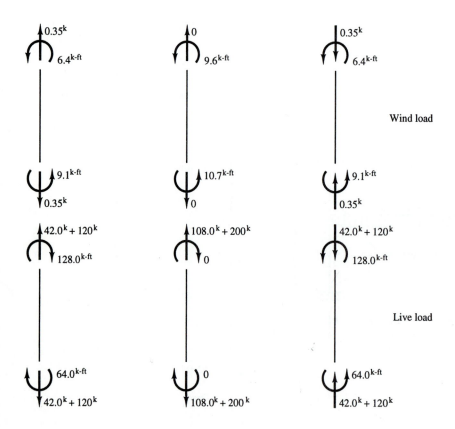

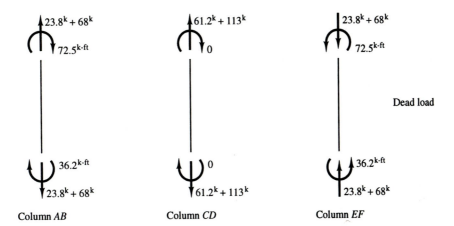

Column *AB*   Column *CD*   Column *EF*

9.13. A nonsymmetric two-bay unbraced frame is required to carry the live and dead loads given. Using the results from the first-order elastic analysis provided, determine whether each column will be adequate. Use A36 steel.

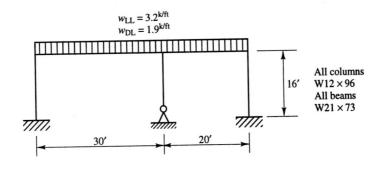

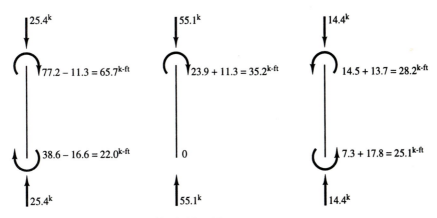

Dead load $M_{nt} + M_{lt}$

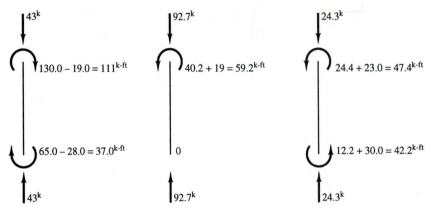

Live load $M_{nt} + M_{lt}$

Assume that the columns described in Problems 9.14–9.17 are pin-ended members of braced frames. Use the methods described in Section 9.7 to select trial sections to carry the indicated loads and check the appropriate LRFD specification equations for the column capacity. Assume A36 steel and use $\beta_1 = 1.0$.

9.14. Select a column with a length of 18 ft to carry a factored load of 510 kips and moment of 250 kip-ft.

9.15. Select a column with a length of 28 ft to carry a factored load of 785 kips and moment of 123 kip-ft.

9.16. Select a column with a length of 14 ft to carry a factored load of 250 kips and moment of 338 kip-ft.

9.17. Select a column with a length of 16 ft to carry a factored load of 900 kips and moment of 324 kip-ft.

9.18. The two-bay unbraced frame shown in the figure contains a single leaning column. The results of an elastic analysis for each of the service loads are given. Determine whether the exterior columns are adequate to provide stability for the frame. Use A36 steel.

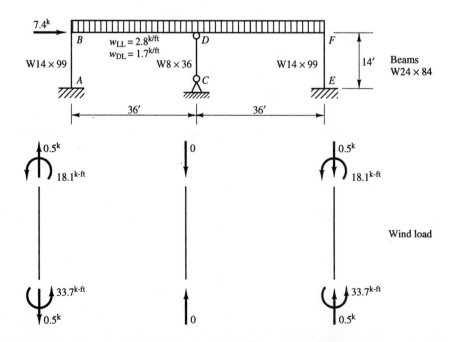

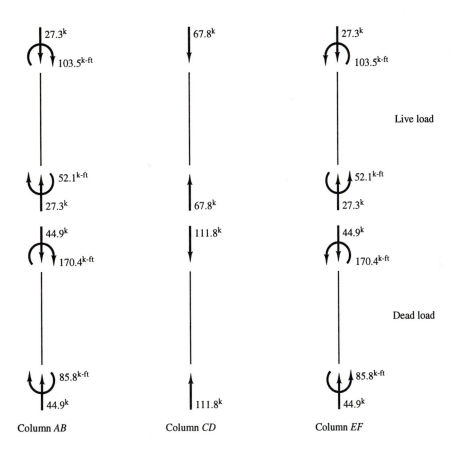

27.3^k

103.5$^{k\text{-}ft}$

52.1$^{k\text{-}ft}$

27.3^k

44.9^k

170.4$^{k\text{-}ft}$

85.8$^{k\text{-}ft}$

44.9^k

Column *AB*

67.8^k

67.8^k

111.8^k

111.8^k

Column *CD*

27.3^k

103.5$^{k\text{-}ft}$

Live load

52.1$^{k\text{-}ft}$

27.3^k

44.9^k

170.4$^{k\text{-}ft}$

Dead load

85.8$^{k\text{-}ft}$

44.9^k

Column *EF*

9.19. The two-story frame shown below relies on the left columns to provide stability. Using the analysis results shown, determine whether the given structure is adequate if $F_y = 50$ ksi.

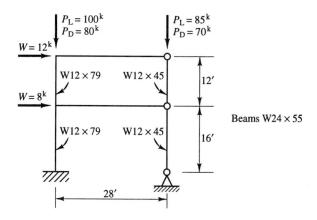

$P_L = 100^k$
$P_D = 80^k$

$P_L = 85^k$
$P_D = 70^k$

$W = 12^k$

W12 × 79   W12 × 45

12′

$W = 8^k$

W12 × 79   W12 × 45

16′

Beams W24 × 55

28′

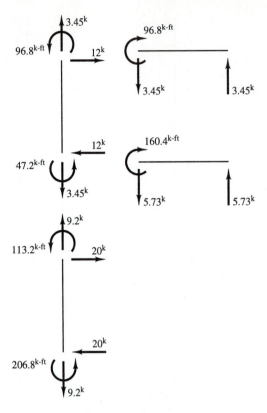

9.20. The two-bay, two-story frame shown below is to be designed. Using the live, dead, snow, and wind loads given, design the columns and beams to resist the imposed loads and provide stability.

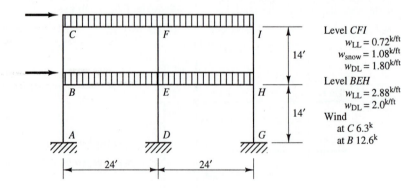

Level *CFI*
$w_{LL} = 0.72^{k/ft}$
$w_{snow} = 1.08^{k/ft}$
$w_{DL} = 1.80^{k/ft}$

Level *BEH*
$w_{LL} = 2.88^{k/ft}$
$w_{DL} = 2.0^{k/ft}$

Wind
at *C* 6.3^k
at *B* 12.6^k

14′

14′

24′ 24′

REFERENCES

[1] Kanchanalai, T., The design and behavior of beam-columns in unbraced steel frames, *AISI Project No. 189*, Report No. 2, Civil Engineering Structures Research Lab, University of Texas, Austin, October 1977.

[2] Galambos, T. V., *Structural Members and Frames*, Prentice-Hall, Englewood Cliffs, NJ, 1968.

[3] Johnson, B. G., Ed., *Guide to Stability Design Criteria for Metal Structures*, 3d ed., SSRC, Wiley, New York, 1976.

[4] Chen, W. F., and Atsuta, T., *Theory of Beam-Columns*, Vol. 1, *In-Plane Behavior and Design*, McGraw-Hill, New York, 1976.

[5] Horn, M. R., The stanchion problem in frame structures designed according to ultimate carrying capacity, *Proceedings, Institute of Civil Engineers*, Vol. 5, No. 1, Part III, April 1956 (pp. 105–146).

[6] Ketter, R. L., Further studies of the strength of beam columns, *Journal of the Structural Division, ASCE*, Vol. 87, No. ST6, 1961.

[7] Uang, C. W., Wattar, S. W., and Leet, K. M., Proposed revision of the equivalent axial load method for LRFD steel and composite beam-column design, *Engineering Journal, AISC*, Vol. 27, 4th Quarter, 1990 (pp. 150–157).

[8] Smith, J. C., *Structural Steel Design: LRFD Approach*, Wiley, New York, 1991.

[9] Yura, J. A., Effective length of columns in unbraced frames, *Engineering Journal, AISC*, Vol. 8, No. 2, 1971 (pp. 37–42).

[10] LeMessurier, W. J., A practical method of second-order analysis, 2: Rigid frames, *Engineering Journal, AISC*, Vol. 14, No. 2, 1977 (pp. 49–67).

[11] Geschwindner, L. F., A simplified look at partially restrained beams, *Engineering Journal, AISC*, Vol. 28, 2nd Quarter, 1991 (pp. 73–78).

[12] Goverdhan, A. V., *A Collection of Experimental Moment Rotation Curves and Evaluation of Prediction Equations for Semi-Rigid Connections*, Master of Science Thesis, Vanderbilt University, Nashville, TN, 1984.

[13] Kishi, N., and Chen, W. F., *Data Base of Steel Beam-to-Column Connections*, CE-STR-86-26, School of Engineering, Purdue University, West Lafayette, IN, 1986.

[14] Gerstle, K. H., and Ackroyd, M. H., Behavior and design of flexibly connected building frames, *Proceedings of the 1989 National Steel Construction Conference*, Nashville, TN, June 1989 (pp. 1.1–1.28).

[15] Deierlein, G. G., Hsieh, S. H., and Shen, Y. J., Computer-aided design of steel structures with flexible connections, *Proceedings of the 1990 National Steel Construction Conference*, Kansas City, MO, March 1990 (pp. 9.1–9.21).

[16] Lui, E. M., and Chen, W. F., Steel frame analysis with flexible joints, *Journal of Constructional Steel Research*, Vol. 8, 1987 (pp. 161–202).

[17] Disque, R. O., Wind connections with simple framing, *Engineering Journal, AISC*, Vol. 1, No. 3, 1975 (pp. 101–103).

[18] Yura, J. A., Summary of bracing recommendations, *Bracing Design Manual*, Lecture Handout, AISC National Steel Construction Conference, 1992.

[19] Milek, W. A., One engineer's opinion, *Engineering Journal, AISC*, Vol. 3, No. 2, 1967 (pp. 88–90).

CHAPTER TEN

Composite Construction

10.1 INTRODUCTION

Any structural member in which two or more materials having different stress–strain relationships are combined and are called upon to work as a single member may be thought of as a composite member. Many different types of members have been used that could be called composite. Such members, as shown in Fig. 10.1, are (a) a reinforced concrete beam, (b) a precast concrete beam and cast-in-place slab, (c) a "flitch" girder combining wood side members and a steel plate, (d) a stressed skin panel where plywood is combined with solid wood members, and (e) a steel shape combined with concrete.

It is this last type of member, and those similar members, that are normally thought of as composite members in building applications. The LRFD specification, Chapter I, provides rules for design of the composite members illustrated in Fig. 10.2. These members are (a) steel beams fully encased in concrete, (b) steel beams with flat soffit concrete slabs, (c) steel beams combined with formed steel deck, (d) steel columns fully encased with concrete, and (e) hollow steel shapes filled with concrete. Encased beams and composite columns, Fig. 10.2a, d, and e, require no special mechanical anchorage between the steel and concrete, other than the natural bond that exists between the two materials; the flexural members shown in Fig. 10.2b and c do require some form of mechanical shear connection.

Regardless of the type of mechanical shear device provided, it must connect the steel and concrete to form a unit and permit them to work together to resist flexure. As a result, the bending strength of the bare steel beam is increased considerably. This form of construction was first used in bridge design in the United States in about 1935 [1]. Until the invention of the shear stud, the concrete floor slab was connected to the stringer beams by means of wire spirals or channels welded to the top flange of the beam, as shown in Fig. 10.3.

In the 1940s the Nelson Stud Company invented the shear stud, a headed rod welded to the steel beam by means of a special device or gun

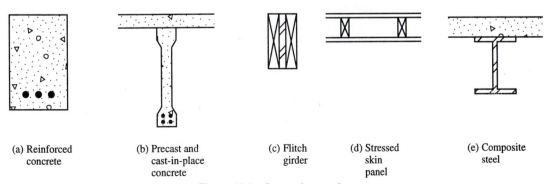

| (a) Reinforced concrete | (b) Precast and cast-in-place concrete | (c) Flitch girder | (d) Stressed skin panel | (e) Composite steel |

Figure 10.1 Composite members.

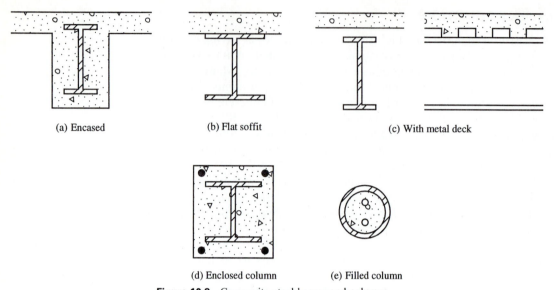

(a) Encased   (b) Flat soffit   (c) With metal deck

(d) Enclosed column   (e) Filled column

Figure 10.2 Composite steel beams and columns.

Figure 10.3 Composite beam using a spiral shear connector. Courtesy McFarland-Johnson, Inc., Binghampton, N.Y.

(Fig. 10.4). With almost unique foresight, Nelson did not enforce its patent. On the contrary, they encouraged nonproprietary use of the system, assuming correctly that the company would get its share of the stud business. In a very short time, studs replaced spirals and channels, so that today, studs are used almost exclusively in composite beam construction.

Figure 10.4   Nelson stud and welder. Courtesy TRW Nelson Stud Welding Division.

In 1952, AISC adopted composite design rules for encased beams in its specification for building design, and in 1956 extended them to beams with flat soffits. Although the design procedure was based on the ultimate strength of the composite section, the rules were written in the form of an allowable stress procedure as was common for the time. As a result, allowable stress design for composite beams has often been criticized as being convoluted and difficult to understand.

In contrast, the LRFD rules, introduced in 1986, for the design of composite beams are straightforward and surprisingly simple. The ultimate flexural strength of the composite member is based on plastic redistribution with the ductile shear connector transferring shear between the steel section and the concrete slab.

The research that forms the basis of the LRFD provisions for composite construction is summarized by Hansell et al. [2]. This chapter discusses the design of both composite beams and composite columns.

10.2 ADVANTAGES AND DISADVANTAGES OF COMPOSITE CONSTRUCTION

One feature of composite construction makes it advantageous for use in building structures. The typical building floor system is composed of two main parts: a deck, which forms the floor surface and carries load to supporting members, usually a concrete slab or slab on metal deck, and the supporting members, which span between girders, usually steel beams or joists. The single advantage of a composite floor system stems from the "double counting" of the already existing concrete slab. All other factors that could be identified as advantages of this type of construction can be traced back to this single feature. A composite beam takes the already existing concrete slab and makes it work with the steel beam to carry the load to the girders. Thus, the resulting system has a greater capacity than would have been available from the bare steel beam alone. The composite beam is stronger and stiffer than the noncomposite beam.

This factor manifests itself in reduced weight and/or shallower depths of members to carry the same loads when compared to the bare steel beam. Since the concrete slab is in compression and the majority of the steel is in tension, both materials are working to their best advantage. In addition, the effective beam depth has been increased from just the depth of the steel to the total distance from the top of the slab to the bottom of the steel, thus increasing the efficiency of the member.

With regard to stiffness, the composite section also has an increased elastic moment of inertia when compared to the bare steel beam. Although actual calculations for the stiffness of the composite section may be somewhat approximate in many cases, the impact of the increased stiffness has a profound effect on static deflection as well as floor vibration [3].

The only disadvantage found with composite construction is the added cost of the required shear connectors. Since the increased capacity, or the reduction in required steel weight, is normally sufficient to offset the added cost of the shear connectors, this is in reality rarely a disadvantage. Current local economic conditions, however, may make it difficult to continue to justify composite construction solely on economic grounds.

10.3 SHORED VS. UNSHORED CONSTRUCTION

Two methods of construction are available for composite beams, referred to as shored or unshored construction. Each has advantages and disadvantages. The difference between these two approaches to the construction of a composite beam is a function of how the self-weight of the wet concrete is carried.

If the steel shape alone is called upon to carry the concrete weight, the beam is considered unshored, stresses will be induced into the steel shape, and the steel will deflect. This is the simplest approach to constructing the composite beam, since the formwork and/or decking is supported directly on the steel beam. Unshored construction, however, may lead to a deflection problem during the construction phase since, as the wet concrete is placed, the steel beam deflects. To obtain a level slab, more concrete will be placed where the beam deflection is greatest. This means that the contractor will be called upon to place more concrete than the initial slab thickness called for and that the designer will need to provide more strength than would have been needed if the slab had remained of uniform thickness.

For shored construction, temporary supports, called shores, are placed under the steel beam to carry the wet concrete weight. In this case, the composite section carries all of the load after the shores are removed. No load is carried by the bare steel beam and, thus, no deflection occurs during concrete placement. Two factors must be considered in the selection of shored construction. First is the additional cost, both in time and money, of placing and removing the temporary shoring, and second is the potential increase in long term, dead load deflection due to creep in the concrete which is called upon to participate in carrying the permanent weight of the slab.

Although elastic stress distribution and deflection under service load conditions will be influenced by whether the composite beam is shored or unshored, research has shown that the ultimate strength of the composite section is independent of the shoring situation. Thus, the use of shoring is entirely a serviceability and constructibility question that must be considered by both the designer and constructor.

10.4 EFFECTIVE FLANGE

A cross section through a series of typical composite beams is shown in Fig. 10.5. Since the concrete slab is normally part of the transverse spanning floor system, its thickness and the spacing of the steel beams will have been established prior to the design of the composite beams. Since the ability of the slab to participate in load carrying decreases as the distance from the beam centerline increases, some limit must be established to determine the portion of the slab that may be used in the calculations to determine the capacity of the composite beam. The LRFD

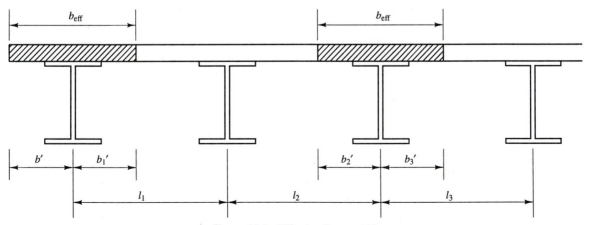

Figure 10.5 Effective flange width.

specification provides two criteria for determining the effective width of the concrete slab for an interior beam and an additional criteria for an edge beam. As shown in Fig. 10.5, the effective width b_{eff} is the sum of b' on each side of the centerline of the steel section. For an interior beam, b' is the least of

$$b' \leq \frac{\text{beam span}}{8}$$

$$b' \leq 1/2 \text{ the distance to the adjacent beam}$$

For an edge beam, the additional criteria is

$$b' \leq \text{the distance to the edge}$$

The depth of the concrete used in calculations is that required to provide sufficient area in compression to balance the force transferred by the shear connectors to the steel shape.

10.5 STRENGTH OF COMPOSITE BEAMS AND SLAB

Flat soffit composite beams, Fig. 10.2b, are constructed using formwork that is at the same elevation as the top of the steel section. The concrete slab is placed directly on the steel section, resulting in a flat surface at the level of the top of the steel. Composite beams with formed steel deck, Fig. 10.2c, are constructed with the steel deck resting on top of the steel beam or girder. The concrete is placed on top of the deck so that the concrete ribs and voids alternate. Provided that the portion of the concrete required to balance the tension force in the steel is available above the tops of the ribs, the ultimate strength of both types of composite beams will be determined in a similar fashion. Although the steel member may be either shored or unshored, the ultimate strength of the composite member is independent of whether or not shores are employed during construction; thus, the design rules are independent of the method of construction.

The nominal flexural strength of the composite beam under positive moment, concrete in compression, is first developed for flat soffit beams. The required modifications to account for the use of metal deck are presented in the next section. For steel sections with a web slenderness ratio h_c/t_w less than $640/\sqrt{F_y}$, the case for all rolled beams, the resistance factor ϕ_b is taken as 0.85 and the nominal moment is determined from the plastic distribution of stress over the composite section.

A composite beam cross section is shown in Fig. 10.6 with three possible plastic stress distributions. Regardless of the stress distribution considered, equilibrium requires that the total tension force must equal the total compression force, $T = C$. In Fig. 10.6a, the PNA is located at the top of the steel shape. The compression force developed using all of the concrete is exactly equal to the tension force developed using all of

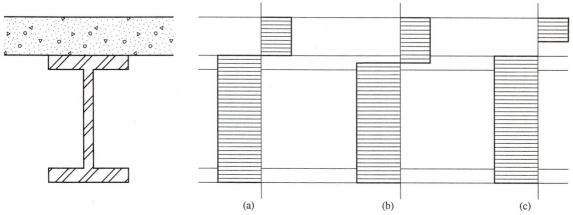

Figure 10.6   Plastic stress distribution.

the steel. For the distribution of Fig. 10.6b, the PNA is located within the steel shape. In this case, all of the concrete is taking compression but this is not sufficient to balance the tension force that the full steel shape could resist. Thus, some of the steel shape will be in compression in order to satisfy $T = C$. The plastic stress distribution shown in Fig. 10.6c is that which occurs when less than the full amount of concrete is needed to balance the force developed in the steel shape. Here the PNA is located within the concrete and that portion of the concrete below the PNA is not used since it would be in tension.

In all three cases, equilibrium of the cross section, when loaded so that the plastic moment is reached, requires that the shear connectors be capable of transferring the force carried by the concrete to the steel. For the cases of Fig. 10.6a and b, this will be the capacity of the concrete. For the case in Fig. 10.6c, this will be the capacity of the steel shape. Since the shear connectors will be carrying the full amount of shear force required to provide equilibrium using the maximum capacity of one of the elements, this is referred to as a "fully composite" beam.

The LRFD specification indicates that the plastic stress distribution in the concrete shall be taken as a uniform stress at a magnitude of $0.85f'_c$. This is the same distribution specified by the ACI code for reinforced concrete beams. In addition, the distribution of stress in the steel is taken as a uniform F_y, as was the case for the plastic moment in a steel shape.

10.5.1 Fully Composite Beams

The stress distribution is determined by calculating the minimum compression force as controlled by three components: the concrete, the steel, and the shear connectors. These are given as

$$C_c = 0.85f'_c \, b_{\text{eff}} \, t \qquad\qquad (10.1)$$

$$C_s = A_s F_y = T_{\text{s-total}} \tag{10.2}$$

and

$$C_q = \Sigma Q_n \tag{10.3}$$

Since full composite action is being assumed at this time, C_q will not control. Thus, if the steel controls, $C_s \leq C_c$, the distribution will be as given in either Fig. 10.6a or c. If the concrete controls, $C_c < C_s$, the distribution of Fig. 10.6b will result. Once the proper stress distribution is known, the corresponding forces may be determined and their point of application found. With this information, the nominal moment M_n may be found by taking moments about some reference point. Since the forces are equivalent to a force couple, any point of reference may be used for taking moments; however, it is convenient to use a consistent reference point, so these calculations will use the top of the steel as the point about which to take moments.

Actual determination of the PNA for the cases of Fig. 10.6a and c is quite straightforward. Since the steel controls in both cases, it is known that the concrete must carry a compression force equal to C_s. Only that portion of the concrete required to resist this force will be used, so that $C = 0.85 f_c' b_{\text{eff}} a$, where a defines the depth of the concrete stressed to its ultimate stress. Setting $T_{\text{s-total}} = C$, or $C_s = C$, and solving for a yields

$$a = \frac{A_s F_y}{0.85 f_c' \, b_{\text{eff}}} \tag{10.4}$$

For the special case where C_s is exactly equal to C_c, the value of a thus obtained will be equal to the actual slab thickness t. This is the case shown in Fig. 10.6a. For all other values of a, the distribution of Fig. 10.6c will result. The nominal flexural strength may then be obtained by taking moments about the top of the steel so that

$$M_n = T_{\text{s-total}} \left(\frac{d}{2} \right) + C \left(t - \frac{a}{2} \right) \tag{10.5}$$

When the concrete controls, $C_c < C_s$, the determination of the PNA is a bit more complex. It is best to consider this case as two separate subcases: the PNA occurring within the steel flange, and the PNA occurring within the web. Once it has been determined that C_c controls, the next step is to determine the force in the steel flange and web from

$$T_f = F_y t_f b_f \tag{10.6}$$

$$T_w = T_{\text{s-total}} - 2T_f \tag{10.7}$$

A comparison between the force in the concrete and the force in the bottom flange plus the web will show whether the PNA is in the top flange or web. Thus, if $C_c > T_w + T_f$, more tension is needed for equilibrium and the PNA is in the top flange. If $C_c < T_w + T_f$, less tension is needed for equilibrium and the PNA will be in the web. In either case, the difference between the concrete force C_c and the available steel force $T_{s\text{-total}}$ must be divided evenly between tension and compression in order to obtain equilibrium. This will allow for the determination of the PNA and the nominal moment capacity. Thus, with

$$A_{s\text{-c}} = \text{area of steel in compression}$$

and

$$A_{s\text{-total}} = \text{total area of steel}$$

equilibrium is given by

$$C_c + F_y A_{s\text{-c}} = T_{s\text{-total}} - F_y A_{s\text{-c}} \tag{10.8}$$

Solving for the area of steel in compression yields

$$A_{s\text{-c}} = \frac{T_{s\text{-total}} - C_c}{2F_y} \tag{10.9}$$

For the case where the PNA is in the flange, the distance from the top of the flange to the PNA is given by x, where

$$x = \frac{A_{s\text{-c}}}{b_f} \tag{10.10}$$

and for the case where the PNA is in the web,

$$x = t_f + \frac{A_{s\text{-c}} - T_f/F_y}{t_w} \tag{10.11}$$

EXAMPLE 10.1

Given: Determine the nominal moment capacity for the interior composite beam shown in Fig. 10.7. The section is a W21 × 44 and supports a 4.5-in. concrete slab. The dimensions are as shown. $F_y = 36$ ksi; $f'_c = 3$ ksi. Assume full composite action.

Solution

Determine the effective flange width:

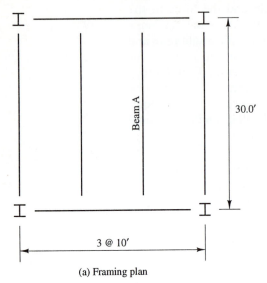

(a) Framing plan

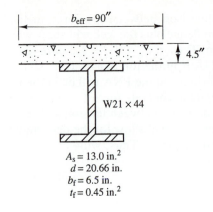

(b) Composite section

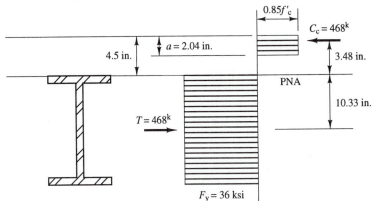

(c) Plastic stress distribution

Figure 10.7   Interior composite beam (Example 10.1).

$$b_{eff} = 30 \text{ ft } (12 \text{ in./ft})/4 \qquad = \quad 90 \text{ in.}$$

or $$b_{eff} = (10 \text{ ft } + \ 10 \text{ ft})(12 \text{ in./ft})/2 = 120 \text{ in.}$$

Therefore use $b_{eff} = 90$ in.
Determine the controlling compression force.

$$C_c = 0.85(3.0)(90)(4.5) = 1033 \text{ kips}$$
$$C_s = 13.0 \ (36) \qquad\qquad = \quad 468 \text{ kips}$$

Assuming full composite action, the shear connectors must carry the smallest of C_c and C_s; thus,

$$C_q = 468 \text{ kips}$$

Since C_s is less than C_c, the PNA is in the concrete.

Determine the PNA location using Eq. (10.4):

$$a = \frac{468}{0.85(3)(90)} = 2.04 \text{ in.}$$

The resulting plastic stress distribution is shown in Fig. 10.6c. Using Eq. 10.5, the nominal moment is determined as

$$M_n = 468\left(\frac{20.66}{2}\right) + 468\left(4.5 - \frac{2.04}{2}\right)$$

$$= 6463 \text{ kip-in.} = 539 \text{ kip-ft}$$

EXAMPLE 10.2

Given: Consider the same arrangement of beams and slab as used in Example 10.1 with a significantly larger steel member, W21 × 111. Determine the nominal moment capacity for the interior composite beam. Use the same materials as in Example 10.1. Assume full composite action.

Solution

The effective flange width will remain the same; thus,

$$b_{\text{eff}} = 90 \text{ in.}$$

Determine the controlling compression force

$$C_c = 0.85 \ (3.0)(90)(4.5) = 1033 \text{ kips}$$
$$C_s = 32.7 \ (36) \qquad\qquad = 1177 \text{ kips}$$

Assuming full composite action,

$$C_q = 1033 \text{ kips}$$

Since C_c is less than C_s, the PNA is in the steel.

Determine whether the PNA is in the flange or web.

$$T_f = 12.34(0.875)(36) = 388.7 \text{ kips}$$
$$T_w = 1177 - 2(388.7) = 399.6 \text{ kips}$$

Thus,

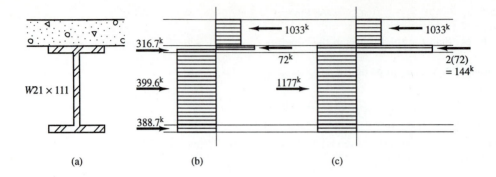

$A_s = 32.7$ in.^{2}
$d = 21.51$ in.
$b_f = 12.34$ in.
$t_f = 0.875$ in.

Figure 10.8   Interior composite beam (Example 10.2).

$$C_c = 1033 > T_f + T_w = 388.7 + 399.6 = 788.3$$

which shows that the PNA is in the flange.

The area of steel in compression is given by Eq. (10.9) as

$$A_{s\text{-}c} = \frac{1177 - 1033}{2(36)} = 2.0 \text{ in.}^2$$

and the PNA is located down from the top of the steel x, given by Eq. (10.10) as

$$x = \frac{2.0}{12.34} = 0.162 \text{ in.}$$

The stress distribution for this PNA is shown in Fig. 10.8b. Moments could be taken about any point to determine the nominal moment; however, a simplified mathematical model is shown in Fig. 10.8c that will make the analysis quicker. In this case, the full area of steel is shown in tension and that portion that is in compression is first removed and then added in compression. This results in only three forces and moment arms entering the moment equation. Thus,

$$M_n = 1177\left(\frac{21.51}{2}\right) + 1033\left(\frac{4.5}{2}\right) - 2(72)\left(\frac{0.162}{2}\right) = 14{,}971 \text{ kip-in.}$$

$$= 1248 \text{ kip-ft}$$

10.5.2 Partially Composite Beams

The composite members investigated thus far have all been fully composite. That means that the shear connectors have been capable of transferring whatever force is required for equilibrium, the force to be carried by the concrete. There are many conditions where the required strength of the composite section is less than that which would result from full composite action. In particular, these are cases where the size of the steel member is dictated by factors other than the strength of the composite section. Since shear connectors make up a significant part of the cost of a composite beam, economies can result if the reduced required strength can be reflected in reduced shear connectors.

If the composite section is viewed under elastic stress distributions, partial composite action may be more easily understood. Figure 10.9 shows elastic stress distributions for three cases of combined steel and concrete. The first case, Fig. 10.9a, results when the concrete simply rests on the steel with no shear transfer. The result is two independent members which may slip past each other at the interface. If the two materials are fully connected, the elastic stress distribution is as shown in Fig. 10.9c and the materials are not permitted to slip at all. If some slip is permitted, the resulting elastic stress distribution is as shown in Fig. 10.9b. This is the partially composite beam in the elastic region.

The plastic moment capacity for a partially composite member is the result of a stress distribution similar to that shown in Fig. 10.10. The PNA will be in the steel and the magnitude of the compression force in the concrete will be controlled by the capacity of the shear connectors C_q.

Regardless of the final location of the PNA, the force in the concrete is limited by the shear connector capacity. Thus, an approach combining those taken for the fully composite sections will be used for the partially composite member. By the definition of partially composite members,

$$C_c = C_q$$

and the depth of the concrete acting in compression is given by

$$a = \frac{C_c}{0.85 f'_c \, b_{\text{eff}}} \tag{10.12}$$

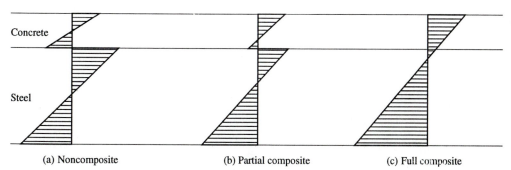

(a) Noncomposite   (b) Partial composite   (c) Full composite

Figure 10.9   Levels of composite action for elastic behavior.

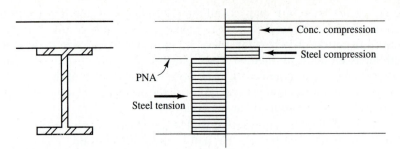

Figure 10.10   Plastic stress distribution for partial composite action.

Equations (10.6) through (10.11) may then be used to determine the location within the steel of the PNA, and the nominal moment may be obtained as before.

EXAMPLE 10.3

Given: Consider the concrete and steel given in Example 10.1 and shown in Fig. 10.7. In this case, assume that the shear connectors are capable of transferring $C_q = 300$ kips. Determine the nominal moment capacity of the resulting composite beam.

Solution

The following values are the same as those determined for Example 10-1:

$$b_{eff} = 90 \text{ in.}$$
$$C_c = 1033 \text{ kips}$$
$$C_s = 468 \text{ kips}$$

From the given data,

$$C_q = 300 \text{ kips}$$

Since the lowest value of the compressive force is given by C_q, this is a partially composite member. From Eq. (10.12),

$$a = \frac{300}{0.85(3)(90)} = 1.31 \text{ in.}$$

From Eq. (10.9), the area of steel in compression is

$$A_{s\text{-}c} = \frac{468 - 300}{2(36)} = 2.33 \text{ in.}^2$$

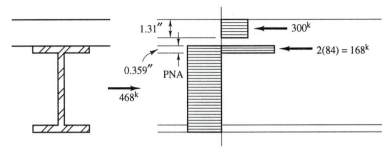

Figure 10.11   Stress distribution and forces used in Example 10.3.

Since this is less than the area of the flange, $6.5(0.45) = 2.93$ in.^{2}, the PNA is in the flange. The location of the PNA is determined from Eq. (10.10) as

$$x = \frac{2.33}{6.5} = 0.359 \text{ in.}$$

Using the three forces shown in Fig. 10.11, the nominal moment is

$$M_\text{n} = 468\left(\frac{20.66}{2}\right) + 300\left(4.5 - \frac{1.31}{2}\right) - 168\left(\frac{0.359}{2}\right) = 5958 \text{ kip-in.}$$

$$= 496 \text{ kip-ft}$$

It should be noted that the nominal moment reduced from 539 kip-ft for full composite action as found in Example 10-1 to 496 kip-ft for this level of partial composite action. This is approximately an 8 percent reduction in strength, corresponding to more than a 35 percent reduction in shear connector strength. The plastic moment capacity of the bare steel beam is $M_\text{p} = 286$ kip-ft.

Usually, the shear connectors provide the governing value for composite member design. Design tables in the *LRFD Manual* are based on this assumption. The authors consider these tables to be an excellent design aid and strongly recommend that the reader become acquainted with them.

10.5.3 Negative Moment Strength

The negative design flexural strength may be taken as that for the bare steel beam or the composite section may be used. For the composite section, the concrete in tension is ignored and reinforcing steel is placed in the tension region. ϕ_b is taken as 0.85, as for the positive moment case, and the nominal flexural strength is calculated assuming a plastic stress distribution similar to that for the positive moment case. The equations of equilibrium are applied, keeping in mind the following provisions:

1. The steel beam must be an adequately braced compact section.
2. Shear connectors are provided in the negative moment region.
3. The required longitudinal reinforcing bars are placed within the effective width of the slab and are properly developed.

10.6 SHEAR STUD CAPACITY

The LRFD specification recognizes two different types of shear connectors: steel studs and channels. The nominal strength of a single steel stud Q_n is given as

$$Q_n = 0.5A_{sc} \sqrt{f'_c E_c} \leq A_{sc}F_u \tag{10.13}$$

where

A_{sc} = cross-sectional area of the shank of the stud (in.^{2})
f'_c = specified compressive strength of the concrete (ksi)
F_u = minimum specified tensile strength of the stud (ksi)
E_c = modulus of elasticity of the concrete (ksi)
  = $w^{1.5} \sqrt{f'_c}$, where w is the unit weight of the concrete in pcf and f'_c is in ksi. Although this is somewhat different than the equation used by ACI, it provides sufficiently accurate results for use in this instance.

The nominal capacity of $\frac{3}{4}$ in. shear studs is given in Table 10.1.
The nominal strength of a channel shear connector Q_n is given as

$$Q_n = 0.3 (t_f + 0.5t_w)L_c\sqrt{f'_c E_c} \tag{10.14}$$

TABLE 10.1 NOMINAL STUD SHEAR STRENGTH Q_n (KIPS) FOR $\frac{3}{4}$-IN. HEADED STUDS.

f'_c (ksi)	w (lb/ft^3)	Q_n (kips)
3.0	115	17.7
3.0	145	21.0
3.5	115	19.8
3.5	145	23.6
4.0	115	21.9
4.0	145	26.1

Source: Reprinted by permission of the American Institute of Steel Construction, Inc., from the 1st edition of the *LRFD Manual.*

where

t_f = thickness of channel flange (in.)
t_w = thickness of channel web (in.)
L_c = length of channel (in.)

10.6.1 Shear Stud Placement

Although the purpose of a shear stud is to transfer load between the steel beam and the concrete slab, it is not necessary to place the studs in accordance with the shear diagram of the loaded beam. Tests have demonstrated that there is sufficient ductility in the studs to redistribute the shear load under the ultimate load condition. Therefore, in design, it is assumed that the studs share the load equally. Thus, the total shear force determined according to Section 10.5 must be transferred over the distance between maximum moment and zero moment. For a uniform load this will result in C_q/Q_n connectors on each side of the center line of the beam span. In the case of concentrated loads placed at the third points of the beam, there would be the same number of studs on each side of the load and a minimum number between the loads, as indicated in Fig. 10.12.

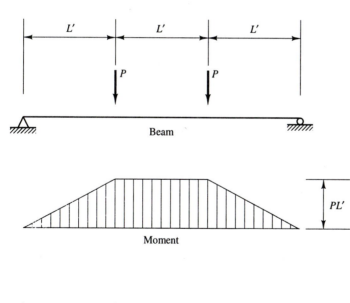

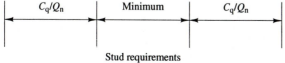

Figure 10.12 Stud placement for concentrated load.

EXAMPLE 10.4

Given: For the fully composite beam of Example 10.1, determine the required number of $\frac{3}{4}$-in. shear studs over the complete beam span. Assume normal-weight concrete (145 lb/ft^3) and the values of Example 10.1.

Solution

From Table 10.1,

$$Q_n = 21.0 \text{ kips}$$

From Example 10.1,

$$C_q = 468 \text{ kips}$$

and

$$\text{Number of studs} = \frac{468}{21} = 22.3 \text{ studs}$$

Therefore, place 23 $\frac{3}{4}$-in. shear studs on each side of the beam span center line.

10.7 COMPOSITE BEAMS WITH FORMED METAL DECK

In the 1960s, manufacturers of formed metal deck developed installation procedures where the stud could be welded to the beam through the metal deck. Research by Grant et al. [6] extended the design procedure of the time to cover the use of formed metal deck. Today, combining metal deck and composite design is considered to be one of the most economical methods of floor construction for office buildings. Cells formed by enclosing the space below the deck and between the ribs may then be utilized for distribution of the electrical and electronic systems of the building, contributing greatly to the economy of the system.

Using the available research results, the LRFD specification provides rules for steel decks with nominal rib heights of up to 3 in. and average rib widths of 2 in. or more. Studs must be $\frac{3}{4}$ in. in diameter and extend at least $1\frac{1}{2}$ in. above the top of the steel deck. The concrete slab thickness above the deck must be at least 2 in.

10.7.1 Deck Ribs Perpendicular to Steel Beam

Beams supporting the steel deck will have the ribs running perpendicular to the beam, as shown in Fig. 10.13. For calculation purposes, concrete below the top of the steel deck is neglected. The design procedure de-

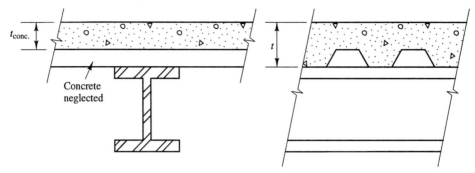

Figure 10.13   Beam with formed metal deck.

scribed for flat soffit beams will apply here as well, provided sufficient concrete is available above the top of the steel deck as determined through Eq. (10.4) or (10.12).

The specified strength of a stud shear connector must be reduced to account for the fact that the force exerted on the stud is applied at a higher point in this application than in a flat soffit beam. The value specified in Eq. (10.13) is multiplied by the following reduction factor:

$$\text{RF} = \left(\frac{0.85}{\sqrt{N_r}} \right) \left(\frac{w_r}{h_r} \right) \left(\frac{H_s}{h_r} - 1.0 \right) \leq 1.0 \qquad (10.15)$$

where N_r is the number of stud connectors in one rib at a beam intersection, not to exceed three in computations, although more than three studs may be installed; w_r is the average width of concrete rib or haunch (in.); h_r is the nominal rib height (in.); and H_s is the length of stud connector after welding (in.), not to exceed the value ($h_r + 3$) in computations, although actual length may be greater. The maximum stud spacing is specified as 36 in., which is convenient because many decks have a rib spacing of 6 in.

10.7.2 Deck Ribs Parallel to Steel Beam

For girders supporting beams that carry steel deck, the ribs will run parallel to the steel section, as shown in Fig. 10.14. Concrete below the top of the deck may be used in calculating the composite section properties and must be used in shear stud calculations.

When the depth of the steel deck is $1\frac{1}{2}$ in. or greater, the average width w_r of the haunch or rib shall be not less than 2 in. for the first stud in the transverse row plus 4 stud diameters for each additional stud. If the deck rib is too narrow, the deck may be split over the beam and spaced in such a way as to allow for the necessary rib width without adversely affecting member strength. The specified strength of a stud shear connect-

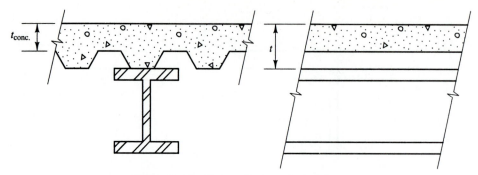

Figure 10.14   Girder with formed metal deck.

or for this arrangement is the value specified in Eq. (10.13) multiplied by the following reduction factor:

$$\text{RF} = 0.6\left(\frac{w_r}{h_r}\right)\left(\frac{H_s}{h_r} - 1.0\right) \leq 1.0 \qquad (10.16)$$

EXAMPLE 10.5

Given: Calculate the design moment and stud requirements for a composite section consisting of a W18 × 35 with a 6-in. slab on a 3-in. metal deck. The beam spacing is 12 ft and the beam span is 40 ft. $f_c' = 3.5$ ksi; $F_y = 36$ ksi. Carry out the calculations for the following three cases:

(a) Full composite action

(b) Partial composite action with $C_q = 279$ kips, which results in the PNA at the center of the top flange of the steel beam

(c) Partial composite action with $C_q = 187.2$ kips, which results in the PNA at the bottom of the top flange of the steel beam

Solution

First determine the effective flange width and the values for compression force using the full concrete and full steel areas:

$$b_{\text{eff}} = \frac{40}{4} \qquad = 10 \text{ ft (governs)}$$

$$b_{\text{eff}} = \text{beam spacing} = 12 \text{ ft}$$

$$C_c = 0.85(3.5)(120)(3.0) = 1071 \quad \text{kips}$$

$$C_s = 10.3 \ (36) \qquad = 370.8 \text{ kips}$$

Case (a) (Fig. 10.15)

For full composite action, C_q is the smallest of C_c and C_s; thus,

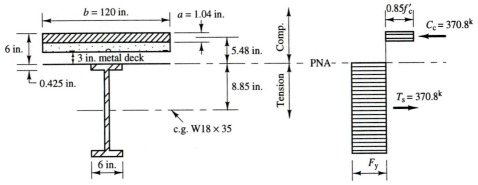

Figure 10.15   Example 10.5, case (a).

$$C_q = 370.8 \text{ kips}$$

Calculate the effective depth of the concrete

$$a = \frac{370.8}{(0.85)(3.5)(120)} = 1.04 \text{ in.}$$

Since a is less than the 3 in. available, the procedures for a flat soffit beam may be used. Thus,

$$M_n = 370.8(8.85) + 370.8(6 - 1.04/2) = 5314 \text{ kip-in.}$$
$$= 443 \text{ kip-ft}$$

Therefore, the design strength is

$$\phi_b M_n = 0.85(443) = 377 \text{ kip-ft}$$

Determine the stud requirements.

The shear that is to be transferred is 370.8 kips. From Table 10.1, the value of a single $\frac{3}{4}$-in. stud, with normal weight concrete $f_c' = 3.5$ ksi, is 23.6 kips.

Since the studs are used in conjunction with metal deck, check for any required reduction. Use $w_r = 5.875$ in., $h_r = 3.0$ in., $H_s = 4.5$ in., and assume one stud per rib.

$$RF = \frac{0.85}{\sqrt{1}} \left(\frac{5.875}{3}\right)\left(\frac{4.5}{3} - 1.0\right) = 0.83$$

Therefore, the capacity of a single stud is

$$Q_n = 0.83(23.6) = 19.6 \text{ kips}$$

and the number of studs required on each side of the maximum moment is

$$\text{Number of studs} = \frac{370.8}{19.6} = 18.9 \text{ studs}$$

Therefore, use 19 studs on each half span or 38 studs.

Case (b) (Fig. 10.16)

Since the value of $C_q = 279$ kips given is less than C_c and C_s, this will be a partially composite beam.

Calculate the effective depth of the concrete as before:

$$a = \frac{279}{(0.85)(3.5)(120)} = 0.782 \text{ in.}$$

Since $a < 3.0$ in., there is sufficient concrete available above the metal deck.

Determine the area of steel in compression and the location of the PNA.

$$A_{\text{s-c}} = \frac{370.8 - 279}{2(36)} = 1.275 \text{ in.}^2$$

$$x = \frac{A_{\text{s-c}}}{b_f} = \frac{1.275}{6.0} = 0.2125 \text{ in.}$$

so the resulting moment is

$$M_n = 370.8(8.85) + 279(6 - 0.782/2) - 2(45.9)(0.2125/2)$$

$$= 4837 \text{ kip-in.}$$

$$= 403 \text{ kip-ft}$$

Design strength $= 0.85 (403) = 343$ kip-ft

Determine the stud requirements.

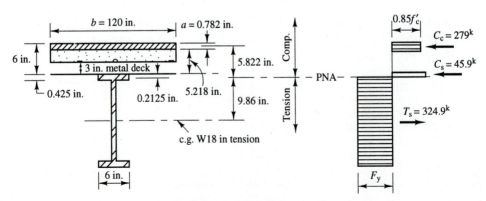

Figure 10.16 Example 10.5, case (b).

The shear to be transferred is 279 kips. From Table 10.1 and the value of RF previously determined,

$$Q_n = 19.6 \text{ kips}$$

The required number of shear studs is

$$\text{Number of studs} = \frac{279}{19.6} = 14.2 \text{ studs}$$

Therefore, use 15 studs in each half span or 30 studs for the beam.

Case (c) (Fig. 10.17)

Again, the value of $C_q = 187.2$ kips, as given, controls. Calculate the effective depth of concrete as before:

$$a = \frac{187.2}{(0.85)(3.5)(120)} = 0.5244 \text{ in.}$$

Determine the area of steel in compression and the location of the PNA:

$$A_{s\text{-}c} = \frac{370.8 - 187.2}{2(36)} = 2.55 \text{ in.}^2$$

$$x = \frac{2.55}{6.0} = 0.425 \text{ in., which is } t_f$$

The nominal moment is then

$$M_n = 370.8(8.85) + 187.2(6 - 0.5244/2) - 2(91.8)(0.425/2)$$
$$= 4317 \text{ kip-in.}$$
$$= 360 \text{ kip-ft}$$

Design strength $= 0.85(360) = 306$ kip-ft.

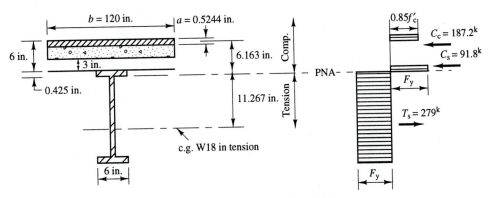

Figure 10.17   Example 10.5, case (c).

Determine the stud requirements.

The shear to be transferred is 187.2 kips. From Table 10.1 and the value of RF previously determined,

$$Q_n = 19.6 \text{ kips}$$

The required number of shear studs is

$$\text{Number of studs} = \frac{187.2}{19.6} = 9.6 \text{ studs}$$

Therefore, use 10 studs in each half span or 20 for the full beam.

10.8 FULLY ENCASED STEEL BEAMS

Steel beams fully encased in concrete that will contribute to the strength of the final member are called encased beams. Such beams may be designed by one of two available procedures. The design flexural strength $\phi_b M_n$ may be calculated from superposition of elastic stresses, considering the effects of shoring, or may be calculated from the plastic strength of the steel section alone. In both cases, ϕ_b is to be taken as 0.9.

For an encased beam to be used as a composite beam, the following must be satisfied:

1. Concrete cover over the steel beam sides and soffit must be at least 2 in.
2. The top of the steel beam must be at least $1\frac{1}{2}$ in. below the top and 2 in. above the bottom of the slab.
3. Concrete encasement must contain adequate mesh or other reinforcing steel to prevent spalling of the concrete.

10.9 SELECTING A SECTION

The design of a composite beam is somewhat of a trial and check procedure, as are numerous other design situations. The material presented thus far in this chapter has been directed toward the determination of section capacity when the shape is known. This section addresses the preliminary selection of the steel shape to go along with an already known concrete slab.

Various rules of thumb have been used over the years to estimate the depth of structural members [7]. The depth recommended through these rules is the depth for the steel section in a noncomposite design or the total depth, including the slab, in composite design. One common rule is the depth of a beam is span/24. Based on this approach, the depth of the steel section in a composite beam is estimated as the given depth less the slab thickness.

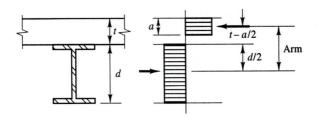

Figure 10.18 Moment arm for preliminary weight determination.

With an estimate of beam depth, the weight of the beam may be estimated. This will be based on the assumption that the PNA is within the concrete so that the full steel section is at yield. The resulting dimensions are given in Fig. 10.18. The moment arm between the tension force in the steel and the compression force in the concrete is given by

$$\text{Arm} = \frac{d}{2} + \left(t - \frac{a}{2} \right) \qquad (10.17)$$

If the required moment capacity is divided by the moment arm, the required tension force will be determined. If that force is divided by the steel yield stress, the required area is determined. Multiplying the required area by 3.4 lb/ft for each square inch will yield an estimate of the beam weight. Thus,

$$\text{Beam wt} = \frac{M_n}{(d/2 + t - a/2)F_y} (3.4) \qquad (10.18)$$

The depth of the beam can be estimated as described above. The thickness of the slab is determined from the design in the transverse direction. Only the effective depth of the concrete is left to be determined. It is generally sufficient to assume that the effective depth of the concrete is 2 in.; therefore, $a/2 = 1.0$ in.

10.10 SERVICEABILITY CONSIDERATIONS

There are several important serviceability considerations associated with the design of composite floor beams: (1) deflection during construction, (2) vibration under service loads and (3) live load deflection under service loads.

10.10.1 Deflection During Construction

As discussed in Section 10.3, the LRFD specification permits either shored or unshored construction. With unshored construction, Section I3.4 of the LRFD specification requires that the steel section alone have adequate strength to support all loads applied prior to the concrete attaining 75% of its specified strength. The bare steel beam, under the weight of these loads, will deflect as an elastic member. Because of this deflection of the beam under the wet concrete, cambering, imposing a permanent upward bend, of the steel beam is often specified. However, predicting the necessary camber is difficult because of the varying methods and sequences of concrete placement employed by different contractors. Also, the end restraint provided by the beam connection reduces deflection and this is difficult to estimate. The usual solution is for the designer to add a little extra concrete load into the dead load and for the contractor to allow for a little extra concrete in the quantity estimate.

With unshored construction, the beam is free to deflect under the wet concrete. Therefore, the calculated extreme fiber stress in the beam is limited by the LRFD specification to $0.90F_y$. This is necessary even though the ultimate strength of the composite section is not dependent on the elastic steel stresses during construction. This rule will prevent excessive permanent deflections after the service load is removed.

In the case of shored construction, deflection during construction is usually not a concern, since the shores are not removed until the concrete has achieved some composite action and the deflection under the wet concrete is a minimum. On the other hand, long-term creep may have to be investigated because the concrete is constantly stressed to a high level under service loads.

10.10.2 Vibration Under Service Loads

Composite construction is usually shallower than comparable noncomposite construction, and therefore may be more susceptible to perceived vibrations, even though, for the same beam size, the composite section is stiffer than the bare steel shape. Problems, if they occur, usually occur with long spans and little damping. For instance, a large area of a department store containing only a light jewelry display might exhibit vibrations that would be perceptible to some customers. On the other hand, an office building, constructed with the same floor system, could contain partitions (full or partial), which would provide sufficient damping to obviate any perceived vibration. Because of the limited research and wide difference in human perception of vibration, the question does not lend itself to simple solutions. The reader is referred to Murray [3] for more information that gives the designer an approach to vibration acceptance criteria, damping, and rational design techniques.

10.10.3 Live Load Deflections

Live load deflections may be a design consideration in many building structures. For instance, in office buildings, an excessive live load floor deflection could cause problems with the proper fit of partitions, doors, and equipment and may also result in an unacceptable appearance. Therefore, a live load deflection calculation should be made. It should be remembered, however, that this calculation is made with service live loads, while the design of the composite beam was actually made with ultimate loads. Live load deflections at ultimate load levels have no design significance.

Deflection calculations are, of course, a function of the moment of inertia of the composite section. Since the true moment of inertia at service loads is not easily determined, the *LRFD Manual* tabulates moments of inertia termed "lower bound." These values are calculated for the case when the composite beam is at the ultimate load condition. This means that the area of concrete required to carry the shear force transferred by the shear connectors is all that is used in a transformed section analysis. With this assumption, the values of I_{lb} determined are smaller than the moment of inertia values for the actual service load condition.

Use of these I_{lb} values, therefore, results in conservative estimates of service load deflections because at service loads, more of the concrete actually participates in resisting deflection. Even for partially composite sections, more of the concrete participates in resisting deflection, since at service load levels, the frictional bond between the top flange of the beam and the concrete has slipped only slightly, if at all, and the real moment of inertia is larger than it is at ultimate load.

10.11 COMPOSITE COLUMNS

Composite columns in building construction have been much slower in development than beams. Rules, published for the first time in 1979 [8], were adopted by AISC in 1986 for the LRFD specification. The use of composite columns in buildings is still quite limited, primarily because connection details are still being developed.

Two types of composite columns are provided for in the LRFD specification. The first is an open shape, usually a wide flange, encased by concrete, while the second is a hollow shape, such as a tube or pipe, filled with concrete. According to the LRFD specification, for a member to qualify as a composite column it must meet the following limitations:

1. The cross-sectional area of the steel member must comprise at least 4 percent of the gross section.
2. The concrete encasement must be reinforced with longitudinal steel as well as lateral ties.
3. The concrete strength must be between 3 and 8 ksi for normal-weight concrete and 4 and 8 ksi for lightweight concrete.
4. The maximum value of F_y to be used in calculations is 55 ksi.
5. Hollow sections must have a minimum wall thickness as specified.

The most important of these limitations is that the cross-sectional area of the steel portion must comprise at least 4% of the total composite cross section. If the steel portion is less than 4%, the member should be designed according to the rules specified by the American Concrete Institute. The other limitations are readily satisfied in a design situation.

The basic column formulas specified by the LRFD specification, Section E2 and discussed in Chapter 6 are used for design of composite columns with the following modifications:

1. A_g is replaced by A_s, the gross area of the steel shape, pipe, or tubing.

2. The radius of gyration r is replaced by the radius of gyration of the steel shape, pipe, or tubing, r_m. For steel shapes, $r_m \geq 0.3$ times the overall thickness of the composite cross section in the plane of buckling (in.).

3. F_y is replaced by a modified yield stress F_{my}, where

$$F_{my} = F_y + c_1 F_{yr} \left(\frac{A_r}{A_s}\right) + c_2 f_c' \left(\frac{A_c}{A_s}\right) \qquad (10.19)$$

where

$c_1 = 1.0$ for concrete filled pipe or tube and 0.7 for concrete encased shapes

$c_2 = 0.85$ for concrete filled pipe or tube and 0.6 for concrete encased shapes

$F_{yr} = $ specified minimum yield stress of reinforcing (ksi)

$f_c' = $ specified compressive strength of concrete (ksi)

$A_r/A_s = $ ratio of the area of the reinforcing to the area of the steel shape

$A_c/A_s = $ ratio of the concrete area to the steel area

4. E is replaced by a modified modulus of elasticity E_m, where

$$E_m = E + c_3 E_c \frac{A_c}{A_s} \qquad (10.20)$$

with

$c_3 = 0.4$ for concrete filled pipe or tube and 0.2 for concrete encased shapes

$$E_c = w^{1.5} \sqrt{f_c'}$$

where w is the unit weight of concrete (lb/ft³) and f_c' is the ultimate strength of concrete (ksi)

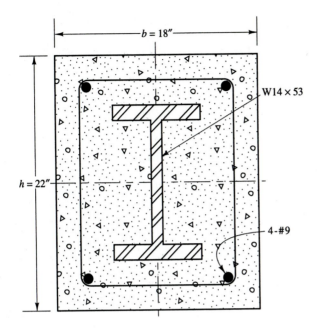

Figure 10.19 Composite column (Example 10.6).

EXAMPLE 10.6

Given: Determine the design strength of the 18×22 composite column shown in Fig. 10.19. Effective length = 15 ft.

Steel shape: W14$\times$53, F_y = 36 ksi
Reinforcing: 4 $-$ #9 bars, Grade 60
Concrete strength: f'_c = 5 ksi
E_c = $(145)^{1.5} \sqrt{5}$ = 3904 ksi
r_{my} = radius of gyration about the y axis = 0.3(18) = 5.40 in.

Solution

$$c_1 = 0.7 \qquad c_2 = 0.6 \qquad c_3 = 0.2$$

$$\frac{A_r}{A_s} = \frac{4 \times 1.0}{15.6} = 0.256$$

$$\frac{A_c}{A_s} = \frac{18 \times 22}{15.6} = 25.4$$

From Eq. (10.19),

$$F_{my} = 36 + 0.7(55)\,(0.256) + 0.6(5)(25.4)$$

$$= 122.1 \text{ ksi}$$

From Eq. (10.20),

$$E_m = 29,000 + 0.2(3904)(25.4) = 48,832 \text{ ksi}$$

Determine the slenderness

$$\lambda_c = \frac{KL}{r_m \ \pi} \sqrt{\frac{F_{my}}{E_m}} = \frac{(15)(12)}{5.40(3.14)} \sqrt{\frac{122.6}{48,832}} = 0.532$$

Thus, since $\lambda_c = 0.532 < 1.5$, use the LRFD specification, Eq. (E2-2):

$$\lambda_c^2 = (0.532)^2 = 0.283$$

and

$$F_{cr} = (0.658)^{0.283} \ (122.6) = 109.0 \text{ ksi}$$

$$\text{Design strength} = \phi_c \ F_{cr} \ A_s$$

$$= 0.85 \ (109.0)(15.6) = 1445 \text{ kips}$$

PROBLEMS

For Problems 10.1–10.6, assume a flat soffit, fully composite beam.

10.1. Determine the location of the plastic neutral axis and the nominal moment capacity for a composite beam composed of a W16×26 spanning 20 ft and spaced 8 ft on center, supporting a 6-in. concrete slab using $f'_c = 4$ ksi and A36 steel.

10.2. For a W16×45 composite beam spanning 22 ft and spaced at 8 ft, determine the location of the plastic neutral axis and the nominal moment capacity if the slab is 5 in. thick. Use $f'_c = 5$ ksi and $F_y = 50$ ksi.

10.3. Determine the location of the plastic neutral axis and the nominal moment capacity for a W18×50 supporting a 5-in. slab. The beams span 20 ft and are spaced 6 ft on center. Use $f'_c = 5$ ksi and $F_y = 50$ ksi.

10.4. A W18×71 is used as a composite beam with a 4-in. concrete slab. The beams span 18 ft and are spaced at 5-ft intervals. Determine the location of the plastic neutral axis and nominal moment capacity using $f'_c = 4$ ksi and A36 steel.

10.5. Determine the location of the plastic neutral axis and the nominal moment capacity for a W14×43 supporting a 4-in. concrete slab with $f'_c = 3$ ksi. The beams have $F_y = 50$ ksi, are spaced 5 ft on center, and span 20 ft.

10.6. For a 24-ft-span W14×61 that supports a 4-in. concrete slab spanning 6 ft, determine the plastic neu-

tral axis and the nominal moment capacity. Use $f'_c = 3$ ksi and $F_y = 50$ ksi.

10.7. Repeat Problem 10.1 with the shear stud capacity limited to $C_q = 250$ kips.

10.8. Repeat Problem 10.2 with the shear stud capacity limited to $C_q = 500$ kips.

10.9. Repeat Problem 10.3 with the shear stud capacity limited to $C_q = 500$ kips.

10.10. Repeat Problem 10.4 with the shear stud capacity limited to $C_q = 400$ kips.

10.11. Repeat Problem 10.5 with the shear stud capacity limited to $C_q = 300$ kips.

10.12. Repeat Problem 10.6 with the shear stud capacity limited to $C_q = 600$ kips.

10.13. A W12 composite beam spaced every 10 ft is to be used to support a uniform dead load of 1.0 k/ft and live load of 0.9 k/ft on a 20 ft span. Using a 4-in. flat soffit slab with $f'_c = 4$ ksi, $\frac{3}{4}$-in. shear studs and A36 steel, determine the least weight shape and the required number of shear connectors to support the load.

10.14. A W14 composite beam is to support a uniform dead load of 1.2 k/ft and live load of 1.2 k/ft. The beam spans 24 ft and is spaced 8 ft from adjacent beams. Using a 5-in. slab and $\frac{3}{4}$-in. shear studs, determine the least weight shape to support the load if $f'_c = 4$ ksi and A36 steel is employed.

10.15. Compare the least weight W16 and W14 members required to support a uniform dead load of 2.4 k/ft and live load of 3.2 k/ft. The beams span 18 ft and are 12 ft on center. They support a 6-in. concrete slab with f'_c = 4 ksi. The steel is A36.

10.16. A series of W16×36 composite beams are spaced at 10 ft on center and span 24 ft. The beams support a 2½-in. formed metal deck, perpendicular to the beam, with a slab whose total thickness is 5 in. Assuming full composite action, determine the nominal moment capacity and the number of ¾-in. shear studs required. The deck has 6-in. wide ribs spaced at 12 in. Use f'_c = 4 ksi and A36 steel.

10.17. Determine the nominal moment capacity of a W18×35 composite beam supporting a slab with total thickness of 5 in. on a 3-in. formed metal deck, perpendicular to the beam. The beam spans 28 ft and is spaced 12 ft from the adjacent beams. Use f'_c = 4 ksi and F_y = 50 ksi.

10.18. A partially composite beam supports a 3-in. metal deck, perpendicular to the beam, and 6-in. total thickness slab. The beam spans 30 ft and is 11 ft from adjacent beams. If the shear stud capacity is 400 kips, f'_c = 5 ksi, and F_y = 50 ksi, determine the nominal moment capacity.

10.19. A composite beam is to span 20 ft and support a 4-in. slab on 1½-in. metal deck spanning 10 ft. The dead load is 60 psf and the live load is 100 psf. The deck has 2-in. ribs spaced 6-in. on center. Determine the steel member to be used and the number of ¾-in. shear studs.

10.20. A composite girder spans 30 ft and supports two concentrated dead loads of 12 kips and live loads of 20 kips at the third points. The 1½-in. metal deck with 2-in. ribs spaced 6 in. on center is parallel to the girder and supports a 5-in. slab. Using f'_c = 4 ksi and A36 steel, determine the required W-shape to carry the load and the number of ¾-in. shear studs required.

10.21. Determine the live load deflection for a W24×76 composite beam with an 8-in. slab on a 3-in. deck. The beam spans 28 ft and is spaced 10 ft. Use a live load of 3.4 k/ft, f'_c = 4 ksi, and A36 steel.

10.22. Determine the live load deflection for a W16×26 composite beam supporting a 6-in. slab on 2½-in. metal deck. The beam spans 24 ft and is spaced at 8 ft on center. The live load is 2.1 k/ft, f'_c = 4 ksi, and F_y = 50 ksi.

10.23. Determine the design strength of an 18 × 18-in. composite column with a W10×68 and 4 - #8, grade 60 bars. Use F_y = 50 ksi, f'_c = 5 ksi, and an unbraced length of 20 ft.

10.24. Determine the design strength of a 22 × 22-in. composite column with a W12×120 and 4 - #9, grade 60 bars. Use A36 steel, f'_c = 4 ksi, and an unbraced length of 16 ft.

10.25. Determine the design strength of a 20 × 22-in. composite column with a W12×136 and 4 - #10 grade 60 bars. Use A36 steel, f'_c = 5 ksi, and an unbraced length of 12 ft.

REFERENCES

[1] Viest, I. M., Fountain, R. S., and Singleton, R. C., *Composite Construction in Steel and Concrete*, McGraw-Hill, New York, 1958.

[2] Hansell, W. C., Galambos, T. V., Ravindra, M. K., and Viest, I. M., Composite beam criteria in LRFD, *Journal of the Structural Division, ASCE*, 104, ST9, September 1978 (pp. 1409–1426).

[3] Murray, T. M., Building floor vibrations, *Proceedings of the 1991 National Steel Construction Conference*, Washington, DC, 5–7 June 1991 (pp. 19.1–19.18).

[4] Ollgaard, J. G., Slutter, R. G., and Fisher, J.W., Shear strength of stud connectors in lightweight and normal-weight concrete, *Engineering Journal, AISC*, Vol. 8, No. 2, 1971 (pp. 55–64).

[5] McGarraugh, J. B., and Baldwin, J.W., Lightweight concrete-on-steel composite beams, *Engineering Journal, AISC*, Vol. 8, No. 3, 1971 (pp. 90–98).

[6] Grant, J. A., Fisher, J. W., and Slutter, R. G., Composite beams with formed steel deck, *Engineering Journal, AISC*, Vol. 14, First Quarter, 1977 (pp. 24–43).

[7] Rice, P. F., and Hoffman, E. S., *Structural Design Guide to AISC Specifications for Buildings*, Van Nostrand Reinhold, New York, 1976.

[8] Task Group 20, Structural Stability Research Council, A specification for the design of steel–concrete composite columns, *Engineering Journal, AISC*, Vol. 16, Fourth Quarter, 1979 (pp. 101–115).

CHAPTER ELEVEN

Connections

11.1 INTRODUCTION

A steel structure is essentially a collection of individual members that are attached to each other to form a stable and serviceable whole, the frame. The assumed behavior of the connection between any two members determines how the structure is analyzed to resist gravity and lateral (wind or seismic) loads. This analysis, in turn, determines the moments, shears, and axial loads for which the beams and columns are designed. It is, therefore, essential that the designer understand the basic behavior of connections.

Members are attached to each other through a variety of connecting elements, such as angles and plates, using mechanical fasteners or fusion joining processes such as welding. The characteristics of these elements and fastening devices must likewise be fully understood, in order to be able to assess the response of the complete connection. Finally, the load transfer mechanism must be understood since it dictates the various limit states of the joint.

11.2 BASIC CONNECTION BEHAVIOR

All beam-to-column connections exhibit certain force–deformation relationships. Typical moment–rotation relationships for three beam-to-column connections are represented in Fig. 11.1. When the connection is

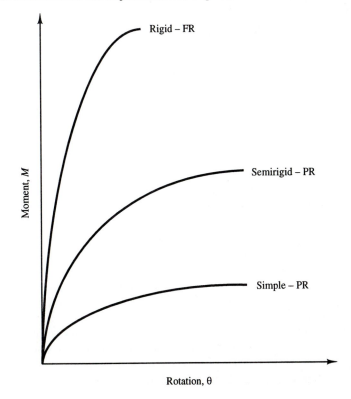

Figure 11.1 Moment–rotation relations for connections.

stiff, it will deform very little even when subjected to large moments. On the other hand, connections can be designed that are quite flexible, will rotate considerably while still carrying a vertical load, and will not develop a significant moment.

For the purposes of design, connections have been assumed to be rigid, simple (pinned), or semirigid. The LRFD specification, however, divides connections into two categories: fully restrained (FR) and partially restrained (PR). FR connections are assumed to be rigid, while PR connections are either simple or semirigid.

It is the designers responsibility to match connection behavior with the appropriate category. Often, this requires experience and judgment; the state of the art is such that it is usually not possible to predict the actual M - θ curve with much accuracy for anything but very simple connections.

11.3 FULLY RESTRAINED CONNECTIONS

The basic assumption for frames with FR connections is that the beams and columns maintain their original relationship during the entire loading history. This is normally referred to as a rigid connection. Figure 11.2 shows examples of beam-to-column connections that are usually treated as FR connections.

Figure 11.2a has a web plate shop-welded to the column and field-bolted to the beam web. The beam flanges have been prepared (chamfered) in the shop and field-welded to the column. Although the beam web is not continuously connected to the column, it has been demonstrated by Huang et al. [1] and others that this connection can adequately transfer the full plastic moment of the beam to the column. Since most of the moment capacity is derived from the flange connections (flange force times beam depth) and the web moment and local strain hardening together help, the connection is able to reach the full plastic moment of the beam.

Figure 11.2b is similar to Fig. 11.2a except that the beam is framing into the web of the column. To ensure that this connection has adequate ductility it is important to extend the flange connecting plates beyond the column flange and also to design these plates a little thicker than the beam flange. Extending the connecting plate reduces the possibility of a triaxial stress condition near the column flange tips. Thickening the plate reduces the average tension stress in the plate. Thickening also facilitates welding to the beam flange, because the beam flange is often slightly "cocked."

Figure 11.2c is an end plate connection. The design procedure is semiempirical and will be discussed later in this chapter. Although an end plate connection is very popular with some fabricators, others tend to avoid it. It must be fabricated with special care so that the end plates are parallel with each other. Also, it is not a very "forgiving" connection and may make erection difficult and expensive.

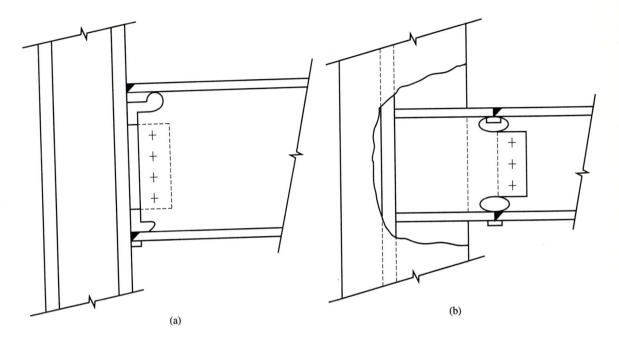

(a)

(b)

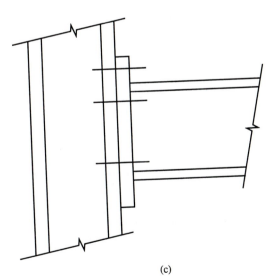

(c)

Figure 11.2   Type FR rigid connections.

11.4 PARTIALLY RESTRAINED CONNECTIONS

A frame with PR connections may be analyzed with either perfectly "simple" or "semirigid" connections or a combination of connections. In the simple case, the analysis assumes that the connections are pinned and free to rotate. The rotation capacity must be sufficient to accommodate the rotation of the beam to which it is connected, when the beam is carrying factored loads.

In choosing the structural system for a frame, the choice of connections depends heavily on how the gravity and lateral loads are to be carried. There are basically two ways in which a simple frame may be designed to resist lateral loads and to provide stability for gravity loads. In one case, a positive bracing system is provided, such as diagonal steel bracing or a masonry shear wall. In the second case, lateral stability is provided by the restraint that is offered by the connections and members themselves. This is done by designing the connections with a limited amount of moment resistance. The connections are flexible enough to rotate under gravity loads so that no gravity moments are transferred to the columns. At the same time it is assumed that the connections have sufficient strength to resist the lateral loads and to provide frame stability [2]. Recent research indicates that this design method should be limited to buildings of less than ten stories to avoid the possibility of excessive drift due to wind load [3].

The second category of PR connections requires that the frame be analyzed considering the true semirigid behavior of the connections. In this case the M - θ curve of the connection must be fairly well defined, which is usually not the case. The analysis also tends to be rather complex, requiring extensive computer analysis, although this obstacle is gradually being removed through advances in computer software development [4, 5].

Figure 11.3 shows simple PR connections that are used with a positive bracing system. Figure 11.3a is a double-angle connection (a pair of clip angles) that has been used extensively over the years. In fact, it is usually considered the standard to which other simple connections are compared. Figure 11.3b shows a single-plate framing connection which is often referred to as a shear tab. These connections were first used in the 1960s but very little research was conducted until Richard et al. [6] and Astaneh [7] in the 1980s.

Figure 11.3c shows a seated connection and Fig. 11.3d shows a stiffened seated connection. Both can be bolted or welded and are usually used to frame a beam into the web of a wide-flange column section. Although they may appear to be stiffer than the standard clip angle connection, they are assumed to rotate sufficiently at factored loads without transferring a significant moment to the column.

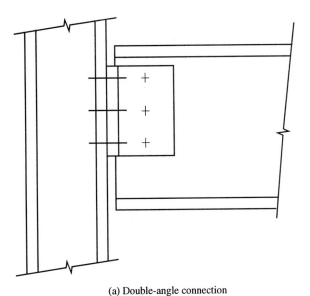

(a) Double-angle connection

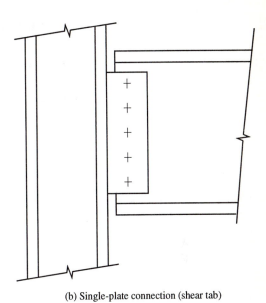

(b) Single-plate connection (shear tab)

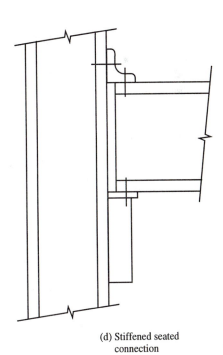

(c) Seated connection

(d) Stiffened seated
connection

Figure 11.3 Type PR connections.

11.5 MECHANICAL FASTENERS

The mechanical fasteners most commonly used are common bolts and high-strength bolts.

11.5.1 Common Bolts

Common bolts are manufactured to ASTM specification A307. They are usually used in PR, simple connections for girts, purlins, light floor beams, bracing, and other applications where the loads are relatively small. They should not be used where the loads are cyclic or vibratory, or where fatigue may be a factor.

Common bolts are also referred to as machine, unfinished, or rough bolts. They may be identified by their square heads and nuts with no marking on the heads. They are available in diameters from $\frac{1}{4}$ to 4 in. These bolts are usually installed using a spud wrench. No specified pretension or torque is required. It is necessary only to tighten the nut sufficiently to prevent it from backing off since no clamping force is assumed. The design shear and tensile strength are specified in LRFD specification, Section J3, Table J3.2.

11.5.2 High-Strength Bolts

Two basic types of high-strength bolts are currently used in steel structures: ASTM A325, manufactured under ASTM specification "High-Strength Bolts for Structural Joints," and ASTM A490, "Quenched and Tempered Alloy Steel Bolts for Structural Joints." Details of the material and other properties of these bolts are described in Chapter 4. The design shear strength of an A490 bolt is about 25% greater than that of an A325 bolt. Both types are used for FR and PR connections and for both static and dynamic loading. Although always very popular in field connections, their use in the shop has increased considerably in recent years with the introduction of new automated equipment and the tension-controlled bolt.

A325 bolts are available in three types. Type 1, made from a medium-carbon steel is the most common and is available in sizes from $\frac{1}{2}$ through $1\frac{1}{2}$ in. diameter. Type 2 is made from low-carbon martensite steel, is less expensive, but is available only from $\frac{1}{2}$ through 1 in. diameter. Type 2 bolts should not be hot-dipped galvanized. Type 3 is a weathering steel bolt with similar corrosion characteristics to that of ASTM A588 and A242 steels. Type 3 bolts are available from $\frac{1}{2}$ through $1\frac{1}{2}$ in. diameter.

A325 Type 1 bolts are identified by the mark "A325" or by three radial lines 120° apart on the bolt head. Type 2 is identified with three radial lines 60° apart, and Type 3 bolts have the designation "A325" underlined. A490 bolts carry the symbol "A490" and radial lines, as shown in Fig. 11.4. All bolts should be marked with the manufacturer's symbol.

Figure 11.5a shows the principle parts and dimensions of a high-strength bolt assembly: head, bolt length, and threaded length. Figure

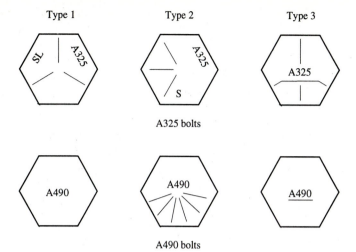

Type 1 | Type 2 | Type 3

A325 bolts

A490 bolts

Figure 11.4 Bolt markings.

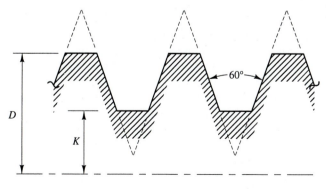

(a)

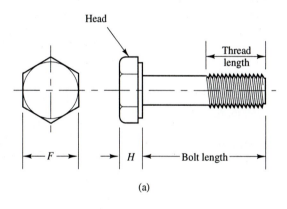

Tensile stress area $= 0.7854 \left[D - \dfrac{0.9743}{n} \right]^2$

n = Number of threads per inch

(b)

Figure 11.5 Bolt definitions.

11.5b shows the basic major diameter D, root diameter K, and tensile stress area. Washers are not always required and will be discussed later in this chapter.

Both A325 and A490 bolts are installed using an impact wrench or, in cases where a clamping force is not necessary, with a spud wrench. Installation requirements are discussed later in this chapter.

11.6 WELDING

Welding is a process of joining steel by melting additional metal into the joint between the two pieces of steel. The ease with which various steels accept being joined by welding, without exhibiting cracks and other flaws is called *weldability,* defined by the American Welding Society (AWS) as "the capacity of a metal to be welded under fabrication conditions imposed, into a specific, suitably designed structure, and to perform satisfactorily in the intended service."

Weldability depends on the chemical composition of the steel, the thickness of the material, and the grain size, to mention the primary factors. Details of weldability are discussed in Chapter 4.

11.6.1 Welding Processes

For structural steel, the four most popular welding processes, along with abbreviations designated by AWS, are as follows:

> Shielded metal arc (SMAW)
> Submerged arc (SAW)
> Gas shielded metal arc (GMAW)
> Flux corded arc (FCAW)

Shielded metal arc welding (SMAW) is the oldest process, often referred to as stick welding. As seen in Fig. 11.6a, an electrode is the source of the metal introduced into the joint; it is the consumable element. A high voltage is set up between the electrode and the metal pieces that are to be joined. When the welding operator strikes an arc, the resulting flow of current melts the electrode and the base metal adjacent to the joint. The electrode is coated with a special ceramic material or flux which protects the molten metal from absorbing hydrogen and other impurities during cooling. When the metal cools, a permanent bond results.

Submerged arc welding (SAW) (Fig. 11.6b) is an automatic or semi-automatic process that is used primarily when long pieces of plate are to be joined. It must be made in the near flat or horizontal position. The flux is a granular material introduced through a flexible tube. Submerged arc welding is an economical process for applications where repetitive and automated fabrication procedures will lend efficiency to the work.

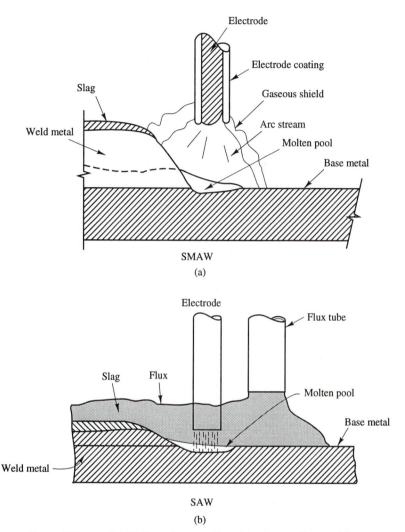

Figure 11.6   (a) Shielded metal arc welding; (b) submerged arc welding.

Gas shielded metal arc welding (GMAW) is a process whereby a continuous wire is fed into the joint to be welded. The molten metal is protected from the atmosphere by gas surrounding the wire. When used in the field, it is necessary to ensure that wind does not blow the gas away from the joint.

Flux cored arc welding (FCAW) is also a continuous wire process, except that the wire is essentially a thin hollow tube, filled with flux that protects the metal as the wire melts. It can be arranged as a semiautomatic process, and exceptionally high production rates can be attained.

11.6.2 Types of Welds

Two basic types of welds are used in steel construction: fillet welds and groove welds, as shown in Fig. 11.7. Figure 11.7a shows a fillet weld. The leg of the weld is measured along the interface between the weld metal and the base metal. The throat of the weld is the shortest dimension of the weld nugget. Since most fillet welds are symmetrical, with a 45° shape, the throat is 0.707 times the leg dimension, as shown. The size of the weld is defined by its leg dimension in increments of $\frac{1}{16}$ in.

A groove weld can be either a full penetration groove weld or a partial penetration groove weld. Both types of groove weld have been prequalified by AWS [8]. A typical complete penetration groove weld is shown in Fig. 11.7b and a partial penetration groove weld is shown in Fig. 11.7c. Prequalification means that certain weld configurations, including

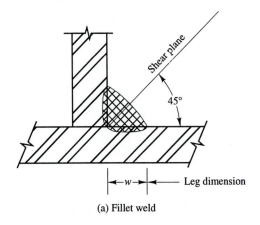

(a) Fillet weld

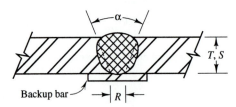

(b) Complete penetration groove weld

Figure 11.7 (a) Fillet weld; (b) complete penetration groove weld; (c) partial penetration groove weld.

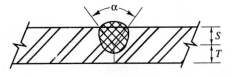

(c) Partial penetration groove weld

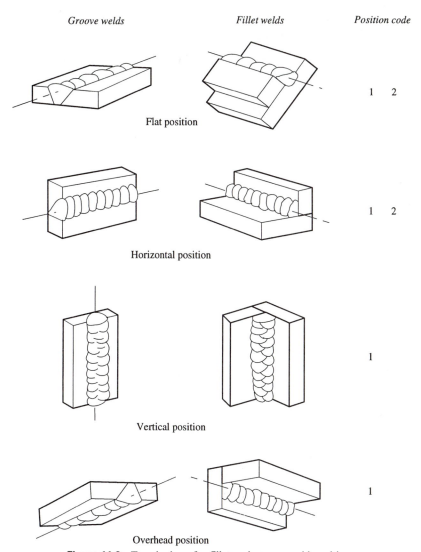

Groove welds   Fillet welds   Position code

Flat position   1   2

Horizontal position   1   2

Vertical position   1

Overhead position   1

Figure 11.8 Terminology for fillet and groove weld positions.

the root opening R, the angle of preparation α, and the effective thickness S, are deemed to be practical to build and will carry the intended load. AWS specifies provisions for prequalifying any weld configuration if circumstances indicate that it would be practical. These prequalified full and partial penetration groove welds are shown in the *LRFD Manual*. The configurations shown in Fig. 11.7 are only representative.

Both fillet welds and groove welds can be made in different positions. The terminology for these positions is shown in Fig.11.8.

11.7 HIGH-STRENGTH BOLTED SHEAR CONNECTIONS: LIMIT STATES

Three basic limit states govern the response of bolted connections loaded in shear: shear through the shank of the bolt, bearing on the material being connected, and block shear on the connected material. In cases where several load reversals are expected and fatigue is a factor, there is also the serviceability limit state of preventing slip at service loads.

Shear through the shank of the bolt is the means whereby the load P in Fig. 11.9 is transferred from the center plate to the two side plates. In the case shown, the bolt is therefore loaded in double shear. If there is only one side plate, the bolt would be in single shear.

Bolt shear failure is prevented by limiting the calculated shear stress to that specified by the LRFD specification, Section J3. When the bolt

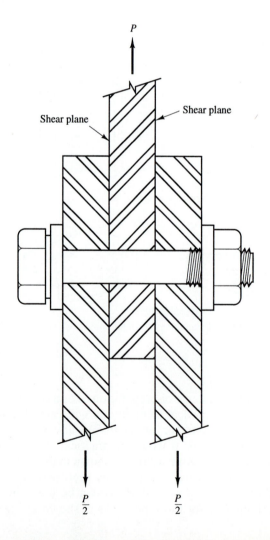

Figure 11.9 Shear through the shank of a bolt.

threads are excluded from the shear plane (X), the specified stress F_v is taken as $0.6F_u$, about 33% greater than when the threads are included in the shear plane (N), where it is taken as $0.45F_u$. In both cases the stress is calculated by dividing the calculated factored bolt force by the cross-sectional area of the bolt shank. When the material plies are relatively thin, it is usually best to design the connection for the worst case of threads included in the shear plane. For both types, $\phi = 0.65$.

Limit states for bearing are also specified in Section J3. Two cases will be considered when bolt spacing is greater than or equal to $3d$. If bolt spacing is less than $3d$, reference should be made to the specification. Case 1 is when the edge distance is less than 1.5 times the bolt diameter. In this case, failure may occur by a piece of material tearing out of the end of the connection, as shown in Fig. 11.10. Resistance to this failure mode ϕV_n is provided by shear along the two planes. From statics,

$$\phi V_n = \phi 0.6 F_u \text{ (2 planes)(edge distance)(material thickness)}$$
$$\phi V_n = \phi 1.2 F_u ED t \tag{11.1}$$

Because the shear plane is actually closer to the edge of the plate by one-half a hole diameter than the distance ED used in the equation, 1.2 in the above equation is reduced to 1.0. Therefore,

$$\phi V_n = \phi F_u t ED \tag{11.2}$$

where

$\phi = 0.75$
$F_u = $ ultimate strength of the connected material (ksi)
$t = $ thickness of the material (in.)
$ED = $ edge distance, measured from the center of the hole to the edge of the material (in.)

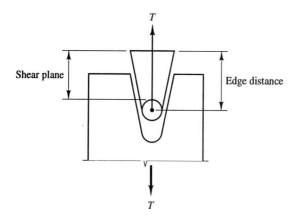

Figure 11.10   Tear out of an end bolt.

Bearing limit state, Case 2, is where the end distance exceeds 1.5 times the bolt diameter. In this case, the limit state is that of hole distortion. If hole distortion is not considered a problem for a particular application, the calculated bolt force may be safely limited to

$$\phi V_n = \phi 3.0 F_u dt \tag{11.3a}$$

where

$$\phi = 0.75$$

If hole distortion is to be limited, 2.4 should be substituted for 3.0 in Eq. (11.3a), yielding

$$\phi V_n = \phi 2.4 F_u dt \tag{11.3b}$$

Figure 11.11 illustrates how these various bearing limit states relate to each other for the case of a 1-in.-diameter bolt. The fabrication limit in the figure refers to the minimum edge distance specified in Table J3.7 in the LRFD specification, Section J3.10. Its purpose is to ensure good workmanship.

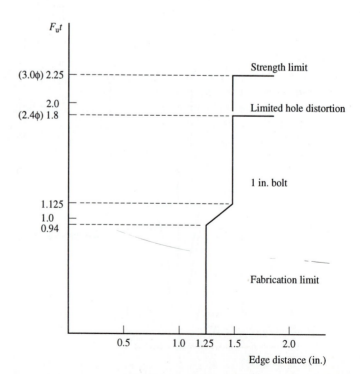

Figure 11.11 Relation of various bearing limit states for a 1-in.-diam bolt.

The third limit state for a bolted shear connection is that of block shear. This limit state is similar to that presented in Chapter 5 and discussed by Birkemoe and Gilmore [9], Ricles and Yura [10], and Hardash and Bjorhovde [11]. If the beam web is thin, it is possible for it to tear out, as indicated in Fig. 11.12. Resistance to this failure mode is provided by the combination of shear on the vertical plane and tension on the horizontal plane. As the reaction R increases, it is initially resisted by yielding in both planes. The ultimate resistance occurs when one of the planes fractures. As with tension members, the design resistance is a combination of the ultimate fracture stress times the area of the fracture plane plus the yield stress times the area of the other plane. As explained in Chapter 5, the larger of the two failure modes controls. For design purposes, it is easiest to check one or the other and, if necessary, use the larger.

In connections where service load slip is not a consideration, high-strength bolted connections can be installed with or without the specified pretension. In either case the design strength is the same. However, in the case where pretension is not specified, and the bolt is installed as snug-tight, the connection should be limited to applications specified in the LRFD specification, Section J1.9.

In some cases a bolted connection transferring shear must be designed as slip-critical. This is a serviceability limit state and is confined to loading conditions where frequent unfactored service load reversals are expected or when fatigue is a factor. This service load design strength ϕV_n is

$$\phi V_n = F_s A_b N_b N_s \qquad (11.4)$$

where

F_s = allowable slip load specified by the Research Council on Structural Connections (ksi) [12]

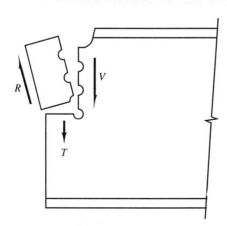

Figure 11.12 Block shear.

A_b = cross-sectional area of the bolt (in.^{2})
N_b = number of bolts in the joint
N_s = number of slip planes

A connection designed as slip critical must also be checked at factored loads for both bolt shear and material bearing. The slip load ϕV_n must be reduced if the friction between the plies is reduced due to a simultaneously applied tension force. This is a straight-line reduction which yields the reduced slip strength $\phi V'_n$ as

$$\phi V'_n = \phi V_n \left(1 - \frac{T}{T_b}\right) \qquad (11.5)$$

where

T = applied tension (kips)
T_b = pretension (kips)

11.8 WELD LIMIT STATES

The only limit state of weld metal is that of fracture. This is true for both fillet welds and groove welds. Yielding of weld metal is not a factor because any slight deformation prior to fracture takes place over such a small distance (gage length) that the performance of the structure is not affected.

11.8.1 Fillet Weld Strength

For a fillet weld, as shown in Fig. 11.7a, the load is transferred by shear through the throat of the weld nugget. The fracture strength is a function of the properties of the electrode. AWS classifies electrodes according to the tensile strength of the weld metal. Chapter 4 describes the classifications and the nomenclature used in identifying the various electrode properties.

AWS and AISC specify that for a particular grade of structural steel, as indicated by yield strength, there is a matching electrode. Table 11.1 shows the matching electrodes for commonly used steels. Both organizations further specify that the steel can be joined only by welding with the matching electrode or one grade higher. The purpose of this is to ensure that the failure of the fillet welded joint occurs along a plane through the weld nugget and not along the plane of the fillet weld leg w. If this rule is followed, tests by Higgins and Preece [13] have indicated that the desired failure mode is ensured.

Since the limit state of all fillet welds is one of shear fracture through the throat, the nominal shear strength V_n of the weld is the shear fracture

TABLE 11.1   STEEL/ELECTRODE MATCH [8]

| | Steel Specification Requirements | | | | | Filler Metal Requirements | | | | |
| Steel Specification | Minimum Yield Point | | Tensile Strength Range | | Electrode Specification | Minimum Yield Point | | Tensile Strength Range | |
	ksi	MPa	ksi	MPa		ksi	MPa	ksi	MPa
ASTM A36	36	250	58–80	345–550	SMAW AWS A5.1 or A5.5 E60XX or E70XX	50 / 60	345 / 415	62 min / 72 min	460 / 495
ASTM A529	42	290	60–85	415–495	GMAW AWS A5.18 E70S-X	60	415	72 min	495
ASTM A242 ASTM A441	42–50 42–50	290–345 290–345	63–70 min 63–70 min	435–485 450–530	SMAW AWS A5.1 or A5.5 E7015, E7016 E7018, E7028	60	415	72 min	495
ASTM A588 (4 in. and under)	50	345	70 min	485 min	SAW AWS A5.17 or A5.23 F7X-EXXX or F7X-EXXXX-X	60	415	70–95	482–620
ASTM 514 (over 2½ in. [63 mm])	90	620	100–130	690–895	SAW AWS A5.23 F10X-EXXX-X	88	605	100–130	690–895
ASTM A514 (2½ in.)	100	690	110–130	760–895	SMAW AWS A5.23 F11X-EXXX-X	98	675	110 min	760

stress F_v times the length of the weld at the throat $0.707w$ times the length of the weld L, expressed as

$$V_n = F_v(0.707wL) \qquad (11.6)$$

The determination of a fillet weld fracture stress F_v has been the subject of considerable research. The most recent is that of Butler et al. [14]. It has been demonstrated that this stress is about 50% larger when the weld is loaded normal to its longitudinal axis than it is when it is loaded parallel to its axis. However, the ductility of the weld is considerably less in the first case, with the actual strength and ductility a function of the angle of loading with respect to the weld axis.

Present AWS and AISC specifications do not consider the effect of the loading angle. With the present state of the art of design, to do so is considered far too complicated. Rather, these specifications take a conservative approach by basing the nominal strength on the least strength case, whereby the load is parallel to the weld axis.

Based on tests reported by Butler et al. [14], the nominal shear stress of a fillet weld is taken as 0.6 times the nominal tensile stress or classification strength of the electrode E_{xx} as discussed in Chapter 4. Therefore,

$$F_v = 0.6F_{EXX} \qquad (11.7)$$

and

$$V_n = 0.6F_{EXX}\,(0.707wL) \qquad (11.8)$$

With ϕ taken as 0.75, the design shear strength is

$$\phi V_n = 0.45F_{EXX}\,(0.707wL) \qquad (11.9)$$

For the most commonly used electrode, $F_{EXX} = 70$ ksi, the design strength can be calculated as

$$\phi V_n = 0.45(70.0)\,(0.707wL)$$
$$= 22.27wL \text{ kips}$$

It is convenient in design to use the fillet weld strength for a fillet weld with a $\frac{1}{16}$-in. leg, which gives

$$22.27\,(\tfrac{1}{16}) = 1.392 \text{ kips per } \tfrac{1}{16} \text{ in. of weld per inch of length}$$

Therefore, a $\frac{1}{4}$-in. fillet weld has a design strength of 1.392×4 (sixteenths) $= 5.57$ kips per inch of length.

11.8.2 Groove Weld Strength

A groove weld may be either a full or partial penetration weld, as shown in Fig. 11.7b and c. In either case the weld is generally used for tension joints. It is not "designed" in the usual sense because the weld metal is always stronger than the base metal. Therefore, it is the design strength of the base metal that controls the design.

In the case of a full penetration groove weld, the nominal strength of the tension joint is the product of the yield strength of the base material and the cross-sectional area of the smallest piece joined. With ϕ of 0.9, the design strength ϕP_n is then

$$\phi P_n = \phi F_y A_{\text{gross}} \tag{11.10}$$

The design strength of a partial penetration groove weld in a tension joint is similar except that the full cross-sectional area of the joined pieces is not effective. In this case, AWS defines an effective throat dimension S, which is a function of the configuration of the bevel, as shown in Fig. 11.7c. The design strength of the joint in tension is

$$\phi P_n = \phi F_y S \text{ (plate width)} \tag{11.11}$$

11.9 SHEAR CONNECTIONS: DESIGN

This section illustrates the design of four of the most commonly used PR, simple connections: double angles, the shear tab, unstiffened seated connections, and stiffened seated connections. The *LRFD Manual* includes many tables that make connection design simple. The examples in this chapter make some use of these tables, but where actual calculations may prove more instructive, these calculations are presented.

11.9.1 Behavior Requirements

There are two basic demands on a shear connection in a structural steel frame:

1. The connection must be able to resist the factored reaction force of the beam it is connecting without failing. Because of the highly redundant nature of these connections, this capacity is determined by testing or, in some cases, a history of satisfactory performance.
2. The connection must have enough rotation capacity and flexibility at factored loads to accommodate the simple beam end rotation of the beam it is connecting. Without this, even though the connection may not fail, it will attempt to transfer moment into the member to which it is attached. It is likely that that

member will not have been designed to carry this newly imposed moment.

Design of shear connections is accomplished by assuming a model and the associated components (bolts, welds, angles, plates) which are proportioned to satisfy statics. It is important for the designer to realize that the static model that is employed may not reflect the actual force distribution in the connection at service load. However, static equilibrium will be satisfied at failure.

11.9.2 Double-Angle Connection: Bolted

A double-angle shear connection (Fig. 11.13) must be checked for the following limit states:

1. *Shear on the bolts*. The total reaction divided by the number of bolts must not exceed the specified design strength of the particular bolt used.
2. *Capacity of the beam web material for bolt bearing*. The design bearing strength of the material is determined for each bolt and added together. All bolts will not necessarily have the same capacity, since the location of the bolt in the connection plays a role in determining the bolt strength. This is very often the case when a beam is coped: In this case, the bolt at the top may have a smaller capacity dictated by tear out, while the other bolts have the larger value resulting from hole distortion.
3. *Capacity of the beam web in block shear*. This check is required only when the beam is coped.

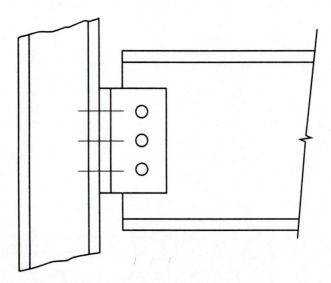

Figure 11.13 Double-angle shear connection.

4. *Capacity of angle material for bolt bearing.* This check is similar to the check of the beam web.
5. *Capacity of the angles for shear fracture through the net section.*

It is possible for a beam to be subjected to an axial load in addition to shear. In this case the connection must be designed to transfer both the shear and the tension. The conservative procedure suggested here is to use a straight-line interaction equation. The capacity of the connection and beam web is calculated for shear alone and for tension alone. The interaction equation is then used to determine the connection capacity:

$$\frac{V_u}{\phi V_n} + \frac{P_u}{\phi P_n} \leq 1.0 \qquad (11.12)$$

where

$$\begin{aligned}
V_u &= \text{required shear strength (kips)} \\
\phi V_n &= \text{design shear strength (kips)} \\
P_u &= \text{required tensile strength (kips)} \\
\phi P_n &= \text{design tensile strength (kips)}
\end{aligned}$$

EXAMPLE 11.1

Given: Design a double-angle shear connection for a W18×50 beam with a factored load of 85 kips. The steel is Grade 50 and the flange is coped 2 in. Use $\frac{7}{8}$-in. A325-N bolts in standard holes ($\frac{15}{16}$ in. diam). See Fig. 11.14.

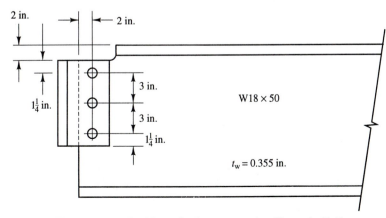

Figure 11.14   Double-angle shear connection (Example 11.1).

Solution

Step 1. Determine number of bolts required (LRFD specification, Section J3.3).

$$\text{Design strength per bolt} = \phi V_n = \phi F_v A$$

$$\phi V_n = 0.65(54.0)(0.6013) = 21.1 \text{ kips}$$

Bolts are in double shear so total number of bolts required is

$$N = 85/(2)(21.1) = 2.01$$

Use 3 bolts.

Step 2. Check web bearing strength at bolt holes (LRFD specification, Section J3.6).

$$\text{Edge distance at top bolt} = 1.25 \text{ in.} < 1.5 \; (\tfrac{7}{8})$$

$$= 1.313. \text{ Use Eq. (11.2).}$$

$$\phi F_u t ED = 0.75 \; (65)(0.355)(1.25) = 21.6 \text{ kips}$$

For bearing due to other bolts, limiting hole distortion, use Eq. (11.3b).

$$\phi(2.4F_u dt) = 0.75 \; (2.4)(65)(\tfrac{7}{8})(0.355)$$

$$= 36.3 \text{ kips}$$

$$\text{Web capacity} = 1 \; (21.6) + 2 \; (36.3)$$

$$= 94.2 \text{ kips} > 85 \text{ kips} \qquad \text{OK}$$

A similar calculation should be made for the bolts on the outstanding legs of the angles that connect to the supporting member. In this case the bolts are in single shear but since there are twice as many bolts, the load per bolt shear plane is the same as the beam web bolts.

Step 3. Check web for block shear.

Using the equations developed in Chapter 5, Eqs. (5.21) and (5.22), first try the shear yield and tension fracture condition. The gross shear area is the distance from the top of the cope to the bottom of the bottom hole times the web thickness.

$$(1.25 + 2 \; (3.0) + \tfrac{15}{32}) \; (0.355) = 2.74 \text{ in.}^2$$

$$\phi V_n = \phi(0.6F_y)A_{vg}$$

$$= 0.75 \; (0.6)(50)(2.74) = 61.7 \text{ kips}$$

The net tension area is the horizontal edge distance less half the hole diameter times the web thickness:

$$(2.0 - \tfrac{15}{32})\,(0.355) = 0.544 \text{ in.}^2$$

$$\phi P_n = \phi F_u A_{nt}$$

$$= 0.75(65)(0.544) = 26.5 \text{ kips}$$

$$R_{bs} = 61.7 + 26.5 = 88.2 \text{ kips}$$

Next check the tension yield and shear fracture condition. The gross tension area is the full horizontal edge distance:

$$2.0\,(0.355) = 0.71 \text{ in.}^2$$

$$\phi P_n = \phi F_y A_{tg} = 0.75(50)(0.71) = 26.6 \text{ kips}$$

Net shear area is the distance from the top of the cope to the bottom of the bottom hole minus the hole diameters times the web thickness.

$$[1.25 + 2\,(3) - 2.5\,(\tfrac{15}{16})]\,(0.355) = 1.742 \text{ in.}^2$$

$$\phi V_n = \phi(0.6)F_u A_{ns}$$

$$= 0.75(0.6)(65)(1.742) = 51.0 \text{ kips}$$

$$R_{bs} = 26.6 + 51.0 = 77.6 \text{ kips}$$

Selecting the larger value yields

$$R_{bs} = 88.2 \text{ kips} > 85 \text{ kips applied.} \qquad \text{OK}$$

Step 4. Check A36 angle material for bolt bearing.
  Assume a $\tfrac{3}{8}$-in. angle.
  Top bolt, use Eq. (11.2):

$$0.75(58)(\tfrac{3}{8})(1.25) = 20.4 \text{ kips}$$

Remaining bolts. Use Eq. (11.3b) with a limit on hole distortion:

$$0.75(2.4)(58)(\tfrac{7}{8})(0.375) = 34.3 \text{ kips}$$

$$20.4 + 2\,(34.3) = 89.0 > \tfrac{1}{2}\,(85) = 42.5 \qquad \text{OK}$$

Step 5. Check angles for shear fracture:

$$A_{sf} = [2(3) + 2\,(1.25) - 3\,(\tfrac{15}{16})]\,(\tfrac{3}{8})$$

$$= 2.133 \text{ in.}^2$$

$$\phi V_n = 0.75(0.6)(58)(2.133) = 55.7 > 42.5 \qquad \text{OK}$$

Conclusion.   The three-bolt connection shown in Fig. 11.14 will adequately carry the imposed load.

EXAMPLE 11.2

Given: A W18×35 beam has a vertical reaction of 24 kips and an axial load of 50 kips. Design a double-angle connection with the beam coped. Steel is A36 with $F_y = 36$ ksi. See Fig. 11.15.

Solution

Step 1. Assume four $\frac{7}{8}$-in. A325–N bolts. As seen in Example 11.1, the capacity in double shear of a $\frac{7}{8}$-in. A325–N bolt is 2(21.1) = 42.2 kips. Thus, the total double-shear capacity is $4 \times 42.2 = 168$ kips. The applied force, the resultant of the shear and tension, is

$$\sqrt{(50^2 + 24^2)} = 55 \text{ kips} < 168 \text{ kips} \qquad \text{OK}$$

Step 2. Calculate web resistance to the reaction in bolt bearing.

$$1.5 \times \text{bolt diam} = 1.31 \text{ in.} > 1\tfrac{1}{4}$$

Therefore use Eq. (11.2) for the top bolt, which yields

$$\text{One bolt @ 16.3 kips} = 16.3$$

For the remaining bolts use Eq. (11.3), which yields

$$\text{Three bolts @ 27.4 kips} = 82.2$$

$$\text{Total} = 98.5 \text{ kips} > 24 \text{ kips} \qquad \text{OK}$$

Step 3. Calculate web resistance to axial load in bolt bearing. Note that the only difference between this calculation and the calculations of step 2 is the end distance, which is now greater than 1.5 times the bolt diameter.

$$1.5 \times \text{bolt diam.} = 1.31 \text{ in.} < 2 \text{ in.}$$

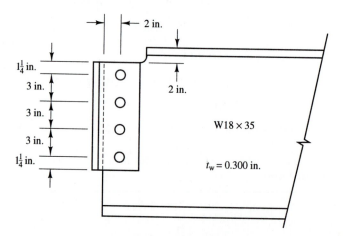

Figure 11.15 Double-angle connection for shear and tension (Example 11.2).

Thus,

$$\text{four bolts @ } 27.4 \text{ kips} = 109.6 \text{ kips}$$

Step 4. Calculate web resistance to the reaction in block shear.

Shear yield, tension fracture:

$$A_{vg} = (1.25 + 3(3) + \tfrac{15}{32})(0.30) = 3.22$$

$$A_{nt} = (2 - \tfrac{15}{32})\,(0.30) = 0.459$$

$$R_{bs} = 0.75[0.6(36)(3.22) + 58(0.459)] = 72.1 \text{ kips}$$

Tension yield, shear fracture:

$$A_{tg} = 2\,(0.30) = 0.60$$

$$A_{ns} = [1.25 + 3(3) - 3.5(\tfrac{15}{16})](0.30) = 2.09$$

$$R_{bs} = 0.75[0.6(58)(2.09) + 36(0.6)] = 70.8 \text{ kips}$$

Select the larger; thus,

$$R_{bs} = 72.1 \text{ kips}$$

Step 5. Calculate web resistance to axial load in block shear.

Shear yield, tension fracture:

$$\text{Shear yield} = \phi(0.6)F_y A_{vg}(2 \text{ planes})$$
$$= 0.75\,(0.6)(36)(2.0)(0.300)(2.0)$$
$$= 19.4 \text{ kips}$$

$$\text{Tension fracture} = \phi F_u A_{nt}$$
$$= 0.75(58)[9 - 3\,(\tfrac{15}{16})]\,(0.300)$$
$$= 80.7 \text{ kips}$$

Thus,

$$\text{Axial}_{bs} = 19.4 + 80.7 = 100.1 \text{ kips}$$

Tension yield, shear fracture:

$$\text{Tension yield} = \phi F_y A_{tg}$$
$$= 0.75(36)(9)(0.300) = 72.9 \text{ kips}$$

$$\text{Shear fracture} = \phi 0.6 F_u A_{ns}$$
$$= 0.75(0.6)(58)[(2 - \tfrac{1}{2})\,(0.300)]$$
$$= 11.7 \text{ kips}$$

Thus,

$$\text{Axial}_{bs} = 72.9 + 11.7 = 84.6 \text{ kips}$$

Select the larger.

$$\text{Axial}_{bs} = A_{bs} = 100.1 \text{ kips}$$

Step 6. Check beam web with interaction equation.

$$\frac{24}{72.1} + \frac{50}{100.1} = 0.83 < 1.0 \qquad \text{OK}$$

11.9.3 Double-Angle Connection: Welded

Table IV of the *LRFD Manual* tabulates capacities for framed welded
angle connections of the type shown in Fig. 11.16 ranging in length from 4
to 32 in. The angles are 4×3 for the larger connections and 3×3 for the
smaller ones. Thickness are $\frac{1}{2}$, $\frac{3}{8}$, and $\frac{5}{8}$ in. Values are tabulated for the
weld on the beam flange (weld A) and for the weld against the support
(weld B).

　　　　Weld A is calculated by using the channel-shaped weld of Table
XXII of the "Eccentric Loads on Weld Groups" in the *LRFD Manual*
with a load angle of attack of 0°. These tables are based on the instantaneous center (IC) method, sometimes called the center of rotation

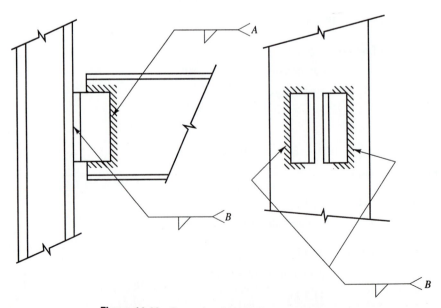

Figure 11.16　Framed welded angle connections.

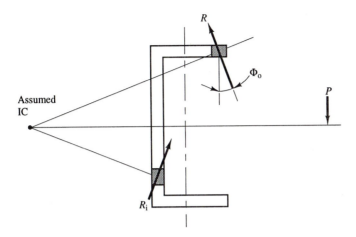

Figure 11.17   Instantaneous center method.

method. A step-by-step description of this procedure follows. Figure 11.17 is used to develop the procedure for weld A.

Step 1. A location of the IC is initially assumed as shown.

Step 2. The load on the heaviest loaded inch of weld segment R (usually farthest from the center of gravity) is assigned the value of the nominal strength of an inch of fillet weld. A deformation Δ for this segment is assumed to be 0.34 in.

Step 3. The load on each of the lesser loaded weld segments R_i is calculated on the basis of a linear calculation of Δ, using the distance from the assumed IC. Then the force is determined from the force–deformation relationship shown in Fig. 11.18.

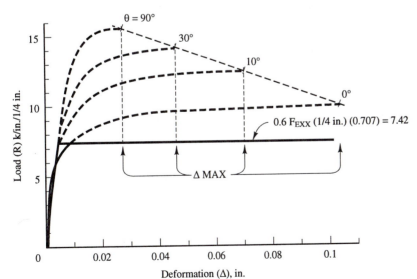

Figure 11.18   Force–deformation relationship. Reprinted by permission of the American Institute of Steel Construction, Inc., from the 1st edition of the *AISC – LRFD Manual.*

Step 4. All forces and moments are summed and statics are checked.

Step 5. The location of the IC is adjusted and rechecked until equilibrium is achieved.

The minimum web thickness t for each weld size is listed in Table IV in the *LRFD Manual*. These thicknesses were calculated by "matching" the capacity of the web in shear yielding to that of the weld in fracture, where

$$\text{Web shear yield capacity} = 0.56F_y t$$

$$\text{Weld fracture capacity} = w(1.392) \ (2 \text{ welds})$$

These two equations are set equal, the weld leg size w is substituted, and the minimum web thickness t is determined. With coped welded framed connections, block shear would probably govern the minimum web thickness. The web thicknesses presented in Table IV of the *LRFD Manual* have also been checked for this limit state and were found to be adequate in all cases.

The origin of the static model for weld B goes back to the 1940s when it was first suggested by LaMotte Grover in the *Manual of Design for Arc Welded Steel Structures* published by Air Reduction. Figure 11.19 shows the basic model. It is assumed that the two angles rotate in opposite directions pressing toward each other at the top and spreading at the

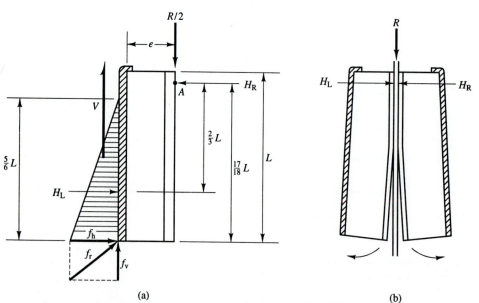

(a)   (b)

Figure 11.19   Static model for determining minimum weld size.

bottom as indicated. The location of the neutral axis and the stress distribution can certainly be debated, but the fact is that this model has served well over the years. Many engineers, including the authors, feel that this model is probably conservative.

The original discussion was based on an allowable fillet weld stress of 13.6 ksi. The LRFD design strength of 31.5 ksi (1.392 kips per inch per $\frac{1}{16}$ in. of the fillet weld leg size w) was used to create the *LRFD Manual* tables. Referring to Fig. 11.19 for horizontal equilibrium, the force on the left angle must equal the force on the right angle:

$$H = H_L = H_R$$

Summing moments about point A in Fig. 11.19a yields

$$\Sigma M = 0$$

$$H(\tfrac{2}{3})L = \left(\frac{R}{2}\right)(e)$$

Solving for H yields

$$H = \frac{3Re}{4L}$$

Using the stress distribution and dimensions of Fig. 11.19 yields

$$H = \left(\frac{f_h}{2}\right)\left(\frac{5L}{6}\right)$$

Setting the above two equations equal and solving for f_h yields

$$f_h = \left(\frac{3Re}{4L}\right)\frac{12}{5L}$$

$$= \frac{9Re}{5L^2}$$

If the shear stress is assumed to be distributed linearly over the length L, then

$$f_v = \frac{R}{2L}$$

Combining the stresses at the bottom of the weld and setting the result equal to the weld strength yields

$$f_r = \sqrt{f_h^2 + f_v^2} = 1.392w \qquad (11.13)$$

The weld leg size *w* (number of sixteenths) may then be determined from Eq. (11.13). Because of the empirical nature of the strength calculations for welds A and B, reliance on tables found in the *LRFD Manual* is deemed to be the most appropriate approach to weld design.

11.9.4 Single-Plate Connections

The single-plate shear connection or ''shear tab,'' as shown in Fig. 11.20, is also a ''pinned'' connection. It consists of a plate shop-welded to the support and field-bolted to the beam.

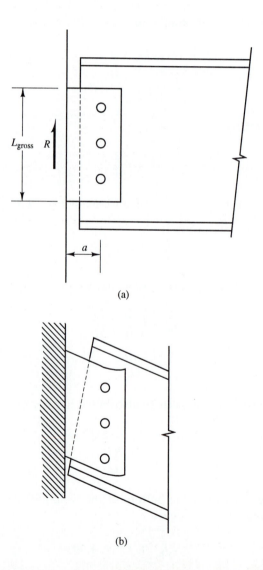

(a)

(b)

Figure 11.20 Single-plate shear connection.

The behavior of this connection is similar to that of a double-angle connection, except that it achieves its rotation capacity in a different way. The double-angle connection achieves its ductility and rotation capacity primarily through the bending of the outstanding legs. A shear tab does not have these legs, so when the load is first applied, the tab is quite stiff. This results in the development of some moment in the connection and the beam, which essentially produces a load at an eccentricity on the bolts and welds. This eccentricity is identified as the location of the point of inflection. Astaneh et al. [15] have shown that as the load on the beam is increased and the force in the connection increases, the shear tab yields. This yielding causes the inflection point to migrate toward the support and, at failure, the eccentricity is quite small. This research has resulted in a design procedure that has been adopted by AISC for the shear tab connection.

Five limit states are investigated for the shear tab connection to determine the capacity of the connection V_n.

1. Shear yielding of the gross area of the plate:

$$\phi V_n = 0.9(0.6)F_y L_{\text{gross}} t$$

2. Shear fracture of the net area of the plate:

$$\phi V_n = 0.75(0.6)F_u L_{\text{net}} t$$

3. Bearing failure of the shear plate or beam web:

$$\phi V_n = C(0.75)(2.4)F_u dt$$

where C is the center of rotation coefficient as a function of eccentricity, *LRFD Manual*, Table X with an angle of 0°. The eccentricity is in inches.

$$\text{Eccentricity} = (n - 1)(1.0 \text{ in.}) - a \geq a \qquad (11.14)$$

where a is the distance from the bolt center line to the face of support and n is the number of bolts.

4. Shear failure of the bolts, considering the eccentricity that exists at ultimate load of the connection:

$$\phi V_n = C(0.65)F_v$$

Where F_v is the nominal strength for the type of bolts in shear, depending on the connection—bolts with threads in the shear plane (*N*) or excluded from the shear plane (*X*).

5. Fracture of the welds, considering the eccentricity that exists at ultimate load of the connection:

$$\phi V_n = CDL$$

where C is the center of rotation coefficient as a function of eccentricity, *LRFD Manual,* Table XVIII with an angle of 0°. The eccentricity of the weld in inches is defined as

$$\text{Eccentricity} = n\,(1.0\ \text{in.}) \geq a \qquad (11.15)$$

The equations for eccentricity of the bolts (11.14) and the welds (11.15) are dimensionally inconsistent. These equations result from tests and are purely empirical. Also, these eccentricities are for the case of an infinitely rigid support to which the tab would be welded. This is conservative since most real supports would be columns or girders, which would have some rotational flexibility.

The required ductility of the connection, designed by this procedure, is obtained through shear yielding of the plate, as illustrated in Fig. 11.20b and described above as the first limit state. Accordingly, the procedure requires that the fracture strength of the welds should be greater than the shear yield capacity but that the fillet weld size need not exceed $\frac{3}{4}$ of the plate thickness.

The procedure also establishes other general requirements to insure proper connection behavior and to be consistent with the research results:

1. Material of the plate is $F_y = 36$ ksi.
2. The ratio of the length of the plate L to the distance to the bolt line a is greater than two.
3. The bolts must be located in one vertical row of between two and seven bolts. Bolts may be ASTM A307, A325, and A490 and may be used fully tightened or snug-tight. Holes may be standard or short slotted.
4. Welds are to be fillets made with E60XX or E70XX electrodes.
5. The vertical pitch = 3 in.
6. Plate length L is at least as long as one-half of the T-dimension of the beam.

EXAMPLE 11.3

Given: Design an A36 shear tab for a factored reaction of 140 kips. The beam is a W27×114, $F_y = 50$ ksi.

Solution

Try a $\frac{3}{8}$-in.-thick plate with $7 - A490-N$, $\frac{7}{8}$-in. bolts. Plate length is 21 in., $a = 3$ in.

1. Shear yielding of the plate:

$$\phi V_n = 0.9(0.6) \, F_y L_{gross} t$$
$$\phi V_n = 0.54(36)(21)(0.375)$$
$$\phi V_n = 153 \text{ kips} > 140 \text{ kips} \quad \text{OK}$$

2. Shear fracture of the plate:

$$\phi V_n = 0.75(0.6)F_u L_{net} t$$
$$\phi V_n = 0.45(58)[(21 - 7(0.875 + 0.0625)](0.375)$$
$$\phi V_n = 141 \text{ kips} > 140 \text{ kips} \quad \text{OK}$$

3. Bearing failure of plate or beam web. Since the plate is thinner than beam web, it will control bearing.

$$\phi V_n = C \, (0.75)(2.4)F_u dt$$
$$\text{Ecc.} = (n - 1 - a) = (7 - 1 - 3) = 3 \text{ in.}$$

From Table X of the *LRFD Manual*, $C = 6.07$

$$\phi V_n = 6.07(1.8)(58)(0.875)(0.375)$$
$$\phi V_n = 208 \text{ kips} > 140 \text{ kips} \quad \text{OK}$$

4. Shear failure of bolts:

$$\phi V_n = C \, (0.65)F_v = C(0.65)(0.45)F_u A_{bolt}$$
$$\text{From step 3, } C = 6.07$$
$$\phi V_n = 6.07(0.65)(0.45)(150)(0.6)$$
$$= 6.07(26.4)$$
$$\phi V_n = 160 \text{ kips} > 140 \text{ kips} \quad \text{OK}$$

5. Weld size. To insure that shear yielding is more critical than weld fracture, design the weld for $\phi V_n = 153$ kips from step 1:

$$w = \frac{\phi V_n(\text{shear yielding, from step 1})}{CL}$$

$$\text{Ecc.} = n = 7 > 3 \quad \text{Use 7 in.}$$

$$\frac{\text{Ecc.}}{\text{Plate length}} = \frac{7}{21} = 0.33$$

From Table XVIII, $C = 1.766$

$$w = \frac{153}{(1.766)(21)} = 4.1 \text{ sixteenths}$$

With a $\frac{3}{8}$-in.-thick plate, the weld size need not exceed

$$\tfrac{3}{4}(\tfrac{3}{8}) = 0.28$$

Therefore, use a $\frac{5}{16}$-in. fillet weld

11.9.5 Seated Connection

A seated connection is the third most commonly used Type PR, simple connection. It may be either unstiffened, as shown in Fig. 11.21, or stiffened as shown in Fig. 11.22. In either case, the connection may be welded or bolted. Its historical performance as a simple ''pinned'' connection has been excellent. It can rotate sufficiently at ultimate load, without fracturing or imposing a significant moment on its abutment, to be considered a simple connection. The seated connection is used mainly to frame a beam into the weak axis of a column. In fact, it was developed for this purpose because it is the safest way to erect a beam that frames into a column web.

     Design loads for these connections are listed in the *LRFD Manual*. The tabulated loads for the unstiffened case are based on the rather complicated procedure discussed below [*16*].

     Two limit states must be checked in calculating the design strength of an unstiffened seated connection: (1) bending of the seat angle and (2) yielding of the beam web. In addition, for beams with slender webs, design strength due to web crippling (LRFD specification, Eq. K1-5) may be less than the capacity of the unstiffened seat as determined by beam web yielding. This condition must also be checked before the connection design may be considered complete.

     Using the notation shown in Fig. 11.23, for an angle of length L, the bending capacity of the angle is

$$M_{\mathrm{p}} = \frac{F_{\mathrm{y,angle}}\, L t_{\mathrm{a}}^2}{4} \tag{11.16}$$

Taking $\phi_{\mathrm{b}} = 0.9$ and expressing the moment in terms of the reaction R and the moment arm e yields

$$\phi_{\mathrm{b}} R = \frac{0.225 F_{\mathrm{y,angle}}\, L t_{\mathrm{a}}^2}{e} \tag{11.17}$$

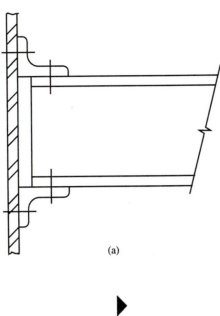

(a)

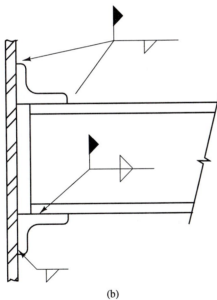

(b)

Figure 11.21 Unstiffened seated connection.

For the beam web, the LRFD specification, Section K1.3b, states that yielding will occur over a length of $(N + 2.5k)$. Thus, if web yielding is to occur under the same load as that which causes the plastic moment to occur on the angle

$$\phi_b R = F_{y,\text{beam}}\, t_w\, (N + 2.5k) \qquad (11.18)$$

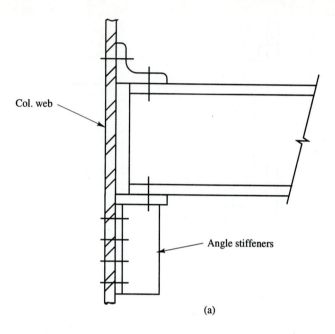

(a)

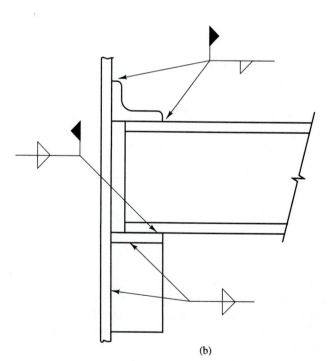

Figure 11.22   Stiffened seated
connection.

(b)

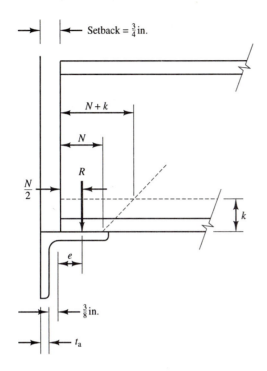

Figure 11.23 Unstiffened seated connection.

The final capacity of the connection is dependent on the assumed location of the reaction in relation to the back of the angle, which is a function of the actual bearing length. For the basic case, the assumed bearing N will be greater than $2.5k$ and less than the available length of the angle, so that the reaction may be assumed to act at the center of bearing, as shown in Fig. 11.23. Based on this geometry, the eccentricity may be expressed as

$$e = \frac{N}{2} + \text{setback} - k_{\text{angle}} \tag{11.19}$$

The minimum clearance between the beam end and the face of the abutment is $\frac{1}{2}$ in.; however, to account for any slight irregularity in the provided beam length, it is normal to provide a setback of $\frac{3}{4}$ in. The critical plane for bending in the projecting leg of the angle is taken at the edge of the fillet as shown. To simplify design, this point is normally defined by assuming a fillet radius of $\frac{3}{8}$ in. Substitution of these values for setback and the location of the critical plane yields

$$e = \frac{N}{2} + \frac{3}{4} - t_a - \frac{3}{8}$$

or

$$e = \frac{N}{2} + \frac{3}{8} - t_a \qquad (11.20)$$

If the limit states for bending of the seat angle and yielding of the beam web are assumed to occur simultaneously, Eqs. (11.17), (11.18), and (11.20) may be combined to determine N, e, and the capacity of the seat. Solving Eq. (11.18), limit state for web yielding, for N and substitution into Eq. (11.20) results in an equation for eccentricity that is linked to the web yielding failure criteria so that

$$e = \frac{\phi_b R}{2 F_{yb} \, t_w} - \frac{5k}{4} + \frac{3}{8} - t_a \qquad (11.21)$$

Substitution of this equation for eccentricity into Eq. (11.17), bending of the seat angle, and rearranging into the standard quadratic form yields

$$\frac{(\phi_b R)^2}{2 t_w \, F_{yb}} + \phi_b R \left(\frac{3}{8} - t_a - \frac{5k}{4} \right) - 0.225 F_{ya} \, L t_a^2 = 0 \qquad (11.22)$$

The solution to Eq. (11.22) is the reaction that will result in simultaneously reaching the two limit states. This value is then used in Eq. (11.18) to determine the resulting bearing length N. Three possibilities may result from this determination of N.

Case 1: If N is between $2.5k$ and the available bearing (for a 4-in. leg angle this is taken as 3.25 in.), the solution is correct and the capacity of the seat has been determined.

Case 2: If N is less than $2.5k$, the assumption that the reaction is applied at $N/2$ is in error and the values obtained for $\phi_b R$ are unrealistic. In this case, the reaction should be placed at $(N + 2.5k)/4$ and the derivation that led to Eq. (11.22) redone. This will result in a more realistic value for the reaction such that

$$\frac{(\phi_b R)^2}{4 t_w \, F_{yb}} + \phi_b R (\tfrac{3}{8} - t_a) - 0.225 F_{ya} \, L t_a^2 = 0 \qquad (11.23)$$

Case 3: If N is greater than the available bearing, the actual bearing length must be used. Since this shorter length would result in lower bending stresses than that from the calculated N, bending cannot be the critical factor and yielding of the web will be the governing limit state. Thus, Eq. (11.18) will give the seat capacity when the available bearing is used for N.

EXAMPLE 11.4

Given: Design an unstiffened bolted seated connection framing into a $W10 \times 33$ column. The beam is a $W14 \times 22$ with a span of 23 ft and a reaction of 15 kips. All material is A36.

Solution

Step 1. The capacity of the beam web is checked for crippling through the Beam Uniform Load Table from the *LRFD Manual*. For the $W14 \times 22$

$$R(N = 3\tfrac{1}{2} \text{ in.}) = 28.0 \text{ kips} > 15 \text{ kips applied} \qquad \text{OK}$$

Step 2. For the $W14 \times 22$ beam, $k = \tfrac{7}{8}$ and $t_w = 0.23$. Assume a $4 \times 4 \times \tfrac{3}{8} \times 6$-in. angle. Substituting into Eq. (11.22) yields

$$\frac{(\phi_b R)^2}{(2)(0.23)(36)} + \phi_b R \left[\tfrac{3}{8} - \tfrac{3}{8} - (\tfrac{5}{4})(\tfrac{7}{8})\right] -0.225(36)(6)(\tfrac{3}{8})^2 = 0$$

which simplifies to

$$(\phi_b R)^2 - 18.1125(\phi_b R) - 113.18 = 0$$

The meaningful root to the quadratic equation is $\phi_b R = 23.02$ kips. Substitution into Eq. (11.18) yields $N = 0.59$, which is less than $2.5k$; thus, Eq. (11.23) must be used. Substituting into Eq (11.23) and solving for the reaction yields

$$\phi_b R = 15.04 \text{ kips} > 15 \text{ kips} \qquad \text{OK}$$

The data presented in Table V of the *LRFD Manual* were developed from these equations and provide a less complex approach to the design of the unstiffened bolted seated connections.

A stiffened seated connection is used when the loads are too large to be supported by an unstiffened seat. For the welded case shown in Fig. 11.24, the loads tabulated in the *LRFD Manual* are calculated by an elastic center of gravity procedure applied to the weld group shown in Fig. 11.24b.

The area of the weld group of unit width is

$$A = 2.4L \tag{11.24}$$

The distance from the center of gravity of the group to the top weld element in tension is

$$c = 0.4167L \tag{11.25}$$

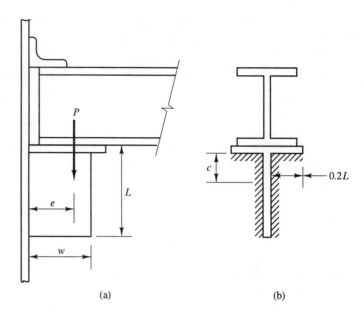

Figure 11.24 Stiffened seated connection.

The moment of inertia of the weld group is

$$I = 0.2501L^3 \tag{11.26}$$

Therefore, the section modulus is

$$S = \frac{I}{c} = 0.6002L^2 \tag{11.27}$$

The vector sum of the vertical stress on the weld, P/A, and the horizontal stress, M/S, is calculated and set equal to the design strength of the fillet weld element ($1.392w$ kips/in. for E70XX electrodes). The design strength of the connection P can then be calculated.

The minimum thickness for the stiffener plate should be kept to the web thickness of the supported beam for unstiffened beam webs. When the stiffener and beam are not of the same material, the plate thickness should be adjusted by the ratio of yield strengths. The top plate should not be less than $\frac{3}{8}$ in. In addition, plates should be of sufficient thickness to accept the required welds.

In determining the moment M that the load P transmits to the weld group, AISC assumes a value of 0.8 (width of seat) for the eccentricity e in the development of the tables in *LRFD Manual*. Therefore, $M = 0.8$ *PW*. The origin of this assumption of eccentricity is unknown to the authors but the connection has had a long history of economical and safe performance and can be used with confidence.

It is common practice to ignore the moment used to calculate the fillet weld size when designing the column to which the connection is framed. There are several reasons why this is safe practice. First, as the load approaches ultimate, the eccentricity decreases. Second, the connection and the beam provide rotational restraint to the column, so that a column design based on an effective length factor of unity is conservative. Thirdly, the shape factor of the column shape about its weak axis is 1.5, but AISC, nevertheless, restricts the nominal strength in this case to a very conservative $F_y S_x$.

11.10 HIGH-STRENGTH BOLTS IN TENSION

11.10.1 Behavior

High-strength bolts are normally used in applications where they will be installed with an initial pretension, thus enabling them to act as a clamp, holding the two connected elements together. Figure 11.25 shows a typical tension hanger where the bolts are expected to carry the applied tension load. The pretension from the bolt actually causes a compressive force to develop between the connected parts. Application of the applied service load will reduce the contact force, but will have little effect on the bolt tension as long as contact is maintained between the plates. Once the plates are separated, the initial conditions have no influence and the bolt force must equal the applied load.

If the attached plate is permitted to deform, as shown in Fig. 11.26, additional forces will develop at the tips of the flange. These additional forces Q are the result of prying action and are called the prying forces. Figure 11.27 shows a typical example of the relation between the applied force and the bolt force for a connection where these prying forces may occur. Note that for applied loads below the service load level, there is very little increase in bolt force with increasing load. However, when the load exceeds the service load by an appreciable amount, the bolt force is increased by the prying force.

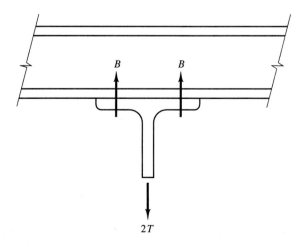

Figure 11.25 Typical tension hanger.

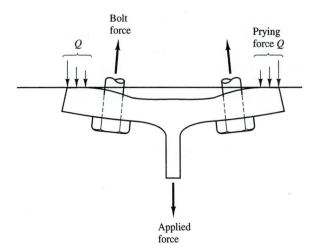

Figure 11.26 Deformation of attached plate in tension hanger.

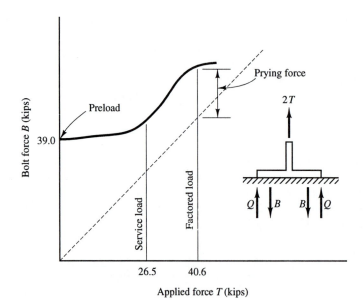

Figure 11.27 Relation between applied force and bolt force when prying forces may occur.

Research reported by Struik and de Back [*17*] has demonstrated that there is a relationship between the thickness of the flange and the prying force, as shown in Fig. 11.28. When *t* is large, the plate does not bend and no prying action takes place. When *t* is small, bending of the plate may be extensive and the prying force will be large. Obviously, prying action may be completely eliminated in a design by selecting a sufficiently thick plate, but this is normally not a practical solution. Prying action may also be avoided if washers are used to keep the flange from coming in contact with the abutment; this, too, is normally not desirable. A better approach

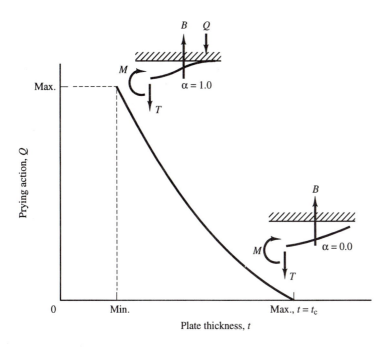

Figure 11.28 Relation between thickness of flange and prying force.

is to select a flange thickness and bolt strength that combine to yield a sufficiently strong and economical connection.

11.10.2 Design

The design of a tension connection, of the type shown in Fig. 11.25, as presented in the *LRFD Manual,* is based on the presentation by Thornton [18]. A somewhat more direct approach will be presented here. The T-shaped element shown in Fig. 11.29a is attached to an abutment with two bolts. It must carry a tensile load of $2T$, resisted by the force B in each of the two bolts and influenced by the prying force Q at each extreme of the flange. The dimensions used to develop the design equations are shown in the figure. Three possible limit states must be investigated for this assembly. They are (1) bending of the flange, (2) tension in the bolt without prying action, and (3) tension in the bolt with prying action.

(1) The beam model shown in Fig. 11.29b is used to develop the bending mechanism. To account for some of the bending in the plate and to provide a better fit between the mathematical model and experimental data, the bolt force B is applied at the inner face of the bolt as shown. The plastic moment capacity of the plate is defined as M_p. At the plane of the bolt, the plastic moment capacity is reduced, due to the hole, to δM_p as shown, where δ is defined as the ratio of net area to gross area given as $\delta = (1 - d'/p)$, p is the width of the plate per hole, and d' is the width of the hole. From equilibrium at formation of the mechanism,

$$(1 + \delta)M_p = T_n\, b' \tag{11.28}$$

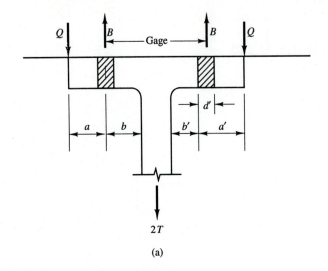

(a)

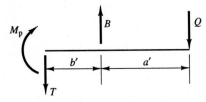

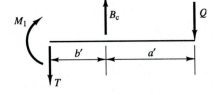

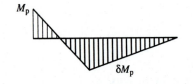

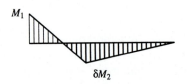

Beam mechanism

Bolt failure with prying action

(b)

(c)

Figure 11.29 Elements in design including prying action.

Recognizing that the flange plate is a rectangular element with

$$M_p = F_y \, t^2 \frac{p}{4}$$

and solving for design strength $T = \phi T_n$ yields

$$T = \frac{\phi F_y \, t^2 \, p}{4b'} (1 + \delta)$$ (11.28a)

(2) If there is no prying action, failure of the bolt will occur when

$$T = B_c \text{ (actual bolt capacity)} \qquad (11.28b)$$

(3) For failure in the bolt prior to formation of the beam mechanism, the moment diagram shown in Fig. 11.29c will be used. Applying the equations of equilibrium yields

$$M_1 = B_c b' - Q(a' + b') \qquad (11.28c)$$

and

$$Q = B_c - T \qquad (11.28d)$$

Substituting Eq. (11.28d) into Eq. (11.28c) and equating M_1 to the plastic moment yields

$$T = \frac{\phi M_p + B_c a'}{a' + b'} \qquad (11.28e)$$

To check the capacity of a hanger connection, T from each limit state (Eqs. (11.28a), (11.28b), and (11.28e)) is determined. The connection is then limited to the lowest value obtained.

It is often desirable to determine the required flange thickness for a given design situation. Again, three possible limit states are involved. Using the forces and moment diagram defined in Fig. 11.29c and summing moments about the force B_c yields

$$M_1 - Tb' + Qa' = 0 \qquad (11.29a)$$

Summation of moments to the right of the bolt force B_c yields

$$\delta M_2 = Qa' \qquad (11.29b)$$

Defining $\alpha = M_2/M_1$ and substituting Eq. (11.29b) into (11.29a) yields

$$M_1 = \frac{Tb'}{1 + \alpha\delta} \qquad (11.29c)$$

Recognizing from Eq. (11.29b) and the definition of α that

$$Q = \frac{\alpha\delta M_1}{a'}$$

and substituting Eq. (11.29c) for M_1 yields

$$Q = \frac{\alpha\delta Tb'}{(1 + \alpha\delta)a'} \qquad (11.29d)$$

If Eq. (11.29d) is substituted into Eq. (11.28d), the force equilibrium equation, the resulting equation may be solved for T, yielding

$$T = \frac{B_c}{1 + [\alpha\delta/(1 + \alpha\delta)](b'/a')} \qquad (11.29e)$$

For the beam mechanism to occur, the moments M_1 and M_2 must be equal; therefore, $\alpha = 1$. This is designated as the balanced condition since the beam mechanism and bolt failure will occur simultaneously. The resulting balanced tension force T_o is given by

$$T_o = \frac{B_c}{1 + [\delta/(1 + \delta)](b'/a')} \qquad (11.29f)$$

If $T \leq T_o$, failure of the plate will occur and the plate thickness may be determined from Eq. (11.28a) as

$$t = \sqrt{\frac{4T\, b'}{\phi F_y\, p\, (1 + \delta)}} \qquad (11.30a)$$

If $T > T_o$, failure will be initiated by the bolt. Summation of moments about the tip of the flange yields

$$M_1 = T(a' + b') - B_c\, a' \qquad (11.30b)$$

Setting M_1 equal to the plastic moment and solving for t yields

$$t = \sqrt{\frac{4[T(a' + b') - B_c\, a']}{\phi F_y\, p}} \qquad (11.30c)$$

Thus, for the determination of the flange thickness, Eqs. (11.29f), (11.30a), and (11.30c) will be used. Both approaches are illustrated in the following examples.

EXAMPLE 11.8a

Given: A WT9×48.5 section, A36 steel, is used as shown in Fig. 11.29 to carry a factored load of 120 kips. Four $\frac{7}{8}$-in.-diameter A325 bolts are used in a 9-in.-long fitting. Determine whether the connection is adequate or inadequate.

Dimensions $t_f = 0.87$ in., $t_w = 0.535$ in., $b_f = 11.145$ in., gage = 4 in.

Solution

$b = (4.0 - 0.535)/2 = 1.73$ in. > 1.375 (wrench clearance)

$a = (11.145 - 4.0)/2 = 3.57$ in. $> 1.25b$ (recommended by AISC)

Thus, take

$$a = 1.25(1.73) = 2.16 \text{ in.}$$

$$b' = 1.73 - (\tfrac{7}{8})(\tfrac{1}{2}) = 1.29 \text{ in.}$$

$$a' = 2.16 + (\tfrac{7}{8})(\tfrac{1}{2}) = 2.60 \text{ in.}$$

$$p = \frac{9}{2} = 4.5 \text{ in. per bolt}$$

$$d' = \tfrac{7}{8} + \tfrac{1}{16} = 0.938 \text{ in.}$$

$$\delta = 1 - (0.938/4.5) = 0.792$$

$$T = 120 \text{ kips/4 bolts} = 30 \text{ kips per bolt}$$

$$B_c = 40.6 \text{ kips (bolt capacity from } LRFD \text{ } Manual)$$

Plate failure limit state, Eq. (11.28a):

$$T = \frac{0.9(36)(0.87)^2 \ (4.5)}{4(1.29)}(1 + 0.792) = 38.3 \text{ kips} > 30 \text{ kips}$$

Bolt failure limit state, Eq. (11.28b):

$$T = B_c = 40.6 \text{ kips} > 30 \text{ kips}$$

Bolt with prying action limit state, Eq. (11.28e):

$$\phi M_p = \frac{0.9(36)(0.87)^2 \ (4.5)}{4} = 27.59 \text{ kip-in.}$$

$$T = \frac{27.59 + 40.6(2.6)}{2.6 + 1.29} = 34.2 \text{ kips} > 30 \text{ kips}$$

Since the capacity of the connection for all three limit states exceeds the applied load, the connection is adequate.

EXAMPLE 11.8b

Given: For the same conditions as in the previous example, determine the minimum thickness required for the flange of the T section provided all of the other dimensions remain unchanged.

Determine the balanced tension T_o, Eq. (11.29f):

$$T_o = \frac{40.6}{1 + (0.792/1.792)(1.29/2.6)} = 33.3 \text{ kips}$$

Since the applied load per bolt is less than T_o, failure will occur in the plate; therefore, use Eq. (11.30a) to determine flange thickness:

$$t = \sqrt{\frac{4(30)(1.29)}{0.9(36)(4.5)(1 + 0.792)}} = 0.77 \text{ in.}$$

Thus, use a WT9×43 with $t_f = 0.77$ in. Since this section has a web thickness somewhat less than that used in calculating a' and b', the designer should recheck the capacity of the connection using updated values.

11.11 HIGH-STRENGTH BOLTS IN SHEAR AND TENSION

11.11.1 Behavior

A bolt loaded in combined tension and shear will have a reduced capacity to resist shear in a bearing type connection. In a slip critical connection, the tension will reduce the contact force and, thus, lower the shear required to cause the connection to slip. This could be important in an application where fatigue is critical.

Tests [19] have shown that the interaction of shear and tension in a bearing-type connection can be fairly well predicted through an elliptical interaction curve. For design, the LRFD specification, Section J3.4 has adopted three straight lines to approximate the ellipse. Figure 11.30 shows the design interaction equation for A325 bolts. When the threads are included in the shear plane,

$$F_t = 85 - 1.8F_v \le 68 \tag{11.31}$$

A similar ellipse predicts the interaction of shear and tension for the case of threads excluded from the shear plane. The straight line equation for A325 bolts in this case is

$$F_t = 85 - 1.4F_v \le 68 \tag{11.32}$$

Similar equations are specified for A490 bolts in bearing type connections.

In a slip-critical connection, the shear–tension interaction equation serves a different purpose. In this case, shear is assumed to be transferred by friction between the plies. The capacity of the connection, therefore, is a linear function of the force compressing the plies. This force is the initial pretension T_b minus the applied load T. The specified slip-critical shear value is, therefore, reduced by the factor $1 - T/T_b$. The resulting interaction equation is

$$F_v' = F_v\left(\frac{1 - T}{T_b}\right) \tag{11.33}$$

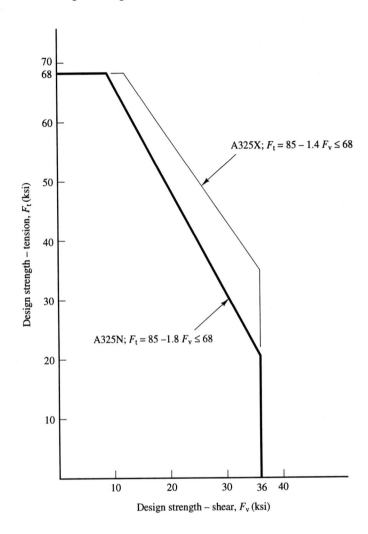

Figure 11.30 Shear tension interaction for A325 bolts.

A325X; $F_t = 85 - 1.4 F_v \le 68$

A325N; $F_t = 85 - 1.8 F_v \le 68$

where

   F_v = the slip-critical design shear
   F'_v = reduced slip-critical design shear due to concurrent applied
        tension

EXAMPLE 11.9

Given: Figure 11.31 shows an inclined hanger supporting a service dead load of
10 kips and a vibrating live load of 50 kips. Steel is A36. Use four A325-N slip-
critical bolts with a pretension of 51 kips (LRFD specification, Table J3.1).

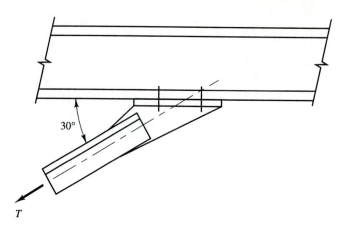

Figure 11.31 An inclined
hanger with bolts
in tension.

Solution

Step 1. A serviceability check is made with unfactored loads and slip-critical bolt values. Try 1-in.-diam. bolts.

$$\text{Bolt tension} = \sin 30° \left(\frac{60}{4}\right) = 7.5 \text{ kips}$$

$$\text{Bolt shear} = \cos 30° \left(\frac{60}{4}\right) = 13.0 \text{ kips}$$

From *LRFD Manual,* Table I-D,

Design shear strength for slip-critical 1-in.-diam. bolt
$$= 13.4 \text{ kips}$$

Reduced design shear strength

$$13.4 \left(1 - \frac{7.5}{51}\right) = 11.4 \text{ kips} > 10.8 \text{ kips} \qquad \text{OK}$$

Step 2. A strength check is made at factored loads using the bearing-type interaction equation with threads included in the shear plane.

$$\text{Bolt tension} = \sin 30° \frac{1.2(10) + 1.6(50)}{4}$$

$$= 11.5 \text{ kips}$$

$$\text{Bolt shear} = \cos 30° \frac{1.2(10) + 1.6(50)}{4}$$

$$= 19.9 \text{ kips}$$

$$\text{Bolt tensile stress} = \frac{11.5}{0.7854} = 14.6 \text{ ksi}$$

$$\text{Bolt shear stress} = \frac{19.9}{0.7854} = 25.3 \text{ ksi}$$

Reduced design tensile stress

$$85 - 1.8(25.3) = 39.5 \text{ ksi} > 14.6 \text{ ksi} \qquad \text{OK}$$

11.12 MOMENT CONNECTIONS

Moment connections are required to transmit moment while keeping the relative rotation of the attached members at zero. They are used for Type FR frames, sometimes for Type PR, and for any situation where moment transfer is required with no relative rotation. In Type FR frames, the connection is designed to resist the moments that result from frame analysis using the worst-case loading condition. The analysis procedure assumes that the beams and columns maintain their original geometric relationship during the entire loading history.

As discussed in Chapter 9, moment connections may be designed to resist the wind load only. The assumption is that under gravity loads, the connection is flexible enough to accommodate the rotation of the beam while retaining sufficient strength to carry the wind load moment.

Four types of moment connections that are in common use when the beam frames into the strong axis of the column are illustrated in Fig. 11.32. The connection shown in Fig. 11.32a is usually considered a PR connection, depending on the size of the plate. A shear plate is shop-welded to the column so the beam can be easily erected by field-bolting. The bottom flange plate is shop-welded to the column flange and field-welded to the beam flange, while the top plate is shipped to the field loose and field-welded to both the column and beam.

Figure 11.32b shows a field-bolted version of the connection shown in Fig. 11.32a. Note that the top flange plate is shop-welded to the column with a clearance that exceeds the theoretical beam depth, usually about $\frac{3}{8}$ in., to account for beam depth overrun and to facilitate erection. Shims are used when it is found that clearance is not necessary.

Figure 11.32c shows a direct field-welded beam-to-column connection. It is used when the full capacity of the beam is to be transferred to the column.

For these three connections, the design assumption is that the moment is transferred by the flanges and the shear is transferred by the web plate. Since the web does participate to some extent in the moment resistance of the beam, this assumption is not completely accurate, but it is acceptable for design. The behavior of these connections was discussed in Section 11.3.

The end plate moment connection shown in Fig. 11.32d is quite different from the other three. The design procedure suggested by AISC is empirical and is the result of research by Krishnamurthy [20]. This connection is not universally popular with fabricators and erectors because it

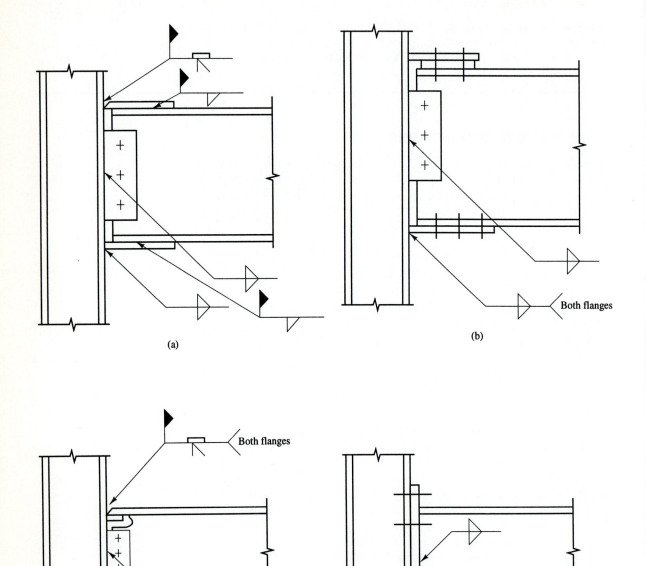

Figure 11.32 Typical moment connections for beams framing into column flange.

is difficult to fabricate the end plates parallel with each other. If they are not parallel, erection is difficult if not impossible. This problem is illustrated in Fig. 11.33. If the plates are welded to the beam square with the flanges (Fig. 11.33a), the natural camber of the beam will cause the plates to be cocked. To avoid this, the plates must be connected square with the theoretical centerline of the beam, as seen in Fig. 11.33b. Not all fabricators are able to do this successfully or are willing to go to this trouble in order to use the end plate connection.

The design of an end plate connection involves selecting the bolt size and the end plate thickness. Using connections with the variable defined as shown in Fig. 11.34, it is assumed that the tension at the top of the connection due to the moment is resisted uniformly by the two bolts above the beam flange and the two bolts below the flange. Because of the presence of the web, it is obvious that no prying action would exist on the lower two bolts. The design procedure does not consider prying action on the top two bolts, since research has indicated that the limit state of the connection behavior is excessive deflection, while these bolts remain intact.

The force transmitted to the upper four bolts is determined by dividing the beam end moment by the moment arm between the flange centerlines. Thus, the flange force is defined as

$$F_f = \frac{M}{d - t_f}$$

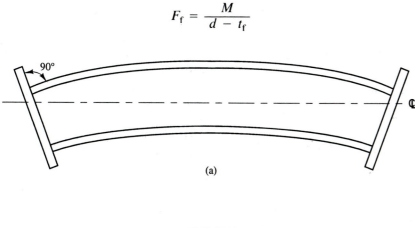

(a)

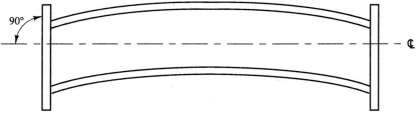

(b)

Figure 11.33   End plates on beams with natural camber.

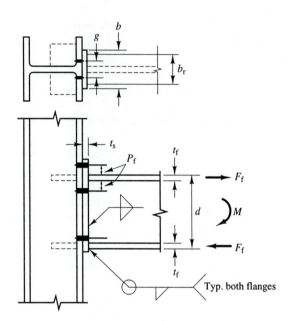

Figure 11.34 Typical end plate connection.

The end plate thickness is determined by calculating its strength in single curvature bending. The effective moment arm P_e is defined as

$$P_e = P_f - \frac{d_b}{4} - w \qquad (11.34)$$

where

d_b = nominal bolt diameter (in.)

P_f = distance from center line of bolt to surface of the tension flange (in.). $d_b + \frac{1}{2}$ in. is generally sufficient to provide wrench clearance

w = fillet weld size or reinforcement of groove weld (in.)

The moment M_e that the end plate must resist is

$$M_e = \frac{\alpha_m F_f P_e}{4} \qquad (11.35)$$

where

$$\alpha_m = C_a C_b \left(\frac{A_f}{A_w}\right)^{1/3} \left(\frac{P_e}{d_b}\right)^{1/4}$$

C_a = constant dependent on bolt type and beam yield stress; see Table A of the *LRFD Manual*, Part 5

$$C_b = b_f/b_s$$
$$A_f = \text{area of tension flange (in.}^2)$$
$$A_w = \text{web area, clear of flanges (in.}^2)$$

The end plate width b_s must be less than or equal to 1.15 times the beam flange width.

The resisting moment M of the plate is the plastic modulus times the design stress $0.9F_y$:

$$M = (\tfrac{1}{4}b_s t^2)(0.9F_y) \tag{11.36}$$

Solving for the plate thickness t yields

$$t = \sqrt{\frac{4M_e}{b_s\,0.9F_y}} \tag{11.37}$$

11.13 COLUMN REINFORCEMENT

For beam-to-column moment connections, as shown in Fig. 11.32, it is sometimes necessary to reinforce the column by adding a doubler plate to the web, as shown in Fig. 11.35, or stiffeners between the flanges, as shown in Fig. 11.36.

Doubler plates are required when the horizontal shear in the web exceeds the design shear. The shear is calculated by considering a free body of the web on a plane immediately below the top beam flange, as shown in Fig. 11.37. The forces in the beam flanges result from the beam moment and are calculated by dividing this moment by 95% of the depth of the beam. The beam moment used in this calculation is the largest possible moment considering both gravity and lateral loading. When beams frame into both sides of the column, the wind moments on the windward side of the column are additive to the gravity moments, while on the leeward side of the column they act in opposite directions and only the net moment should be used. When the moments are additive, it is conservative to use the full dead and live gravity loads. When the moments are subtractive, the designer may elect to assume that only the gravity dead loads are available. The story shear is then subtracted from the net force due to the beam moments to determine the shear within the column web.

The column web shear, calculated as described above, is compared to the design shear which is the product of the design stress, $0.9(0.6)F_y$, multiplied by the area of the column web, $d_c t_w$. If the calculated shear exceeds this value, a doubler plate is necessary. The thickness of this plate is added to the column web thickness so that equilibrium is achieved.

Columns may also have to be reinforced by adding stiffeners, because of the force delivered to the column by the beam flange. This force

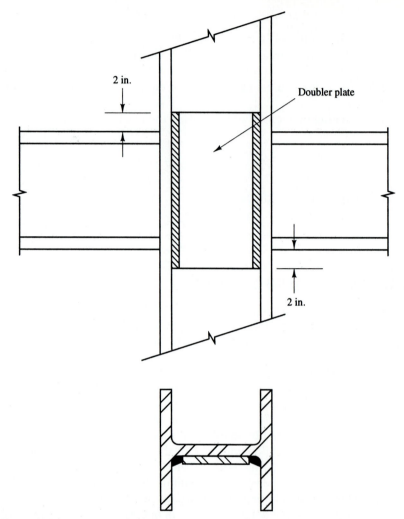

Figure 11.35 Beam-to-column moment connection with doubler plate added to the web.

is calculated by dividing the beam moment by the beam depth. There are three cases to be considered. The following discussion applies to moment connections shown in Figures 11.32a–c.

Case 1: Column Web Yielding. The column web must be prevented from yielding as a result of the force (tension or compression) in the beam flange, as seen in Fig. 11.38. The LRFD specification, Section K1.3, stipulates that this force is resisted by the column web by an area defined by distributing the force on a $2\frac{1}{2}$ to 1 slope through the column flange on a plane at the k distance. The resisting force R_n, therefore, is

$$\phi R_n = (5k + t_f) F_y t_w \tag{11.38}$$

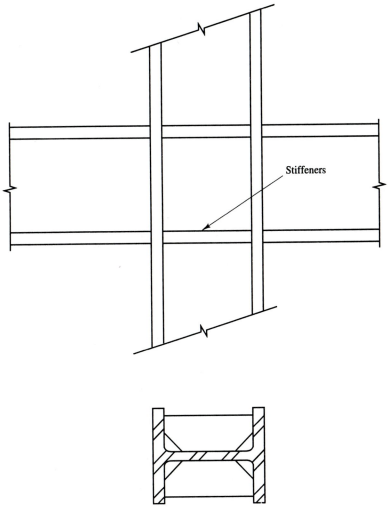

Figure 11.36 Beam-to-column moment connection with stiffeners between the flanges.

where t_f is the beam flange thickness and t_w is the column web thickness. The factor ϕ is taken as 1.0.

Equation 11.38 can be rearranged as

$$\phi R_n = 5kF_y\, t_w + t_f F_y t_w \qquad (11.39)$$

The quantity $5kF_y t_w$ is a constant for each column shape and is tabulated in the *LRFD Manual* as P_{wo}. The second term in Eq. (11.39) is also tabulated with $P_{wi} = F_y\, t_w$. This value is multiplied by the beam flange thickness t_f and added to P_{wo} to evaluate the column web resistance to yielding.

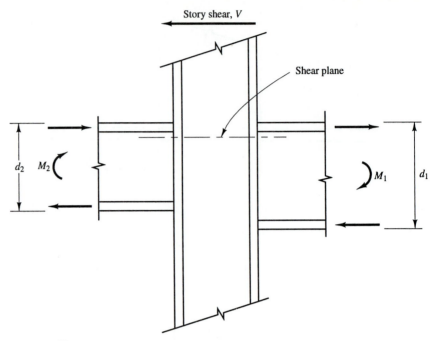

Figure 11.37 Shear plane for determination of column web shear.

Case 2: Column Web Buckling. Where the beam flange delivers a compression force to the column, the column web has the potential to buckle, as shown in Fig. 11.39, if its slenderness is large. The LRFD specification, Section K1.6, defines $\phi = 0.9$ and the resisting force as

$$\phi R_n = \frac{3690 t_w^3 \sqrt{F_{yw}}}{d_c} \qquad (11.40)$$

This value is also unique for every shape and is tabulated in the *LRFD Manual* as P_{wb}.

Case 3: Column Flange Bending. Where the beam delivers a tension force, the column flange tends to bend, as shown in Fig. 11.40. If the column flange is too flexible, shear lag develops and the weld at the flange–web juncture is subjected to an overload. To prevent this, the LRFD specification, Section K1.2, defines $\phi = 0.9$ and the resisting force as

$$\phi R_n = 5.625 t_f^2 F_{yf} \qquad (11.41)$$

The *LRFD Manual* lists this value as P_{fb}.

It can be seen that column stiffener requirements are a "go – no go" situation. A stiffener may be required by a very small margin. Unfor-

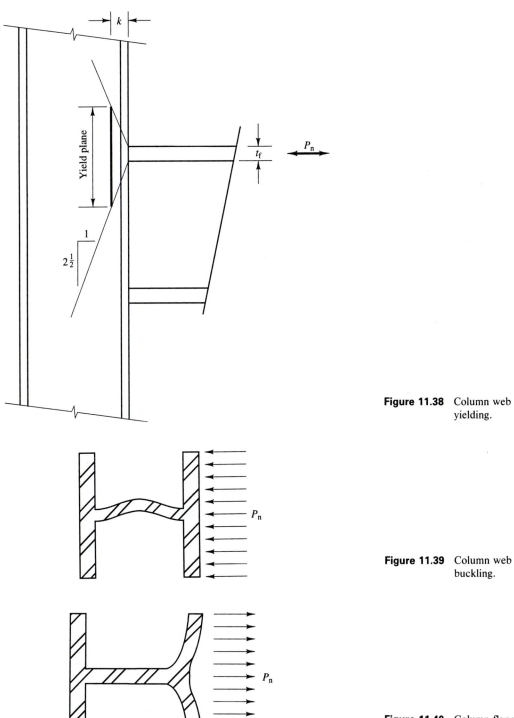

Figure 11.38 Column web yielding.

Figure 11.39 Column web buckling.

Figure 11.40 Column flange bending.

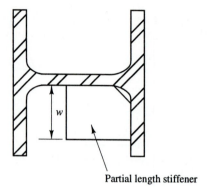

Partial length stiffener

Figure 11.41 Partial length column web stiffeners.

tunately, stiffeners are very expensive, especially if they must be fitted between the column flanges. In many cases, it would be much more economical for the design engineer to select a column section that may be larger than required for the axial load but would not require stiffeners. Stiffener requirements should not be left for the detailing stage; rather, they should provide a significant piece of input during the design stage.

If stiffeners cannot be avoided, they are designed to resist a force calculated as the beam flange force minus the resisting force, as defined by Eqs. (11.39) through (11.41). This net force is resisted by the cross section of two stiffeners in yielding wtF_y. If the stiffeners are partial length, as shown in Fig. 11.41, the weld to the column web must be designed for this same force.

AISC suggests the following rules for proportioning stiffeners:

1. The width of each stiffener plus one-half the thickness of the column web shall not be less than one-third the width of the flange or moment connection plate delivering the concentrated force.
2. The thickness of stiffeners shall not be less than one-half the beam flange thickness, $t_b/2$.
3. When the concentrated force is delivered to only one column flange, the stiffener need not extend past the column center line.
4. The weld joining stiffeners to the column web shall be sized to carry the force in the stiffener caused by unbalanced moments on opposite sides of the column.

EXAMPLE 11.10

Given: Design a welded beam-to-column moment connection of the type shown in Fig. 11.32c with beams of the same depth framing in from both sides. The beam is a W24 × 76 and the column is a W14 × 109. Bolts are A325-N and the electrodes are E70XX. All steel is A36. The factored moment is 500 kip-ft (333 kip-ft gravity, 167 kip-ft wind) and the factored shear is 60 kips. The factored story shear is 30 kips.

Solution

Step 1. From the *LRFD Manual*, the beam and column properties are tabulated:

Beam (W24 × 76) d = 23.92 in. b_f = 8.99 in.
                                      t_w = 0.44 in. t_f = 0.68 in.
                                      Z = 200.0 in.^{3}

Column (W14 × 109) d = 14.32 in. b_f = 14.605 in.
                                        t_w = 0.525 in. t_f = 0.86 in.
                                        P_{wo} = 148 P_{wi} = 19
                                        P_{wb} = 281 P_{fb} = 150

A325-N bolts, single shear = 21.1 kips

Step 2. Check beam strength:

$$Z_{req'd} = \frac{500(12)}{0.9(36)} = 185 \text{ in.}^3$$

$$< 200 \text{ in.}^3 \quad \text{OK}$$

Step 3. Design web connection:

$$\frac{60}{21.1} = 2.8 \quad \text{Try 3 bolts}$$

Check bearing on beam web
  From the *LRFD Manual*, Part 5, Table I-E,

$$3(91.3)(0.44) = 121 \text{ kips} > 60 \text{ kips} \quad \text{OK}$$

Design shear plate
  Try $\frac{3}{8}$-in. plate, 4-in. pitch, L = 12 in.

Check fracture:

$$A_{net} = 0.375[12 - 3\,(\tfrac{7}{8} + \tfrac{1}{16})]$$

$$= 3.45 \text{ in.}^2$$

$$\phi V_n = 0.75(0.6)(58)(3.45) = 90.0 \text{ kips}$$

$$> 60 \text{ kips} \quad \text{OK}$$

Check yielding:

$$A_{gross} = 0.375(12) = 4.5 \text{ in.}^2$$

$$\phi V_n = 0.9(0.6)(36)(4.5) = 87.5 \text{ kips} > 60 \text{ kips}$$

Shear plate is OK

Check bearing on shear plate:
From the *LRFD Manual,* Part 5, Table I-E,

$$3(34.3) = 103 \text{ kips} > 60 \text{ kips} \qquad \text{OK}$$

Weld to column flange

$$w_{\min} = 60/(2(1.392)(12)) = 1.8 \text{ sixteenths}$$

Therefore, use $\frac{5}{16}$ fillet. See the LRFD specification, Table J2.5.

Step 4. Check column web for story shear:
Web shear due to beam flange forces from wind moments:

$$\frac{M_{\text{left}}}{0.95 d_{\text{left}}} + \frac{M_{\text{right}}}{0.95 d_{\text{right}}} = \frac{167(12)}{0.95(23.92)} + \frac{167(12)}{0.95(23.92)} = 176 \text{ kips}$$

Net web shear $= 176 - 30 = 146$ kips
Web design shear $= 0.9(0.6)(36)(14.32)(0.525)$
$= 146 \text{ kips} > 146 \text{ kips}$
Web doubler plate not required.

Step 5. Check column web stiffener requirements:

$$\text{Flange force} = \frac{M_{\text{total}}}{d - t_f}$$

$$= \frac{500(12)}{23.92 - 0.68}$$

$$= 258 \text{ kips}$$

Design strength: web yielding

$$P_{\text{wo}} + (P_{\text{wi}} \, t_f) = 148 + (19)(0.68)$$

$$= 161 \text{ kips} < 258 \text{ kips}$$

Therefore, stiffeners are required for web yielding

Design strength: web buckling

$$P_{\text{wb}} = 281 \text{ kips} > 258 \text{ kips}$$

Stiffeners not required for web buckling

Design strength: flange bending

$$P_{\mathrm{fb}} = 150 \text{ kips} < 258 \text{ kips}$$

Thus, stiffeners are required for flange bending

Because stiffeners are required for web yielding, they are necessary for both the tension and compression flange. Based on the design guidelines given above,

$$\text{Minimum stiffener width} = b_f/3 - t_w/2$$

$$= 8.99/3 - 0.525/2$$

$$= 2.73 \text{ in.}$$

$$\text{Stiffener thickness} = 0.68/2 = 0.32 \text{ in.}$$

$$\text{Use } \tfrac{3}{8} \text{ in.}$$

$$\text{Stiffener area req'd} = (258 - 150)/36$$

$$= 3.0 \text{ in.}^2$$

$$\text{Stiffener width req'd} = \tfrac{1}{2}(3.0/\tfrac{3}{8})$$

$$= 4.0 \text{ in.}$$

Step 6. Design stiffener welding:
The force to be transferred to the column flange
from each stiffener $= \tfrac{1}{2}(258 - 150)$

$$= 54 \text{ kips}$$

Fillet weld length is the stiffener width less the clipped portion at the column:

$$\text{Fillet} = 4.0 - 1.0 \text{ (assumed)} = 3.0 \text{ in.}$$

$$2 \text{ welds}(1.392)(w)(3.0) = 54$$

$$w = 6.5 \text{ sixteenths}$$

Use $\tfrac{7}{16}$ fillet weld

PROBLEMS

For Problems 11.1–11.6, use $\tfrac{5}{16}$-in. A36 angles, $\tfrac{3}{4}$-in. A325-N bolts in standard holes, and uncoped beams. Assume that the supporting member is not critical.

11.1. Design a double-angle connection for a $W18 \times 50$, A36 beam to carry a factored load of 85 kips.

11.2. Design a double-angle connection for a $W27 \times 102$, A36 beam to carry a factored load of 180 kips.

11.3. Design a double-angle connection for a $W24 \times 146$, A36 beam to carry a factored load of 150 kips.

11.4. Design a double-angle connection for a $W16 \times 67$, A36 beam to carry a factored load of 124 kips.

11.5. Design a double-angle connection for a $W18 \times 143$, A36 beam to carry a factored load of 152 kips.

11.6. Design a double-angle connection for a $W8 \times 40$, A36 beam to carry a factored load of 55 kips.

For Problems 11.7–11.12, use $\frac{5}{16}$-in. A36 angles, $\frac{3}{4}$-in. A325-N bolts in standard holes, and assume that the beams are coped so that the edge distance is $1\frac{1}{4}$-in. Assume that the supporting member is not critical.

11.7. Design a double-angle connection for a W30×191, A36 beam spanning 40 ft and carrying the maximum permitted uniform load.

11.8. Design a double-angle connection for a W18×76, A36 beam to support a factored load of 100 kips.

11.9. Design a double-angle connection for a W21×68, A36 beam spanning 20 ft and carrying the maximum permitted uniform load.

11.10. Design a double-angle connection for a W24×84, A36 beam to support a factored load of 150 kips.

11.11. Design a double-angle connection for a W12×87, A36 beam to support a factored load of 85 kips.

11.12. Design a double-angle connection for a W16×67, A36 beam spanning 20 ft. and carrying the maximum permitted uniform load.

11.13. Design a double-angle connection to support a W14×74 with a factored shear force of 40 kips and a factored axial force of 35 kips. Use $F_y = 50$ ksi and assume that the beam is not coped.

For Problems 11.14–11.18, use weld data provided in the *LRFD Manual* and design a welded connection in which both the beam and the supporting member are welded.

11.14. Design a double-angle connection for an uncoped W18×50, A36 beam to carry a factored load of 95 kips.

11.15. Design a double-angle connection for an uncoped W16×67, A36 beam to carry a factored load of 110 kips.

11.16. Design a double-angle connection for an uncoped W8×40, A36 beam to carry a factored load of 60 kips.

11.17. Design a double-angle connection for a coped W18×76, A36 beam to carry a factored load of 110 kips.

11.18. Design a double-angle connection for a coped W24×84, A36 beam to carry a factored load of 140 kips.

For Problems 11.19–11.22, use a $\frac{3}{8}$-in. thick shear tab and $\frac{3}{4}$-in. A325-N bolts. Assume that the supporting member is not critical.

11.19. Design a shear tab connection for an uncoped W18×50, A36 beam to carry a factored load of 85 kips.

11.20. Design a shear tab connection for an uncoped W27×102, A36 beam to carry a factored load of 180 kips.

11.21. Design a shear tab connection for a coped W21×68, A36 beam spanning 20 ft and carrying the maximum permitted uniform load. Assume that the edge distance at the cope is $1\frac{1}{4}$ in.

11.22. Design a shear tab connection for a coped W18×76, A36 beam to carry a factored load of 100 kips. Assume an edge distance of $1\frac{1}{4}$ in.

11.23. A W16×26 beam spans 24 ft, carries a uniform load, and frames into the web of a W10×39 column. Design an unstiffened seated connection to carry the maximum capacity of the beam assuming all elements to be A36 and bolts to be A325-N.

11.24. Design a moment resisting connection of the type shown in Fig. 11.32a to carry the full plastic capacity of a W16×31, A36 beam.

11.25. A moment resisting connection is to be designed to transfer the full plastic moment capacity of a W18×71. Using WT sections at the top and bottom flange and double angles on the web, design a bolted connection. Assume A36 steel and A325-N bolts.

REFERENCES

[*1*] Huang, J. S., Chen, W. F., and Beedle, L. S., *Behavior and Design of Steel Beam-to-Column Moment Connections,* Welding Research Council, Bulletin 188, New York, 1973.

[*2*] Disque, R. O., Wind connections with simple framing, *Engineering Journal, AISC,* Vol. 1, No. 3, 1964 (pp.101–103).

[*3*] Ackroyd, M., Simplified frame design of type PR

construction, *Engineering Journal, AISC,* Vol. 24, No. 4, 1987 (pp. 141–146).

[4] Goto, Y., and Chen W. F., On the computer-based design analysis for the flexibly jointed frames, *Journal of Constructional Steel Research*, Vol. 8, 1987 (pp. 203–231).

[5] Lui, E. M., and Chen, W. F., Steel frame analysis with flexible joints, *Journal of Constructional Steel Research*, Vol. 8, 1987 (pp. 161–202).

[6] Richard, R. M., Gillett, P. E., Kriegh, J. D., and Lewis, B. A., The analysis and design of single plate framing connections, *Engineering Journal, AISC,* Vol. 17, No. 2, 1980 (pp. 38–52).

[7] Astaneh, A., New concepts in design of single plate shear connections, *Proceedings, National Steel Construction Conference,* AISC, Nashville, TN, 1989 (pp. 3.1–3.17).

[8] American Welding Society, *Structural Welding Code—Steel(D1.1),* AWS, Miami, FL, 1992.

[9] Birkemoe, P. C., and Gilmore, M. I., Behavior of bearing critical double-angle beam connections, *Engineering Journal, AISC,* Vol. 15, No. 4, 1978 (pp. 109–115).

[10] Ricles, J. M. and Yura, J. A., Strength of double-row bolted web connections, *Journal of the Structural Division, ASCE,* Vol. 109, No. ST1, 1983 (pp. 126–142).

[11] Hardash, S., and Bjorhovde, R., New design criteria for gesset plates in tension, *Engineering Journal, AISC,* Vol. 22, No. 2, 1985 (pp. 77–94).

[12] Research Council on Structural Connections, *Allowable Stress Design Specification for Structural Joints Using ASTM A325 or A490 Bolts,* AISC, 13 November, 1985.

[13] Higgins, T. R. and Preece, F. R., Proposed working stress for fillet welds in building construction, *Welding Journal Research Supplement*, October 1968.

[14] Butler, L. J., Pal, S., and Kulak, G. L., Eccentrically loaded welded connections, *Journal of the Structural Division, ASCE,* Vol. 98, No. ST5, 1972.

[15] Astaneh, A., Call, S. M., and McMullin, K. M., Design of single plate shear connections, *Engineering Journal, AISC,* Vol. 26, No. 1, 1989 (pp. 21–32).

[16] Garrett, J. H., and Brockenbrough, R. L., Design loads for seated-beam connections in LRFD, *Engineering Journal, AISC,* Vol. 23, No. 2, 1986 (pp. 84–88).

[17] Struik, J. H. A., and de Back, J., *Tests on Bolted T-Stubs with Respect to Bolted Beam-to-Column Connections,* Report 6-69-13, Stevin Laboratory, Delft University of Technology, Delft, The Netherlands, 1969.

[18] Thornton, W. A., Prying action: A general treatment, *Engineering Journal, AISC,* Vol. 22, No. 2, 1985 (pp. 67–75).

[19] Chesson, E., Jr., Faustino, N. L., and Munse, W. H., High-strength bolts subjected to tension and shear, *Journal of the Structural Division, ASCE,* Vol. 91, No. ST5, 1965.

[20] Krishnamurthy, N., A fresh look at bolted end-plate behavior and design, *Engineering Journal, AISC,* Vol. 15, No. 2, 1978 (pp. 39–49).

Index

A

ACI, 357, 378
Ackroyd, M., 331
Air Reduction, 410
AISC, 95
AISI, 95, 121
Allowable Stress Design (ASD), 17
Aluminum, 107
AMOCO Building, 15, 90
Analysis factor, 80–81
ANSI, 20, 115
  A58.1, 68–69
Area
  effective net, 153–54, 166
  gross, 153–54, 156–57, 199
  net, 153–54, 157
ASCE, 20
ASCE 7-88, 65, 67–71
Astaneh, A., 386, 413
ASTM, 88, 95–97
  A6, 95, 195
  A7, 11
  A36, 116
  A53, 119–20
  A108, 130
  A242, 116
  A307, 122–23

  A325,122–23, 388, 390
  A354, 122, 124, 134
  A370, 111, 113, 134
  A440, 106, 116
  A441, 116–18
  A449, 122–24
  A490, 106, 122–24, 134, 388, 390
  A500, 119–20
  A501, 119–20
  A514, 106, 116–18
  A529, 116–18
  A570, 121–22
  A572, 106, 115–18
  A588, 106, 117–19, 139
  A606, 121–22
  A607, 121–22
  A618, 120
  A709, 116–18
AWS, 88, 109, 125
Axial forces, 302–3

B

Base metal, 129
Basler, K., 286
Bauschinger effect, 114
Beam bearing, 419

Beam-columns, 302–22
  axial forces, 302–3
  biaxial bending, 303
  C_m, 309–14
  drift, 303–5
  equivalent moment factor, 309–12
  interaction, 302–7
  interaction equations, 305, 314
  lateral deflection, 303–4
  member second-order effects, 310, 314
  moment amplification, 307
  moment gradient, 303–4
  $P\Delta$, 304–5, 315
  $P\delta$, 308–9
  second-order effects, 303–4
  structure second-order effects, 315–19
  sway, 303–5
  trial section, 319–22
Beam design, 262–73
  BF, 266
  laterally unsupported, 265–73
  lateral support, 262
  moment gradient, 270–71
  moment gradient factor, 267–70
  plastic, 262
  unbraced length, 262
Beam failure mechanism, 273
Beam line, 329
Beams, 251–81. *See also* Beam design
  behavior, 253–62
  bending, 253–62
  built-up, 259
  C_b, 267–70
  compact shapes, 254
  composite (*see* Composite beams; Composite columns)
  deflection, 279–81
  double angle, 277–78
  encased beams (*see* Composite beams)
  flange local buckling, 259–60
  noncompact shapes, 254
  plastic analysis, 273–77
  plastic moment, 251–52
  plastic neutral axis (PNA), 252
  plate girder, 285–99
  shear, 257–58, 264–65
  slender shapes, 254
  T-beams, 277–78
  torsion, 254, 273
  web local buckling, 260–62

Bearing wall, 12
Beedle, L., 208
Bell Atlantic Tower, 1
Bending limit states
  lateral torsional buckling, 253–57, 273, 277
  local buckling, 253–54, 258–62, 277
Bending members. *See* Beams
Bessemer, 89–90
Biaxial bending, 303
Biaxial state of stress, 115
Bijlaard, P., 162–63
Birkemoe, P., 169, 397
Bjorhovde, R., 169, 397
Bleich, F., 234
Block shear, 168–78, 397–98
Bolts, 122–25, 388–390
  combined stress, 430–33
  common, 388
  high strength, 388–90
    combined stress, 430–33
    tension, 423–30
  interaction, 430–31
  limit states, 394–98
  tension controlled, 388
Braced frames, 307–14
Bracing
  design, 338–42
  frame support, 339–41
  member support, 338–39
Brittle fracture, 134–35
Brooklyn Bridge, 90
Buckling, 200–213, 253–54
  elastic, 200–204, 223–30
  Euler, 200–204, 215
  flange, 253–54, 259–60
  inelastic, 206–10, 230–33
  plate, 234–36
  torsional, 239–42
  web, 253–54, 260–62
Building codes, 20, 59
Built-up columns, 234–46
  modified slenderness, 244–45
Butler, L., 400

C

C_b, 267–70
C_m, 309–14

Carbon, 105, 107
Carbon equivalent, 109
Charpy V-notch test, 115, 134
Chen, W., 331
Chernenko, D., 216
Chesson, E., 166
Chromium, 107
Chrysler Building, 90, 143
Citicorp Center, 87
Coalbrookdale Bridge, 89
Cochrane, V., 162
Coefficient of variation (COV), 35, 38–43, 51–52
Cold forming, 193
Cold-straightening
 gag straightening, 193
 rotorizing, 193
Colorado State University, 71
Columbium, 106–7
Columns, 184–246
 behavior, 185, 200–213
 buckling, 200–213
 built-up, 234–46
 effective length, 223–34
 Euler buckling, 200–204
 leaning, 227, 322–29
 length, 187, 214–18
 long columns, 216–17
 maximum strength, 211
 short columns, 214–15
 strength, 185, 218–19
  cross section, 187
  end support, 187
  initial out-of-straightness, 187, 204–6, 218
  length, 187, 218
  manufacturing method, 186–87 (see also Production)
  material properties, 186
  residual stress, 188–93, 216, 218
Commerce Square, 23
Components, 5–9
 bending members, 8
 combined force members, 9
 compression members, 7
 tension members, 6
Composite beams, 354–77
 deflection, 375–76
 effective flange width, 355–56
 encased, 374
 flat soffit, 356
 lower bound moment of inertia, 377

 metal deck, 356, 368–74
 negative moment, 365–66
 plastic neutral axis, 356
 plastic stress distribution, 357
 serviceability, 375–77
 shored construction, 354–55
 strength, 356–59
 trial section, 374–75
 unshored construction, 354–55
 vibration, 376
Composite columns, 377–80
Composite construction, 351–80
 advantages, 353–54
 disadvantages, 354
 flitch girder, 351
 full composite action, 357–62
 partial composite action, 362–65
Compression members, 184–246
Connections, 319, 383–445
 behavior, 383–84
 double angle
  bolted, 402–8
  welded, 408–12
 end plate, 433–37
 fully restrained, 383–85
 hanger, 427
 instantaneous center of rotation, 409
 moment (see Moment connections)
 moment rotation, 383
 partially restrained, 383, 386–87
 rigid, 383–85
 seated, 416–23
 semirigid, 383
 shear, 397, 401–23
 simple, 383
 single plate, 412–16
 slip-critical, 397, 430
 static model, 402
Considère, 206
Copper, 107
Cornell, C., 69
Corrosion, 138–40
Cost, 140–41
CSA, 88
 G40.12, 195
 G40.20, 95
 G40.21–44W, 119
 G40.21–50A, 119
 G40.21–100Q, 119
Culver, C., 69

D

de Back, J., 424
Deflection
  beam, 279–81
  lateral, 303–4 (*see also* Drift)
Deierlein, G., 331
Design, 4
  criteria, 16
  phases, 3–5
  philosophy, 16–19
Drift, 279, 303–5, 314

E

Eads Bridge, 90
Eccentric load, 204–6
Effective length factor, 223–33
Effective length nomograph, 227–34
Eiffel Tower, 90
Elastic buckling, 200–204
Elastic moment, 251
Ellingwood, B., 69
Empire State Building, 90, 183
End plate connection, 384–85, 433–37
Engesser, F., 206
Equivalent moment factor, 309–12
Euler, L., 200–204, 215

F

Fabrication factor, 43, 45, 47
Factor of safety, 25–26
Fatigue, 135–38
FCAW, 126, 129
Filler material, 126
First City Tower, 301
First-order second-moment (FOSM), 29
Firth of Fourth Bridge, 90
Flame cutting, 134, 136
Flange local buckling, 259–60
  compact, 259
  noncompact, 259
  slender, 260
Flexible wind connections, 331–33
  moment redistribution, 333–34
  shakedown, 331–33
Formed steel deck, 351
420 5th Avenue, 250

Fracture, 134–38
Frames
  braced, 307–14
  unbraced, 314–19
Friction, 430

G

Galvanizing, 132–34
George Washington Bridge, 90
Gere, J., 234
Gerstle, K., 331
Gilmor, M., 169, 397
Girders. *See* Beams; Plate girder
GMAW, 126, 128–29
Golden Gate Bridge, 90
Grant, J., 368
Grover, L., 410

H

Hanger connection, 427
Hardash, S., 169, 397
Higgins, T., 398
High-rise construction, 14
Holes, 157–66
  oversized, 160
  slotted, 160
  staggered, 161–66
  standard, 158
HSLA, 106, 120, 122
Humber River Bridge, 90

I

Inelastic buckling, 206–10
Influence area, 52, 66–67, 83–85
Initial crookedness, 193–96, 204–6
Interaction, 302–7
Interaction equations, 305, 314
Intermediate-length columns, 215–17
Iron, 89

J

John Hancock Center, 14, 133, 350
Johnston, B., 208

Julian, O., 227
Jumbo shapes, 136

K

Kanchanalai, T., 305
Kavanagh, T., 227
Kelly, 89–90
Kennedy, D., 216
Kings Bridge, 135
Krishnamurthy, N., 433

L

Lateral bracing, 338–41
Lateral deflection, 303–4
Lawrence, L., 227
Leaning column, 227, 322–29
Lehigh University, 208, 286
LeMessurier, W., 322
Liberty ships, 135
Limit states, compression, 218, 239–42
Limit states design (LSD), 27
Live load reduction, 68–69, 83–85
Load and resistance factor design (LRFD), 18
    philosophy, 26–28
    principles, 28–32
Load-deformation, 148–49, 151–52
Load effects, 48–54
Load Factor Design (LFD), 18
Loads
    arbitrary point-in-time, 62
    blast, 64
    combination, 78, 83–85
    dead, 61, 65–66
    factors, 27, 53–54, 81–82
    impact, 64
    live, 61–62, 66–69
    maximum lifetime, 62
    modeling, 59
    reduction (*see* Live load reduction)
    seismic, 63–64, 71–77
        base shear, 75
        dead weight, 75
        importance factor, 75
        overturning, 77
        period, 75–76
        site coefficient, 75–76
        system coefficient, 75

    torsion, 77
    zone factor, 72, 74–75
settlement, 65
snow, 62–63, 69–70
statistical characteristics, 65–71
sustained, 62
thermal, 65
transient, 62
variability, 59
wind, 63, 70–71
Local buckling, 234, 236, 253–54, 258–62
Loma Prieta earthquake, 72–73
Long columns, 216–17
Long-span construction, 13
Lui, E., 331

M

Manganese, 105, 107
Material factor, 42, 44, 46
McCormick Place, 131
McGuire, W., 162
McGuire, R., 69
Member second-order effects, 310, 314
Metal deck, 351
Metal matching, 129
Miner-Palmgren hypothesis, 136
Modulus of elasticity, 112–13
    reduced, 207–8
    tangent, 206–7, 209–10
Molybdenum, 106–8
Moment amplification factors, 307
Moment connections
    column flange bending, 440–42
    column reinforcement, 437–45
    column stiffeners, 438–40
    column web buckling, 440
    column web yielding, 438–39
    doubler plate, 437–38
    fully restrained, 433–37
    partially restrained, 433
Moment gradient, 256–57, 303–4
Munse, W., 166
Murray, T., 376

N

Nelson Stud Company, 351–53
Neutral axis
    determination of, 358–59

Neutral axis (cont.)
  plastic, 252
New River Gorge Bridge, 90
Nickel, 108
Nomenclature, 103–4
Nondimensional slenderness, 215
Normalizing, 133

O

One Mellon Bank Center, 284
Osgood, W., 208
Oversized holes, 160

P

$P\Delta$, 80, 304–5, 315
$P\delta$, 308–9
Partially restrained connections, 383, 386–87
  moment rotation curves, 329–30, 332
Partially restrained frames, 329–37
  analysis, 331
Perry-Robertson formula, 204, 206, 211
Phosphorus, 108
Plastic analysis, 273–77
Plastic design (PD), 18
Plastic moment, 251–52
Plastic neutral axis (PNA), 252
  determination, 358–59
Plastic section modulus, 253
Plate buckling, 234
Plate girder, 285–99
  flange local buckling, 287–88, 291
  homogeneous, 286–87
  hybrid, 285–86
  lateral torsional buckling, 289, 291
  noncompact web, 287
  nontension field, 293–95
  Pratt truss model, 293
  shear, 293–95
  slender web, 290–93
  stiffeners, 293, 296
  tension field, 294–95
  web local buckling, 287–88
Point Pleasant Bridge, 135
Poisson's ratio, 114

Preece, F., 398
Probability, 29
Production, 89, 90–95, 186–91
  Algoma Steel Co., 100
  basic oxygen, 90–91
  Bethlehem Steel Corp., 94
  blast furnace, 91
  casting, 94
  killed steel, 93
  NUCOR, 94
  open-hearth, 90
  pig iron, 91
  rimming, 91
  rolling, 93
  soaking pit, 93
  USX Corp., 94, 101
Professional factor, 43, 45, 47
Prying action, 423–28

Q

Quenching, 134

R

Rankine-Gordon equation, 204
Rayleigh analysis, 75
Reduced modulus of elasticity, 207–8
Reliability, 79
  index of, 33–34, 80
Research Council on Structural Connections, 397
Residual stress, 150–51, 188–93
Resistance, 150
Resistance factor, 28, 37–38, 43–47
Richard, R., 387
Ricles, J., 169, 397
Rivets, 122

S

SAE, 125
Safety, 19–20, 24–26
SAW, 126, 128
Sears Tower, 15, 90, 350
Seated connections, 416–23
Second-order effects, 303–4

Serviceability, 279–81
Serviceability limit states, 27, 56
Shanley, F., 208
Shapes
    angles, 198–99
    bars, 98–99
    C, 96–97
    HP, 96–97
    M, 96–97
    MC, 96–97
    S, 96–97
    tube (CF), 98, 121
    tube (HF), 97–98, 121
    Universal Mill Plate (UM), 98–99
    W, 96–97
    WT, 96–97
Shear connectors, 351, 365. *See also* Shear stud
Shear lag, 166–68
Shear limit states, 257
Shear modulus, 114
Shear stud, 130, 351–70
    capacity, 366–67
    placement, 367–68
    reduction factor, 369–70
Shear tab, 412
Shear yield, 114, 257
Short columns, 214–15
Siemens, W., 90
Silicon, 108
Single-plate framing connection, 386–87
Single-story construction, 15–16
Slender elements, 234–39
Slenderness, 258–59
    ratio, 204
Slip-critical connections, 397, 430
Slotted holes, 160
SMAW, 126–28
Smith, T., 162
Society Tower, 58
Staggered holes, 161–66
Standard holes, 158
Standard Oil Building, 14, 90
Steel grades, 115–25. *See also* ASTM; CSA
Steel Structures Painting Council, 139
Stiffened seat, 386–87
Stiffness reduction factor, 230–34
Strain hardening, 113–14
Stress-strain, 110–15, 148–49
Structure second-order effects, 315–19

Struik, J., 424
St. Venant torsion, 254, 273
Sulfur, 108
Sustained live load, 62
Sway, 303–5, 314
Sydney Harbor Bridge, 90

T

Tangent modulus of elasticity, 206–7, 209–10
Teal's approach, 75
Temperature, 105, 130–35, 188–91
    coefficient of thermal expansion, 115
Tension
    angles, 146–47
    braces, 144
    eye-bars, 146, 178–79
    hangers, 144
    pin-connected links, 146, 178–80
    sag rods, 144
    tie rods, 144
    tubes, 146–47
Tension connections
    design, 425–30
    limit states, 425
Tetmajer equation, 204
Thornton, W., 425
312 Walnut, 382
Timoshenko, S., 234
Transient live load, 62
Tresca failure criterion, 162
Triaxial state of stress, 115
Tributary area, 52, 66–67
Turkstra's rule, 79

U

Uang, C., 321
Ultimate limit states (ULS), 27, 55
Unbraced frames, 314–19
Uniform Building Code, 72–77
University of Western Ontario, 71

V

Vanadium, 106, 108

Verrazano Narrows Bridge, 90
Vibration, 279, 376
von Kármán, T., 208
von Mises criterion, 114, 162, 257

W

Warping torsion, 254, 273
Weathering steel, 139. *See also* ASTM, A242, A588
Web local buckling
  compact, 260
  noncompact, 260–62
  slender, 261 (*see also* Plate girder)
Weldability, 105, 108–9
Welding, 390–93
  electrode matching, 399
  electrodes, 125
  FCAW, 390–91
  fillet, 392

fillet weld strength, 398–400
GMAW, 390–91
groove, 392
groove weld strength, 401
limit states, 398–401
preheat, 132
prequalified, 392
process, 390–91
SAW, 390–91
SMAW, 390–91
Woolworth Building, 90
World Trade Center, 14, 90
Wrought iron, 89

Y

Yang, H., 208
Yield point, 112–13
Yura, J., 169, 322, 397